Springer Theses

Recognizing Outstanding Ph.D. Research

Aims and Scope

The series "Springer Theses" brings together a selection of the very best Ph.D. theses from around the world and across the physical sciences. Each volume has been selected for its scientific excellence by an internationally renowned institute or department. For greater accessibility to non-specialists, the published versions include an extended introduction, as well as a foreword by the student's supervisor explaining the special relevance of the work for the field. The series as a whole provides a valuable resource both for newcomers to the research fields, and for other scientists seeking detailed background information on special questions. Finally, it serves as an accredited documentation of the valuable contributions made by today's younger generation of scientists.

How to Apply

Theses prepared at internationally leading universities and institutes may be nominated for publication by supervisors with the endorsement of the head of department. Each such department is entitled to nominate one thesis per year. Enquiries and requests for a nomination form should be sent, together with the thesis, to Ms. Nathalie Berg (see link on right).

Eligibility Criteria

To qualify for consideration, theses should fulfil the following criteria:

- They must be written in good English.
- The topic should fall within the confines of Chemistry, Physics, Earth Sciences, Engineering and related interdisciplinary fields such as Materials, Nanoscience, Chemical Engineering, Complex Systems and Biophysics.
- The work reported in the thesis must represent a significant scientific advance.
- A maximum of 30% of the text may be a verbatim reproduction from the author's previous publications.
- They must have been examined and passed during the 12 months prior to nomination.
- The theses should have a clearly defined structure including an introduction accessible to new PhD students and scientists not expert in the relevant field.
- After acceptance, theses should be supplemented by a "Supervisor's Foreword" outlining the wider scientific relevance of the work.

Indexed by zbMATH.

Tyler Johnson

The First Indication of Neutrino-Induced Nuclear Fission

Doctoral Thesis accepted by
Duke University, Durham, USA

Springer

Author
Dr. Tyler Johnson
Yale University
New Haven, CT, USA

Supervisor
Prof. Phillip Barbeau
Duke University
Durham, NC, USA

ISSN 2190-5053 ISSN 2190-5061 (electronic)
Springer Theses
ISBN 978-3-032-01935-6 ISBN 978-3-032-01936-3 (eBook)
https://doi.org/10.1007/978-3-032-01936-3

This Springer imprint is published by the registered company Springer Nature Switzerland AG
The registered company address is: Gewerbestrasse 11, 6330 Cham, Switzerland

For my parents, Sharon and Markl Johnson,
and siblings, Lexi and Blake.

Supervisor's Foreword

It is not the critic who counts; not the man who points out how the strong man stumbles, or where the doer of deeds could have done them better. The credit belongs to the man who is actually in the arena, whose face is marred by dust and sweat and blood; who strives valiantly; who errs, who comes short again and again, because there is no effort without error and shortcoming; but who does actually strive to do the deeds; who knows great enthusiasms, the great devotions; who spends himself in a worthy cause; who at the best knows in the end the triumph of high achievement, and who at worst, if he fails, at least fails while daring greatly, so that his place shall never be with those cold and timid souls who neither know victory or defeat.

—Theodore Roosevelt

It is the nature of the human condition that there are always those daring souls *in the arena* who push the boundaries of the known universe; who explore; seek adventure; and attack the hard problems set before us. But in many ways today, we are too often held back. In our scientific endeavors, the demand is often "why"? Why is *this* problem the one we should work on. Why is it important, and who advocates for it? How will this help our project, my program, our group, this lab, or that facility? And the answer "because it is there" is never sufficient to those *timid* souls.

It is the exploration of the unknown that delivers paradigm-shifting progress. It teaches us what is possible. What are the limits of our knowledge and the human condition. So much more so if the cost of this adventure is contained; if the effort is resourceful; and the risk is limited to a small group.

This thesis, on the search for neutrino-induced fission, describes such an adventure. It is the diary and autobiography of one student's absurd idea to measure something impossible. To measure the interaction strength of a neutrino, one of the most enigmatic and ephemeral particles, on a radioactive nucleus of Thorium that regularly undergoes nuclear fission. This courageous and creative endeavor tells the story of the first neutrino cross-sectional measurement on a radioactive nucleus, and the first measurement of neutrino-induced fission, and all of it for less than the price of the average family vehicle. It is a story of an unassailable challenge, of grit, and of the thrill of a new discovery, and should serve as a model for graduate students and senior scientists in fundamental physics research. We celebrate these triumphs of those whose *faces are marred by sweat and blood* for the achievement, but also for the risk they assumed and their sacrifice. This thesis is one of those human triumphs.

Durham, NC, USA
July 2025

Prof. Phillip Barbeau

Acknowledgements

I owe all that I am to my parents, Sharon and Markl Johnson. This doctorate would not be possible without their unwavering support throughout the entirety of my life. My love of physics was nurtured and allowed to grow because they gave me every opportunity to pursue this passion. I owe so much to my little brother and sister, Blake and Lexi. When I needed a mental or emotional reset, you all gave me that and more.

There is nothing more valuable to me than my family and with such a large one, I can't reasonably fit everyone here. It goes without saying that the constant and unconditional support of everyone in Kansas City and St. Louis sustained me. I firmly believe that all the important facets of me that allowed me to complete this program were instilled in me by my family. My grandma Dorothy imbued me with the habit of challenging my beliefs and to pulling apart the fabric of my own arguments in order to grow and make them stronger. My grandma Nonny engendered me with a grit and fortitude that only someone with such a rich and layered life could embody. My partner, Choiselle, sets the standard for excellence in our household. I can honestly say that I wouldn't have made it past first year if it weren't for her relentless encouragement.

My time in graduate school was enriched and made worthwhile thanks to many people. My adviser, Phil Barbeau, was there for me even before day 1—from calling me while I was an undergrad while I was in his old stomping ground, the Hallowed Grounds Coffeeshop at the University of Chicago, to allowing me to embark on the wild journey of building NuThor. My philosophy of "why not?" for doing the crazy sounding ideas for experiments was directly instilled from the many years of whiteboard detector design sessions and tin foil hat club chats. Diane Markoff continually and unwaveringly supported me and NuThor throughout the years. My thesis work is only possible because of that support.

My time at Duke was lush with numerous mentors in the faculty, student-body, staff, and university administrators. Professor Calvin Howell was someone who really convinced me that Duke was a place I could thrive and succeed. His work the Sloan Foundation (among other things) demonstrated to me how physicists can still make a material difference in our committee. Professor Roxanne Springer made sure I found my bearings when I first arrived to Duke and routinely checked-in with me, which went a long way toward assuring me Duke could be home. Dean Looney was a life-changing source of confidence and comfort. Along with Phil, she called me on the phone when I was an undergrad to explain to me how Duke would take care of me and how I could, in turn, be impactful at Duke. One of the first things Dean Looney ever told me was "grad school is bad for your health!" and as ever, she was right, but she also gave me the tools to thrive despite this. Deans Alan Kendrick, Cisco Ramos, and Johnna Frierson are some of the kindest human beings that I've ever met and I'm so glad they were the first faces I meet when I first arrived for Sloan early start—I can think of no greater beginning.

I had the pleasure of being mentored by my fellow graduate students as well. None more than Sam Hedges who taught me more than I could ever list on a page. Connor Awe brought me into the fold of the Barbeau group and taught me everything from operating the tandem to how to design a detector. I am so grateful to all of my lab mates who have kept me company and helped me get across the finish line when I inevitably take on too much on my own. Jay Runge was my office mate for years and years and I was always so impressed by the depth of his curiosity and zeal for figuring out tough problems. Jay helped me build NuThor and it could only have gone so smoothly because he was there MacGyvering solutions to every issue I overlooked. Ryan Bouabid is someone I'd like to collaborate with for the span of our research careers. Thanks for putting up with me annually pitching the same, bad idea for detecting relic neutrinos (but also prepare yourself for the next time I pitch the same thing). Adryanna Major is a singular force. I owe so much to her for making the impossible happen while we deployed to Neutrino Alley. One of the most helpful and productive time in my graduate career was from our weekly analysis meetings. Ana Zaalishvili was extremely helpful in getting through the gargantuan NuThor datasets. There isn't a better person at removing people from our office when I needed to get work done. Charlie Prior would always brighten my day with his colorful phrases. I'm so grateful to the entire lab and want to thank Natalie Jones, Nixon Ogoi, Jane O'Reilly, Aitor Bracho, Godwin Smart, Erem Ujah, Long Li, Peibo An, Emma van Nieuwenhuizen, and Jose Colon Rivera.

My colleagues in the COHERENT Collaboration also have my eternal gratitude for paving the way for my work to be possible. Without tireless advocates like Jason Newby, I cannot imagine NuThor being built—let alone getting over one hundred pounds of thorium safety installed in a neutrino detector inside an accelerator facility.

It is because I had the continual support from the NNSA Consortium for Monitoring, Technology and Verification that I was able to freely pursue this line of inquiry. I am extremely gratful for Profs. Sara Pozzi and Shaun Clarke for allowing me to be a part of the MTV fellowship. I must thank the Sloan Foundation for building a local support network for me in this new part of the country. Additionally, the Mellon-Mays Undergraduate Fellowship and Elise Larose at the University of Chicago really set me up to successfully apply to prepare for graduate school.

Contents

1 Introduction 1
1.1 Neutrinos 2
1.1.1 The First Observation of the Neutrino 3
1.1.2 Neutrino Flavors 3
1.1.3 Neutrino Oscillations 4
1.1.4 Neutrino Sources 6
1.1.5 Neutrinos as a Probe of Nuclear Physics 11
1.2 Nuclear Fission 11
1.2.1 Induction by Neutrons 12
1.2.2 Induction by Other Particles 13
1.3 Neutrino-Induced Nuclear Fission 13
References 17

2 Theoretical Background 21
2.1 Neutrino Interaction Theory 21
2.1.1 Charged-Current Interactions 22
2.1.2 Neutral-Current Interactions 31
2.2 Nuclear Theory 33
2.2.1 Nuclear Structure 33
2.2.2 The Liquid Drop Model 34
2.2.3 The Shell Model 35
2.2.4 The Theory of Nuclear Fission 35
2.2.5 Fission Barriers 36
2.2.6 Barrier Lowering Processes 36
2.2.7 Final State Particles 38
References 46

3 Neutrinos at the SNS 49
3.1 The COHERENT Experiment 49
3.2 Neutrino Alley 51
3.2.1 The CEvNS Program 52
3.2.2 Continuing the CEvNS Legacy 52

3.2.3 NaI ... 54
3.2.4 INCOHERENT Program ... 54
3.2.5 Eljen Cell ... 55
3.2.6 Neutrino Cubes ... 56
3.2.7 NaIvE-185 ... 58
3.2.8 D_2O ... 59
3.3 Neutrino Production at the SNS ... 60
3.3.1 Beam Power and Energy ... 64
3.3.2 Neutrino Oscillations in the Standard Model ... 67
References ... 68

4 Neutrino-Induced Fission on Thorium ... 71
4.1 Target Nucleus Selection ... 71
4.2 Charged-Current Neutrino-Nucleus Interaction on Thorium ... 73
4.2.1 Final State Lepton Energy Distribution ... 79
4.2.2 Cross Check with Fermi-Dirac Supernova Flux-Averaged Cross Section ... 79
4.2.3 Cross Check with Uranium-238 Cross Section ... 81
4.3 De-Excitation Via Nuclear Fission ... 81
4.3.1 Isobaric Analog State Decay Cross Check ... 84
4.3.2 Neutron Energy Modeling ... 84
4.4 Neutrino-Induced Nuclear Fission Cross Section ... 88
4.5 Neutral-Current Neutrino-Nucleus Interactions on Thorium ... 89
4.5.1 Event Rate at Nuclear Reactors ... 93
References ... 94

5 The NuThor Detector ... 97
5.1 The Thorium ... 98
5.1.1 Radioactivity Survey ... 100
5.1.2 Double Escape Peak ... 102
5.2 Overview of the NuThor Design ... 103
5.2.1 NaI Scintillators ... 107
5.2.2 Neutron Multiplicity Meter ... 108
5.3 Deployment ... 115
5.3.1 NuThor Phase 0 ... 116
5.3.2 NuThor Phase 1 ... 118
5.3.3 NuThor Phase 2 and On ... 118
5.4 Electronics and Data Acquisition System ... 119
5.4.1 Trigger Scheme ... 123
5.5 Signal and Backgrounds ... 125
5.5.1 Signal ... 125
5.5.2 Backgrounds ... 129
References ... 139

6 Signal Processing 141
6.1 Timing Signals 143
6.2 Detector Signals 144
6.3 Savitzky-Golay Smoothing 149
6.4 Conditional Moving Average Filter 151
6.5 Frequency-Domain Matched Filter 157
6.6 Integration Optimization 159
6.7 Differential Evolution Optimization 161
6.8 Data Cleaning 164
6.9 Energy Calibration 168
6.10 Health Checks 173
6.11 Event Building 174
6.11.1 Coincidence Window Selection 176
6.11.2 Correlation with SNS Beam Pulses 180
References 181

7 Simulation of Signal and Backgrounds 183
7.1 Motivation Behind DICEBOX 186
7.2 Gadget 190
7.2.1 MCNP Neutron Simulation 191
7.2.2 Photon Simulation 192
7.3 Simulation Post-processing 194
7.3.1 The Truth Level 194
7.3.2 Energy Smearing 194
7.3.3 Time Smearing 196
7.3.4 Mock Data Stream 198
7.4 NuFission Signal Expectations 198
7.5 Background Expectations 205
7.5.1 Neutrino-Induced Neutrons on ^{232}Th 205
7.5.2 Neutrino-Induced Neutrons on ^{208}Pb 205
7.5.3 Spontaneous Fission of ^{232}Th 205
7.5.4 Beam-Related Neutrons 208
References 210

8 Event Selection 211
8.1 Preselection Cut 212
8.1.1 Maximum Energy Cut 214
8.2 Building the BDT Model 215
8.2.1 Training and Testing Samples 215
8.2.2 Neutron Tagging 216
8.2.3 Neutron Model Performance 240
8.2.4 BDT Validation with Cosmic Muon-Induced Neutrons 246

8.2.5 Fission Tagging ... 262
8.2.6 Fission Model Performance ... 268
8.2.7 Fission-Tag Score—Primary Cut ... 275
References ... 279

9 Maximum Likelihood Fitting ... 281
9.1 PDF Construction ... 281
9.1.1 Steady-State Backgrounds ... 282
9.1.2 Neutrino-Induced Nuclear Fission Neutrons ... 282
9.1.3 Neutrino-Induced Neutrons ... 284
9.1.4 Beam-Related Neutrons ... 284
9.2 Binned Maximum Likelihood Fitter ... 288
9.3 Unblinded Results ... 293
References ... 302

10 Conclusion ... 303
References ... 304

Appendix A: Alternative NuThor Designs ... 305

Appendix B: Seismic Stability ... 309

Appendix C: Nuclear Transitions in Thorium ... 315

Curriculum Vitae ... 319

Chapter 1
Introduction

Abstract The introduction to this thesis takes a perfunctory look at several pivotal aspects about the neutrino and nuclear fission. These concepts about both the fundamental nature of the neutrino and complex modeling of the nucleus undergird the work embarked upon in this thesis and informs most of the experimental choices made to conduct this measurement campaign for neutrino-induced nuclear fission.

The most successful scientific theory in history is arguably the Standard Model of Particle Physics. The Standard Model neatly boils the cacophonous cosmos down to a small cadre of fundamental particles whose interactions define the universe. The most fundamental particles in the theory are divided into two categories: fermions (half-integer spin particles) and bosons (integer spin particles). The fermions are further portioned into quarks and leptons. There are (likely) six quarks: up, down, charm, strange, top, and bottom. There are (likely) six leptons: electrons, muons, tau, electron neutrinos, muon neutrinos, and tau neutrinos. These elementary fermions can be grouped into "families" or "generations"—first, second, or third [1]. However, all of the stable matter that we experience belongs to the first generation. Whether there exists a fourth generation of particles in the Standard Model is a topic of much vigorous study and discussion that is beyond the scope of this work (Fig. 1.1).

The connective tissue that joins together the diverse constituents of the Standard Model is the four fundamental forces mediated by the bosons. The fundamental forces are electromagnetism, the strong interaction, the weak interaction, and gravity. The strong nuclear force, which holds the nucleus together, is a by-product of the strong interaction. These forces differ greatly in their relative strengths (See Table 1.1). This disparity in strength is encapsulated in their coupling constants.

This study explores a connection between these fundamental forces by examining the interplay of neutrinos and nuclear physics. The two primary aspects of this thesis are neutrinos and nuclear fission, and how they influence each other. The introductory portion of this work will lay out the rudimentary concepts for both neutrinos and nuclear fission. The second part concludes with a discussion of neutrino-induced nuclear fission, and the NuThor experiment.

T. Johnson, *The First Indication of Neutrino-Induced Nuclear Fission*,
https://doi.org/10.1007/978-3-032-01936-3_1

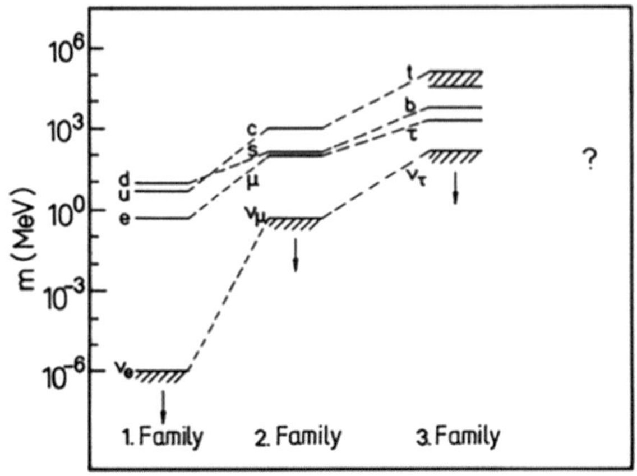

Fig. 1.1 The standard model of particle physics illustrating the three generations of particle families [1]

Table 1.1 Phenomenology of the fundamental interactions [1]. The four fundamental forces are enumerated as well as the residual strong interaction holding the nuclei together. Note that the strength of the weak interaction is given in units of $m_p \frac{\hbar^3}{c}$ where m_p is the proton mass and used typically in older literature due to the familiarity of the proton mass. For gravitation, M_{Pl} is the Planck energy given by $\sqrt{\hbar c^5/G}$

Interaction	Strength	Range	Mediator
Strong (color)	$\alpha_s \approx 1$	Confinement	g (gluon)
Strong (nuclear)	$\frac{g_\pi^2}{4\pi} \approx 14$	$\approx m_\pi^{-1} \approx 1.5$ fm	π (pion)
Electromagnetic	$\alpha = \frac{1}{137.036}$	∞	γ (photon)
Weak	$G_F = 1.166 \times 10^{-5}\,\text{GeV}^{-2}$	$\approx M_W^{-1} \approx 10^{-3}$ fm	$W^\pm$, Z^0
Gravitation	$G_N = M_{\text{Pl}}^{-2}$	∞	Graviton (hypothetical)
	$M_{\text{Pl}} = 1.22 \times 10^{19}$ GeV		

1.1 Neutrinos

"Neutrinos are as close to nothing that something can be."
— Isaac Asimov

Imagine the fundamental particles that make up the Standard Model as a jigsaw puzzle. When the puzzler gets to the neutrino pieces, they discover that the pieces are ever so slightly misaligned, they don't quite fit as expected. The neutrinos are quirkier than quarks, more bizarre than the bosons, and despite nearly a century to study them, they remain the least understood of all known fundamental particles.

The induction of the neutrino into the Standard Model brought with it unmatched opportunity to break beyond history's most successful scientific framework. For as many questions that the neutrino answered, it generated at least as many new, challenging questions that still beguile the field. Relatively little is known about the diminutive particle compared to its more massive and electrically charged subatomic neighbors.

The neutrino's ghostly nature arises from its lack of electric charge and extremely small mass, causing it to interact exclusively through the weak interaction. This raises the question: why is the neutrino so much lighter than the other fundamental particles? One explanation is that its mass is generated from a mechanism other than the Higgs mechanism, or that there is a heavy counterpart that balances the light neutrinos observed [2, 3].

The ubiquity of massive neutrinos in the universe has vast and pivotal implications to physics far beyond the auspices of particle physics. The neutrino mass washes out cosmic structure [4].[1] A massive neutrino implies that there should exist both a left and a right handed neutrino [6]—yet only left handed neutrinos have ever been detected [7]. Alternatively, one could expand the Standard Model to allow for Majorana neutrinos, abandoning lepton number conservation.[2]

1.1.1 The First Observation of the Neutrino

The experimental history of the neutrino begins at the P Reactor site at Savannah River National Laboratory. In 1956, Frederick Reines and Clyde Cowan became the first observers of the neutrino using a detector composed of a cadmium-loaded scintillator[3] to exploit the inverse beta decay interaction of the electron anti-neutrino on hydrogen nuclei [8].

1.1.2 Neutrino Flavors

Neutrinos come in different forms or "flavors." These flavors correspond to each of the leptons in doublets: electron, muon, and tau. However, one could just as easily represent neutrinos as a set of mass states. In fact, the flavor states, ν_l, can be represented as a linear combination of mass states according to Eq. 1.1.

---

[1] Conversely, cosmological structure provides a handle to model dependently constrain the sum of the neutrino mass eigenstates, $\sum_i m_{\nu_i}$ [5].

[2] Majorana neutrinos would act as their own anti-particles thus allowing rare, nuclear reactions like neutrino-less double beta decay to occur and violate lepton number conservation.

[3] Inverse beta decay emits a neutron that scatters for several microseconds before finally capturing on either a cadmium or hydrogen nucleus. Requiring two, coincident signals from the positron annihilation and neutron capture drastically diminishes backgrounds.

$$\nu_l = \sum_{m=1}^{N=3} U_{lm}\nu_m, l = e, \mu, \tau \tag{1.1}$$

where U_{lm} is the Pontecorvo-Maki-Nakagawa-Sakata (PMNS) mixing matrix, ν_m are the mass eigenstates, l denotes the three flavor eigenstates. The PMNS matrix can be written as Eq. 1.2:

$$\begin{bmatrix} c_{12}c_{13} & s_{12}c_{13}e^{i\phi_{12}} & s_{13}e^{i\phi_{13}} \\ -s_{12}c_{23}e^{-i\phi_{12}} - c_{12}s_{13}s_{23}e^{i(\phi_{23}-\phi_{13})} & c_{12}c_{23} - s_{12}s_{13}s_{23}e^{i(\phi_{12}+\phi_{23}-\phi_{13})} & c_{13}s_{23}e^{i\phi_{23}} \\ s_{12}s_{23}e^{-i(\phi_{12}+\phi_{23})} - c_{12}s_{13}c_{23}e^{-i\phi_{13}} & -c_{12}s_{23}e^{-i\phi_{23}} - s_{12}s_{13}c_{23}e^{i(\phi_{12}-\phi_{13})} & c_{13}c_{23} \end{bmatrix} \tag{1.2}$$

where θ_{ij} is the mixing angle, $c_{ij} \equiv \cos\theta_{ij}$ and $s_{ij} \equiv \sin\theta_{ij}$, and ϕ_{ij} are the CP-violating phases [9]. This formalism assumes the paradigm of three mass states and three flavor states. If the neutrino is Dirac, then only the ϕ_{13} phase is non-zero. Alternatively, if the neutrino is Majorana, the ϕ_{12} and ϕ_{23} phases may be non-zero.

This work will not spend considerable time on the topic of neutrino flavors; however, the flavor of neutrinos emitted from the various sources can be critically important for experimenters. Research focused on low-energy neutrinos (<100 MeV) is primarily concerned with the electron neutrino, because at these energies, the kinematic threshold for charged-current interactions involving the muon and tau neutrino flavors cannot be met.[4] Consequently, the study of compound nuclear reactions induced by neutrinos is largely limited to interactions with electron neutrinos.

1.1.3 Neutrino Oscillations

Neutrinos have the peculiarity that their can flavor continuously changes as they propagate through space or "oscillate." A more simplistic two-flavor neutrino oscillation framework yields an oscillation probability expressed as Eq. 1.3

$$P(\nu_\alpha \rightarrow \nu_\beta) = \sin^2(2\theta)\sin^2\left(1.27\frac{\Delta m^2 L[\text{km}]}{E[\text{GeV}]}\right) \quad (\alpha \neq \beta) \tag{1.3}$$

where θ is the mixing angle, Δm^2 is the squared-mass differences ($\Delta m^2_{ij} \equiv m^2_j - m^2_i$), L is the distance traveled by the neutrino, and E is the neutrino's kinetic energy. However, we know that there are, in fact, three neutrino flavors. Thus, the oscillation probability must be adjusted. A convenient set of expressions for oscillation probabilities is provided for typical use cases below. In the case of small L/E, the oscillation probabilities are:

---

[4] There is a range of kinematic thresholds set by the mass differences of the target nucleus and its subsequently created daughter nucleus. This can be optimized to reduce the thresholds faced by experiments. However, the rest mass of the electron poses a final, insurmountable barrier to reducing the threshold to zero.

$$P(\nu_\mu \to \nu_\tau) = \cos^4(\theta_{13})\sin^2(2\theta_{23})\sin^2\left(1.27\Delta m^2_{23}\frac{L}{E}\right) \tag{1.4}$$

$$P(\nu_e \to \nu_\mu) = \sin^2(\theta_{13})\sin^2(2\theta_{23})\sin^2\left(1.27\Delta m^2_{23}\frac{L}{E}\right) \tag{1.5}$$

$$P(\nu_e \to \nu_\tau) = \sin^2(\theta_{13})\cos^2(2\theta_{23})\sin^2\left(1.27\Delta m^2_{23}\frac{L}{E}\right) \tag{1.6}$$

The current neutrino oscillation data yields the following values for the mass splittings and mixing angles according to the Particle Data Group (PDG) [10]:

$$\left|\Delta m^2_{21}\right| \simeq 7.6 \times 10^{-5}\text{eV}^2 \tag{1.7}$$

$$\left|\Delta m^2_{31}\right| \simeq 2.4 \times 10^{-3}\text{eV}^2 \tag{1.8}$$

$$\frac{\left|\Delta m^2_{21}\right|}{\left|\Delta m^2_{31}\right|} \simeq 0.032 \tag{1.9}$$

$$\sin^2\theta_{12} = 3.10 \times 10^{-1} \tag{1.10}$$

$$\sin^2\theta_{23} = 33.82 \times 10^{-1} \tag{1.11}$$

$$\sin^2\theta_{13} = 2.241 \times 10^{-2} \tag{1.12}$$

There are now many experiments around the world that have reported the oscillatory nature of the neutrino at a variety of neutrino sources and baselines. One example of an observation of neutrino oscillation is the KamLAND experiment's results shown in Fig. 1.2.

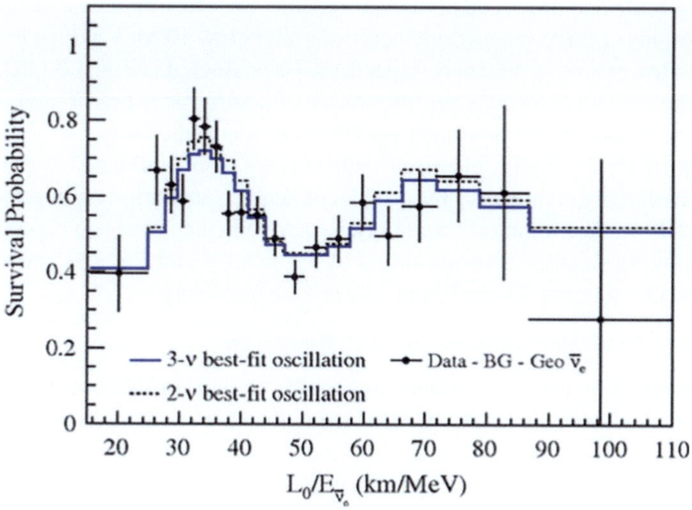

Fig. 1.2 An example of reactor neutrinos appearing to oscillate in the KamLAND experiment [11]

Experiments aimed at studying neutrino oscillations from low-energy sources, like the Sun, or core-collapse supernovae, with neutrino-induced nuclear reactions must be cognizant of oscillations into or out of the observable neutrino flavor.

1.1.4 Neutrino Sources

One of the great ironies of neutrinos is that despite the terrible difficulty involved in measuring them, they are exceptionally, almost shockingly, plentiful. Neutrinos rain down from a myriad of cosmological sources, spanning the entire conceivable range of energy scales. Neutrinos explode out of the bellies of nuclear reactions, out of the particle-laden mess of accelerator targets, and the churning molten core of the Earth. But also from bananas, human bodies, and kitty litter.

A non-exhaustive list of neutrino sources with helpful rules-of-thumb ensues in the subsequent section. Each source offers unique advantages for exploring neutrino-nucleus interactions, which will be discussed.

Nuclear Reactors

Inside the core of nuclear reactors, the uranium- and (to a lesser extent) plutonium-filled fuel rods undergo nuclear fission reactions for which the heavy nuclei split into two large pieces—called fission fragments. There are thousands of possible fission fragments and they are often quite unstable. As a result, they tend to beta decay numerous times before they reach stability. On average, each fission fragment beta decays three times for a total average of six beta decays per fission. With each beta decay, one electron anti-neutrino is emitted. It is worth emphasizing that the vast majority of neutrinos produced at nuclear reactors are electron anti-neutrinos.

The energies of these reactogenic neutrinos are below 10 MeV with a heavy skew towards the lower end of the energy spectrum—as depicted is Fig. 1.3 [12]. Nuclear reactors represent an extremely prolific source of neutrinos: approximately 6×10^{23} electron anti-neutrinos are produced per GWhr of thermal energy.

To a much lesser extent, electron neutrinos are also produced. These electron neutrinos are largely created as a by-product of neutron activation on materials inside and around the nuclear reactor's core [14]. Reactogenic neutrons can capture on ^{51}Cr and ^{55}Fe to yield electron neutrino fluxes of 8.3×10^{-4} and 3×10^{-4} neutrinos per fission with Q-values of 753 keV and 231 keV, respectively.

Relevance of Neutrino-Induced Nuclear Reactions

Neutrino-induced nuclear reactions can provide clearer insight into events occurring within a nuclear reactor's core. This capability is beneficial for applications in national security, such as verifying compliance with nuclear weapons agreements or monitoring the production of special nuclear materials. The typical avenue for reactor neutrino detection is via inverse beta decay (IBD), which uses a hydrogen nucleus as a target. However, IBD has an energy threshold of 1.8 MeV. A 1.8 MeV threshold excludes roughly 70% of the reactor neutrino flux. By carefully selecting

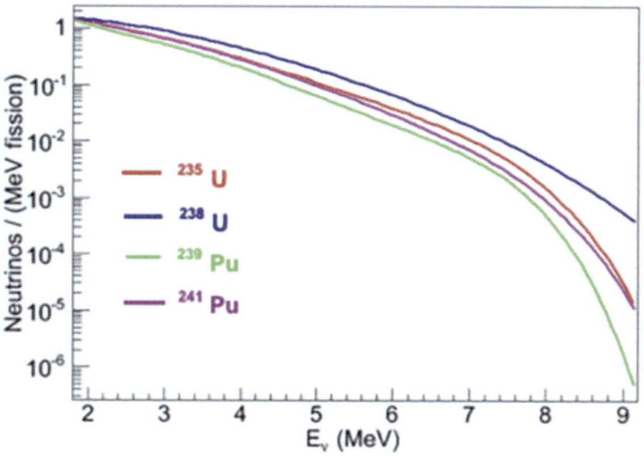

Fig. 1.3 The electron anti-neutrino energy spectrum emitted from nuclear reactors broken down into the four primary fissile isotopes responsible for generating neutrinos via nuclear fission [13]

an alternative nuclear target, lower energy thresholds and greater interaction cross sections can be achieved. For instance, charged-current neutrino-nucleus interactions on ^{79}Br have an energy threshold of 1.17 MeV [15].

In addition to the potential improvements in cross section and energy threshold, these nuclear reactions can produce multiple particles in the final state. The emission of multiple particles, such as several neutrons, enhances background discrimination. However, a significant challenge in utilizing neutrino-induced nuclear reactions for fundamental neutrino physics searches at nuclear reactors, such as oscillation parameter measurements, is the scarcity of measured cross sections. Even in cases where cross sections are available, they often suffer from large uncertainties, hindering the precision of such measurements.

Stopped Meson Decay-at-Rest Sources

There are many avenues for producing neutrinos with a particle accelerator-based source. One key species of accelerator neutrino sources is the Decay-at-Rest source in which high energy primary particles such as protons are collided with a dense target to produce mesons. When the primary proton beam exceeds the rest mass of the pion (139.57039(18) MeV/c^2), these sources are referred to as pion Decay-at-Rest (πDAR) sources. Notable examples of πDARs are the Spallation Neutron Source (SNS) at Oak Ridge National Laboratory, the Lujan Neutron Source at Los Alamos National Laboratory, and Chinese Spallation Neutron Source (CSNS), and in the future, the European Spallation Source (ESS)—these are tabulated in Fig. 1.4 [16]. Both positive and negative pions are created following the proton's collision with the target. A large fraction of the negative pions are captured by the target medium itself while the positive pions will scatter until they come to rest. Once at rest, they will radioactively decay into positive muons and muon neutrinos. Those muons will then decay into positrons, electron neutrinos, and muon anti-neutrinos.

If the particle accelerator's proton beam energy exceeds the rest mass of the kaon (493.677 $\pm$ 0.016 MeV), it is then able to create said kaons in addition to the flurry

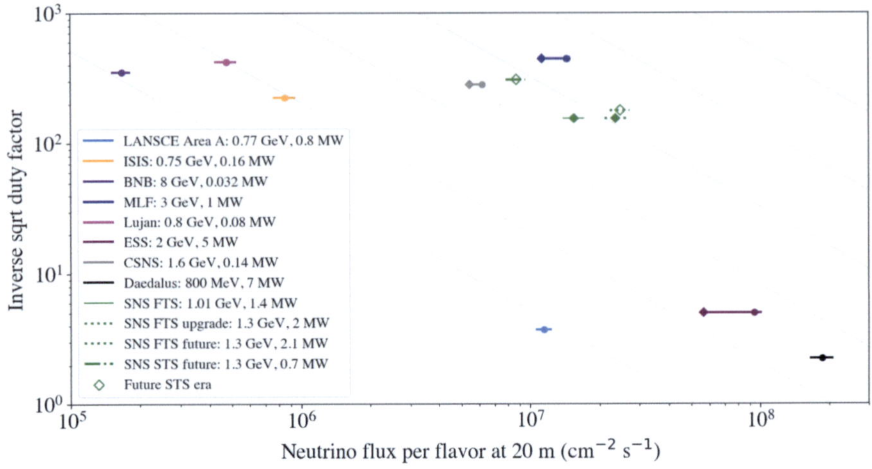

Fig. 1.4 Figure of merit for stopped-pion decay-at-rest sources around the world [16]. It is preferable to be the upper right quadrant of the figure

of pions–these are so called "KDAR" sources. A notable example of a KDAR source is the J-PARC Spallation neutron Source (JSNS).

Relevance of Neutrino-Induced Nuclear Reactions

DAR neutrino sources are effective testbeds for measuring neutrino interaction cross sections. Their electron neutrino energies reach into the tens of MeV. At these energies, kinematic thresholds act as less of a constraint. A significant challenge in utilizing neutrino-induced nuclear reactions for fundamental neutrino physics searches, such as oscillation parameter measurements, is the scarcity of measured cross sections. Even in cases where cross sections are available, they often suffer from large uncertainties, hindering the precision of such measurements. DAR sources add to the body of measured cross sections to mitigate this shortcoming.

Neutrino Beams

Alternatively, particle accelerators can also be utilized in a neutrino beam configuration. Starting from a similar point as the meson decay-at-rest source, a packet of high-energy protons is collided with a dense target, like graphite, to produce charged pions and kaons whose trajectories are bent with magnetic focusing horns towards the forward direction. Often times, the polarity of the current in the magnetic focusing horns can determine whether the beam is mostly neutrinos or anti-neutrinos.

Neutrino beams are typically used for neutrino oscillation studies to allow for sufficient baselines for optimal oscillations or accumulation of matter effects [17]. The largest neutrino beam experiment is the Long Baseline Neutrino Facility (LBNF) supplying neutrinos for the DUNE experiment (Fig. 1.5).

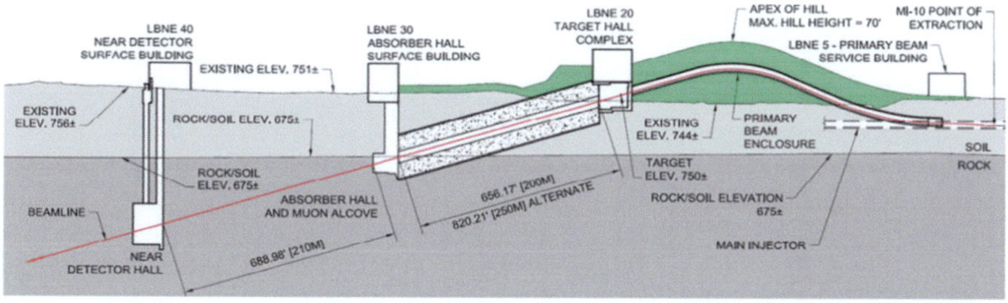

Fig. 1.5 The DUNE Neutrino beam schematic, from Ref. [18]. The energies exhibited from a neutrino beam are in the GeV range

Relevance of Neutrino-Induced Nuclear Reactions

Neutrino beams are often at the vanguard in the pursuit towards physics beyond the Standard Model. Although the neutrinos are substantially higher energy than many other sources, the understanding of neutrino-induced nuclear reactions is pivotal to their science goals. DUNE's campaign to measure the magnitude of CP-violation in the lepton sector (via δ_{CP}) hinges on few-percent precision modeling of the neutrino interaction and subsequent nuclear reaction on argon nuclei [19]. At the medium energy ($\sim$1 GeV) achieved at neutrino beams, interactions can be loosely modeled as scattering off of a single nucleon with the remaining nucleons acting as non-participants. However, contributions from nuclear effects lead to multiple nucleon emission, which must be accounted for to avoid introducing bias in the respective precision physics measurements. The energies exhibited in neutrino beams lead to excitation energies where nuclear reactions are no longer going through an intermediate compound nucleus. For the purposes of studying neutrino-induced nuclear fission, a compound nuclear reaction is preferred, thus, the neutrino projectile energies of a neutrino beam are non-ideal, because the primary de-excitation mechanism is through pre-equilibrium or direct nuclear reactions. The ideal neutrino energies are achievable at neutrino beams by moving to the off-axis where the neutrino energies decrease, but so does the neutrino flux.

Solar Neutrinos

The Sun is a tremendous source of neutrinos as it continually undergoes nuclear fusion processes that result in the emission of neutrinos. The Sun serves as a great complication to next-generation WIMP dark matter direct-detection searches [20]. Because these detectors have reached such stellar sensitivities, they are now able to breach into a region of the phase space where nuclear recoil induced by CEvNS interactions from Solar neutrinos impede the measurement of the sought after dark

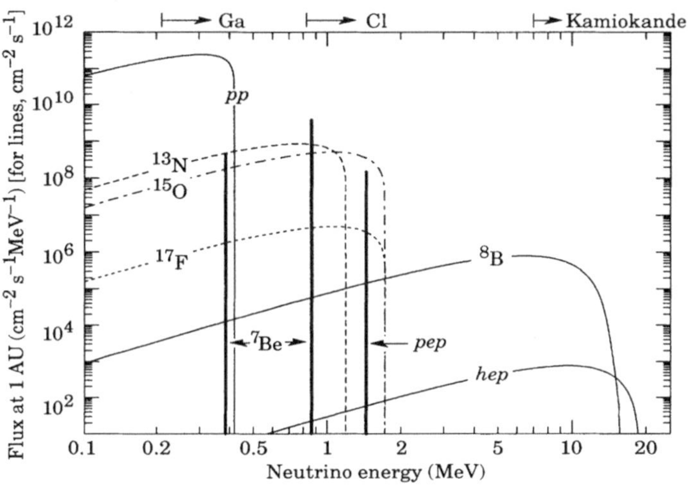

Fig. 1.6 Solar neutrino energy spectrum from various processes inside the Sun [25, 26]

matter [21].[5] Both the PandaX and XENONnT collaborations have recently detected ^{8}B solar neutrinos via CEvNS interactions [22, 23].

Solar neutrinos have historically played an integral role in the evolution of neutrino physics and launched the line of inquiry which ultimately led to the discovery of neutrino oscillations. The first indication of this came from the Homestake experiment led by John Bahcall, which originated the "Solar neutrino problem" [24].

Relevance of Neutrino-Induced Nuclear Reactions

As seen in Fig. 1.6, solar neutrinos are heavily weighted to the lower energies. The presence of neutrinos from the ^{8}B and hep populations, has the potential to complicate terrestrial neutrino-nuclear measurements. Because the flux of the neutrinos greater than $\sim$ 1 MeV is subdominant, and actually quite small, the degree that solar neutrinos interfere with terrestrial neutrino-nucleus interaction measurements is limited. Doubly so for measurements requiring the emission of a nucleon or nuclear fission due to the even greater energy thresholds.

[5] This was colloquially dubbed the "neutrino floor" or "neutrino fog" to symbolize an irreducible background to nuclear recoil-based WIMP detectors. There are approaches to burrowing through this floor that rely on directional detection of nuclear recoils, but that technology has yet to bear fruit.

1.1.5 Neutrinos as a Probe of Nuclear Physics

There are a wide variety of neutrino detector technologies of all sizes dotted across the world. Many of these neutrino detectors are comprised of either water or some form of organic scintillator material. Both of these provide neutrinos with an abundance of hydrogen nuclei as targets.[6] However, some neutrino detectors utilize heavier targets such as lead, carbon, gallium, iron, iodine, etc. Unlike hydrogen, these nuclear targets are quite complex, which makes predictive work incredibly difficult. This often leads to large uncertainties on their calculated cross sections and subsequent disparities between the expected and measured interaction rates.

Thus, before one can truly capitalize on the promise of using neutrinos as a laboratory of physics beyond the Standard Model, one must first grapple with the entwined problem of understanding neutrino interactions on nuclei. Even physics squarely within the auspices of the Standard Model, like the modeling of neutrino transport in astrophysical environments (e.g. core-collapse supernovae and neutron-star mergers), relies heavily on models for neutrino-nuclear interactions—many of the nuclei are heavy, unstable and inaccessible in a terrestrial laboratory setting [27].

Despite the clear importance, measurements of neutrino-nuclear interactions at energies lower than 300 MeV (relevant to πDAR, EC, Solar, and supernova, reactogenic and geo-neutrinos) are almost completely unstudied [28]. In fact, there are only 11 elements with measured cross sections in this energy regime. From the sparse experimentation conducted on heavier nuclear targets, it is becoming increasingly clear that there are complex nuclear effects encumbering the neutrino-nuclear interactions [29]. These nuclear effects remain experimentally understudied.

A complete examination of such effects invites the use of many nuclear targets over a range of masses. As one advances up the periodic table's atomic masses, the nuclei become more and more susceptible to unbound de-excitations like nuclear fission and neutron emission. As nuclei exceed atomic masses of approximately 200 amu, nuclear fission becomes a dominant decay channel for excitation energies in the 10s of MeV. An understanding of unbound nuclear reactions following a neutrino interaction is paramount to surveying the neutrino's relationship with nuclear matter. The remainder of this section is devoted to the unbound de-excitation channel, nuclear fission.

1.2 Nuclear Fission

The discovery of the neutron by James Chadwick [30] was the gateway to a new world of nuclear reactions. Separately, Irene Joliot-Curie and Frederic Joliot-Curie discovered induced-radioactivity of aluminum by alpha particles with the

[6] Additionally, the electrons orbiting the nuclei are also potential targets for the neutrinos to scatter on.

$\alpha + ^{27}_{13}\text{Al} \rightarrow ^{30}_{15}\text{P} + e^+$ reaction [31]. These pivotal pieces of the puzzle enabled Enrico Fermi and his team to build on this alchemical technique by turning to the neutrally charged neutron for greater effectiveness in creating new elements. To that end, they moderated neutrons to thermal energies with paraffin in order to subsequently produce transuranic elements out of readily available neighboring nuclei and a Radium-Beryllium neutron source [32]. Naturally, the experimenters sought to push beyond the heaviest known element at the time, uranium, and produce the Z = 93 element (later named Neptunium). In doing so, they found peculiar results that seemed to suggest something other than just the production of neptunium was occurring.

The radioactivity present after the neutron bombardment of uranium was unlike what was expected by any of the heavy, transuranic elements. After extremely painstaking radiochemical analysis carried out by Hahn and Strassman, the anomalous presence of radioactive barium was confirmed [33]—in addition to other elements of roughly half the mass of the original uranium. The correct interpretation of this astounding finding came from Lisa Meitner and Frisch who claimed that the barium implied that uranium was splitting into two large chunks of medium weight [34]. In fact, they postulated that barium's co-conspiring product would be krypton because its mass would account for the difference between uranium and barium that was missing. Krypton was soon confirmed by Hahn and Strassman in a derivative experiment. Meitner and Frisch minted the new process as *kernspaltung la fission nucléaire* or nuclear fission.

1.2.1 Induction by Neutrons

As established through the discovery of nuclear fission, neutrons are capable of splitting heavy nuclei. Like the discovery experiment suggests, nuclear fission can be set off by thermal neutrons, but that is by no means a requirement. Thermal neutrons are very handy when it comes to absorption on uranium nuclei because the nucleus possesses myriad resonances that greatly enhances the cross section. Using ^{235}U as an example, it is apparent why nuclear fission caught the eyes of the world immediately after its first sighting. The daughter nucleus of thermal neutron capture on ^{235}U is unsurprisingly ^{236}U, but with notable excitation energy given by:

$$
\begin{aligned}
m(^{236}\text{U}^*) &= m(^{235}\text{U}) + m_n \\
&= 236.052589\text{u} \\
E_{\text{ex}} &= (236.052589\text{u} - 236.045563\text{u}) \times 931.502\text{MeV/u} \\
&= 6.5\text{MeV}
\end{aligned}
\tag{1.13}
$$

where m_n is the neutron's mass and m(^{236}U*) is the sum of individual contributions from the ^{235}U and a neutron. The excitation energy, E_{ex} is the difference between the sum of the masses and the rest mass of ^{236}U. ^{236}U's rest mass is 236.045563 u.

Note that this is true when the neutron has negligible kinetic energy. This excitation energy exceeds the energy threshold for undergoing nuclear fission, referred to as the fission barrier. Alternatively, if the nuclear target is the naturally abundant form of uranium, ^{238}U, the excitation energy generated by neutron capture is 4.8 MeV which is lower than the fission barrier for the daughter ^{239}U meaning the remaining 1.8 MeV must be provided by the neutron's kinetic energy.

1.2.2 Induction by Other Particles

There is a small cadre of particles other than neutrons that are able to induce nuclear fission. A non-exhaustive list of these fission mechanisms are listed here:

1. **Photofission**: Discovered in 1941 [35] by R. O. Haxby, W. E. Shoupp, W. E. Stephens, and W. H. Wells on uranium and thorium using 6.3 MeV γ-rays from the $F^{19}(p,\gamma)$ reaction with a high-energy proton beam. Photofission benefits by stringent selection rules which largely leads to the photon exciting the nucleus to the Giant Dipole Resonace (GDR) before undergoing nuclear fission.
2. **Electrofission**: electrofission refers to fission induced by the scatter of a high-energy electron on a fissionable target nucleus. The electromagnetic interaction mediated by a photon can alternatively be thought of as induced by the exchanged virtual photon [36, 37].
3. **Deuteron-Induced Fission**: Discovered in 1939 [38] by D. H. T. Gant who sought out an alternative to neutron-induced nuclear fission.

1.3 Neutrino-Induced Nuclear Fission

Since the discovery of neutrinos, nuclear fission and neutrinos have been inextricably linked. However, this relationship between neutrinos and nuclear fission was always that nuclear fission begat neutrinos. As the title would suggest, neutrino-induced nuclear fission flips that paradigm; neutrinos can also bring about nuclear fission.

Spanning back as far as 50 years, neutrinos were predicted to be able to interact with a nucleus causing it to be critically deformed and subsequently decay via nuclear fission [39]. Despite the long standing predictive work, there has yet to be an experimental confirmation of this process. However, there is considerable theoretical work outlining the process, from where it might naturally occur to the specifics of which neutrino flavors and nuclear targets should be utilized for an observation [40].

Motivation

National Security

The first observation of neutrinos occurred at a nuclear reactor. From that first discovery, it has long been accepted that neutrinos give a perspective into the happenings of nuclear reactors. Among other purposes, neutrino detectors have the ability to verify non-use of reactions [41], sense the transmutation of uranium into weapons-grade plutonium within the core [42], and discover undeclared reactors [28]. Most nuclear reactor neutrino monitoring apparatuses exploit inverse beta decay (IBD) on hydrogen-laden materials, then subsequently observe the final state positron and neutron. Because nuclear reactors emit largely the electron anti-neutrino type of neutrinos, one is largely restricted to the neutral-current (NC) rather than the charged-current (CC). In contrast to IBD, NC-based neutrino-induced nuclear fission has a comparatively low cross section but benefits from an unambiguous and dramatic signal in which over 200 MeV is released in a single nuclear fission.

Nucleosynthesis

Half of the elements heavier than gold (and all of the uranium and thorium) were created by rapid neutron capture nucleosynthesis, also known as the R-Process [43–45]. The remainder of the heavy isotopes were produced by the s-process [46]. There are several possible astrophysical sites where the R-process is believed to take place. The top contender for R-Process locations has changed several times in the past, but the current favorite sites are binary neutron star mergers and core-collapse supernovae. An aspect that both of these sites have in common is their immense flux of neutrons and neutrinos. In these violent stellar environments, a gargantuan flux of neutrons rapidly capture onto seed nuclei to produce greater and greater atomic masses so rapidly that the nuclei do not have time to radioactively decay through beta-decay (See Fig. 1.7 for a depiction of the journey taken by seed nuclei to fantastically heavy products).

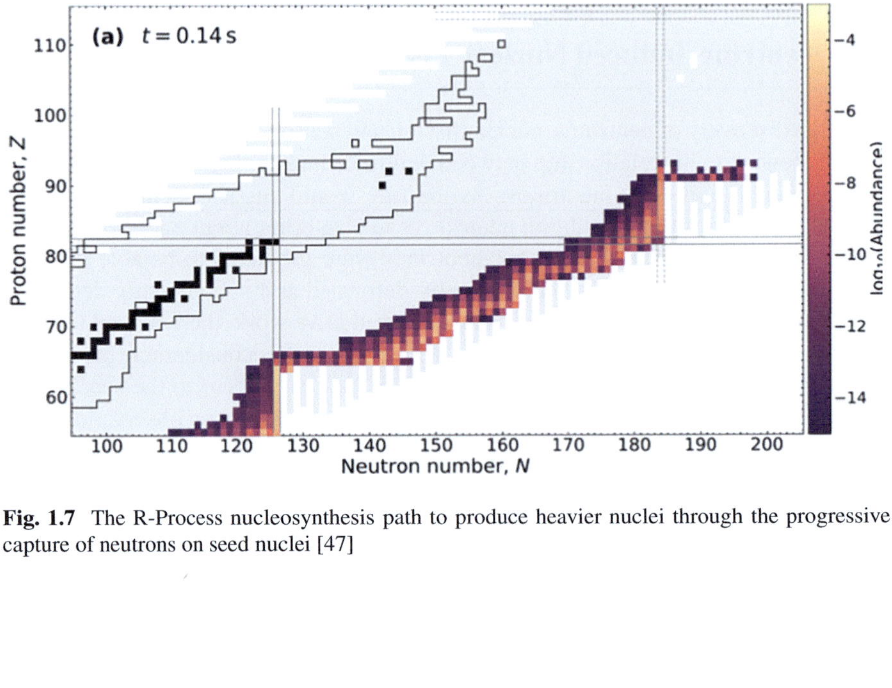

Fig. 1.7 The R-Process nucleosynthesis path to produce heavier nuclei through the progressive capture of neutrons on seed nuclei [47]

The final abundance of isotopes produced by the R-Process is an extremely non-trivial field of inquiry that won't be touched on in great detail here other than to mention a notable discrepancy between observed and predicted R-Process isotopic abundances from ultra-metal-poor (UMP) stars. UMP stars are defined as stars with a logarithmic abundance ratio of approximately [Fe/H] ~ -3. UMP stars are unique, stellar laboratories because they were produced so early in galactic formation that they were formed before the s-process had enough time to produce significant levels of heavy nuclei. Thus, the heavy nuclei present in their volumes is majority due to the R-process. This makes them an interesting probe for R-process abundances.

A discrepancy was observed between observational data of UMP star CS 22892-052 and the predictions made by the Solar R-Process mode—which is computed by taking the difference between the solar inventory and the s-process model [48]. The model curve was adjusted to fit the nuclei around Eu which brings good agreement for that and many other regions, but with the drawback of having tension with observations near Rh, Ag, Pd, and Cd. In Ref. [49], Qian et al suggest that the tension at these isotopes can be explained by neutrino-induced nuclear fission of progenitor nuclei with $190 \leq A < 230$. Because nuclei of this mass range tend to produce fission fragments with mass ratios of 1–1.2, they will likely produce daughter nuclei at $86 \leq A < 125$ which covers the masses of the discrepant isotopes. Additionally, nuclei of this mass range are not sufficiently susceptible to spontaneous fission to account for the deficits observed, but neutrino-stimulated reactions may be up to the task of inducing fission as these progenitor nuclei are decaying towards stability after neutron freeze-out.

Additionally, the neutrino-driven wind inducing nuclear reactions on the surrounding nuclei can have an effect on the nuclear reaction network calculations used to understand the production and destruction of isotopes. This network also helps to account for the availability of neutrons to spur on the R-Process. The following network equation (Eq. 1.14) contains components for neutrino-induced nuclear fission [50].

$$\begin{aligned} Y(Z, A) &= n_n Y(Z, A-1)\sigma_{A-1} + Y(Z, A+1)\lambda_{A+1} \\ &\quad - Y(Z, A)(n_n\sigma_A + n_n\sigma_{nf}(Z, A) + \lambda_A + \sum_n \lambda_{\beta n}^{A} + \lambda_{\beta f}^{Z,A} + \lambda_{\nu f}^{Z,A} \\ &\quad + \sum_n Y(Z-1, A+n)\lambda_{\beta n}^{Z-1,A+n} + \sum_{Z',A'} Y(Z', A')\lambda_{\beta f}^{Z',A'} \\ &\quad + \sum_{Z',A'} n_n\sigma_{nf}(Z', A')Y(Z', A') + \sum_{Z',A'} Y(Z', A')\lambda_{\nu f}^{Z',A'} \end{aligned} \tag{1.14}$$

where Y is the abundance of an isotope with Z protons and A atomic mass number, $\lambda_{\nu f}^{Z',A'}$ and $\lambda_{\nu f}^{Z,A}$ are the terms in the network related to neutrino-induced nuclear fission. The remainder of the network terms are conventional reactions: beta-delayed neutron emission, beta-delayed nuclear fission, and neutron-induced nuclear fission. Even if the contributions from neutrino-induced fission are slight compared to more dominant fixtures of the reaction network, it is better to quantify the impact and incorporate it to the modeling for a more data-driven initiative. It is also worth mentioning that the study of neutrino-induced nuclear fission likely involved nuclear targets of exceptionally large atomic mass.

Supernova Neutrino Detection

When a star reaches the end of its lifespan, its grand finale might be a core-collapse supernova [51]. The death knell for the star is when it begins to fuse heavier nuclei such as iron, which is energetically unprofitable.[7] Thus, the star's own gravity begins to win the battle of tug-of-war with the outward pointing forces. This is the collapse. The pressure exerted on the core's contents eventually exceeds the neutron degeneracy pressure. The electrons capture onto the protons making neutrons in the "neutronization". As the protoneutron-star de-leptonizes, the emission of electron neutrinos is the primary cooling mechanism [53]. The neutrinos are largely emitted over the span of just 10 s.

Some of the principle champions of measuring and exploiting neutrino-induced nuclear fission were Goldhaber and Shrock [54] in which the authors suggest co-locating a neutrino-induced nuclear fission-based detector and an inverse beta decay-based detector for supernova neutrino detection.[8] Further, that work advocates for the use of uranium as a target. Despite the fact that uranium suffers from a substantial spontaneous fission rate, the coincident neutrino detections in the inverse beta decay for the $\mathcal{O}(10)$ seconds of the supernova event was argued to counteract the significant spontaneous fission backgrounds.

The immense flux of electron neutrinos from the neutronization process within the core-collapse supernova propagates to Earth where a constellation of neutrino detectors around the globe experience a sudden influx of interactions. Typically, the detectors involved with supernova detection are massive and are party to the Supernova Early Warning System (SNEWS) [55]. Currently, inverse beta decay-based detectors make up $\sim$98% of the available mass, meaning that electron anti-neutrinos are the primary signal particle. Alternatively, detectors with heavy nuclear targets, like HALO [56], are sensitive to supernova electron neutrinos. Neutrino-induced nuclear fission (and its counterpart, neutrino-induced neutron emission) would add to the inventory of electron neutrino-sensitive detectors and diversify the suite of detection channels for capturing these once-in-a-lifetime supernova events.

---

[7] Nuclear fusion releases energy if the product nucleus has a higher binding energy than the constituent pieces that came together to form it. However, iron has the greatest binding energy per nucleon, thus represents a turning point after which nuclear fusion becomes endothermic. A more detailed explanation can be found in most introductory nuclear physics textbooks, but a particularly helpful one can be found in Ref. [52].

[8] Goldhaber and Shrock suggest using a ^{238}U target for the neutrino-induced nuclear fission reaction which has a large spontaneous fission rate, thus the purpose of the IBD detector is to act as a external trigger to reject that significant spontaneous fission background. Further, they propose looking for "subthreshold" reactions where the neutrino-induced nuclear fission detector may be sensitivity to muon neutrinos despite the average energy of muon neutrinos being 25 MeV from core-collapse supernovae.

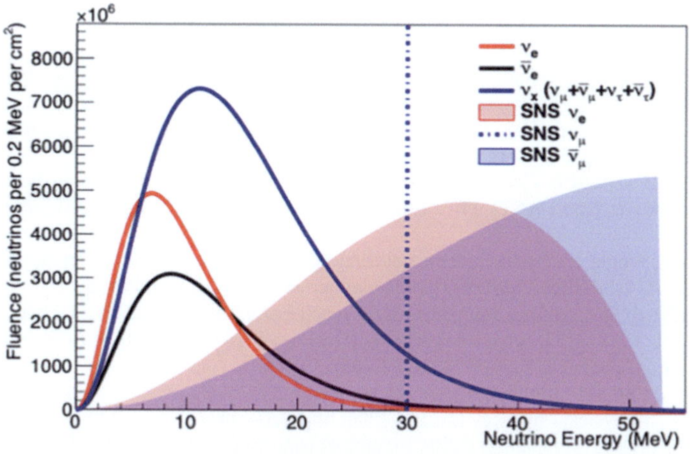

Fig. 1.8 The neutrino energy spectra of core-collapse supernovae and the spallation neutron source broken into their individual neutrino flavor compositions [53]

Interestingly, there is significant overlap of the energy spectra of core-collapse supernova and stopped pion decay-at-rest neutrino sources—Fig. 1.8. For this reason, one can argue that such neutrino production facilities are opportune test beds for supernova neutrino monitoring technologies.

References

1. Grotz K, Klapdor HV (1990) Weak interaction in nuclear particle and astrophysics. IOP Publishing Ltd
2. Akhmedov EK, Branco GC, Rebelo MN (2000) Seesaw mechanism and structure of neutrino mass matrix. Phys Lett B 478(1):215–223. ISSN: 0370-2693. https://doi.org/10.1016/S0370-2693(00)00282-3
3. Adame AG et al (2025) DESI 2024 VI: cosmological constraints from the measurements of baryon acoustic oscillations. J Cosmol Astropart Phys 21. https://doi.org/10.1088/1475-7516/2025/02/021
4. Zeng Z, Yeung S, Chu MC (2019) Effects of neutrino mass and asymmetry on cosmological structure formation. J Cosmol Astropart Phys 15. https://doi.org/10.1088/1475-7516/2019/03/015
5. Chudaykin A, Ivanov MM (2019) Measuring neutrino masses with large-scale structure: euclid forecast with controlled theoretical error. J Cosmol Astropart Phys 34. https://doi.org/10.1088/1475-7516/2019/11/034
6. Goldhaber AS, Goldhaber M (2011) The neutrino's elusive helicity reversal. Phys Today 64(5):40–43. ISSN: 0031-9228. https://doi.org/10.1063/1.3592004, https://pubs.aip.org/physicstoday/article-pdf/64/5/40/10093314/40_1_online.pdf
7. Goldhaber M, Grodzins L, Sunyar AW (1958) Helicity of neutrinos. Phys Rev 109:1015–1017. https://doi.org/10.1103/PhysRev.109.1015

8. Reines F, Cowan CL (1953) Detection of the free neutrino. Phys Rev 92:830–831. https://doi.org/10.1103/PhysRev.92.830
9. Maki Z, Nakagawa M, Sakata S (1962) Remarks on the unified model of elementary p. Prog Theor Phys 28(5):870–880. ISSN: 0033-068X
10. Workman RL, Particle Data Group (2022) Review of particle physics. Prog Theor Exp Phys 2022:083C01. https://doi.org/10.1093/ptep/ptac097
11. Gando A et al (2011) Constraints on θ_{13} from a three-flavor oscillation analysis of reactor antineutrinos at KamLAND. Phys Rev D 83:052002. https://doi.org/10.1103/PhysRevD.83.052002
12. Hayes AC, Vogel P (2016) Reactor neutrino spectra. Annu Rev Nucl Part Sci 66:219–244. ISSN: 1545-4134. https://doi.org/10.1146/annurev-nucl-102115-044826
13. Kim Y (2016) Detection of antineutrinos for reactor monitoring. Nucl Eng Technol 48(2):285–292. ISSN: 1738-5733. https://doi.org/10.1016/j.net.2016.02.001, https://www.sciencedirect.com/science/article/pii/S1738573316000498
14. Xin B et al (2005) Production of electron neutrinos at nuclear power reactors and the prospects for neutrino physics. Phys Rev D 72:012006. https://doi.org/10.1103/PhysRevD.72.012006
15. Hedges SC (2023) Evaluating $CeBr_3$ for antineutrino detection below the IBD threshold. Phys Life Sci
16. Akimov D et al (2022) The COHERENT experimental program. arXiv: 2204.04575 [hep-ex]
17. Wolfenstein L (1978) Neutrino oscillations in matter. Phys Rev D 17:2369–2374. https://doi.org/10.1103/PhysRevD.17.2369
18. Papadimitriou V et al (n.d) Longitudinal section of the LBNE Beamline facility in the MI-10, Shallow beamline configuration. Batavia, IL, USA, https://creativecommons.org/licenses/by/3.0/
19. DUNE Collaboration et al (2022) Snowmass neutrino frontier: DUNE physics summary. arXiv: 2203.06100 [hep-ex]
20. Roszkowski L, Sessolo EM, Trojanowski S (2018) WIMP dark matter candidates and searches–current status and future prospects. Rep Prog Phys 81(6):066201. https://doi.org/10.1088/1361-6633/aab913
21. Billard J, Figueroa-Feliciano E, Strigari L (2014) Implication of neutrino backgrounds on the reach of next generation dark matter direct detection experiments. Phys Rev D 89(2):023524. https://doi.org/10.1103/PhysRevD.89.023524
22. PandaX Collaboration et al (2024) First measurement of solar ^{8}B neutrino flux through coherent elastic neutrino-nucleus scattering in PandaX-4T. arXiv: 2407.10892 [hep-ex]
23. Aprile E et al (2024) First measurement of solar ^{8}*B neutrinos via coherent elastic neutrino-nucleus scattering with XENONnT* arXiv: 2408.02877 [nucl-ex]
24. Haxton WC (1998) The solar neutrino problem: mixing of neutrinos and mixing in the sun. Progr Part Nucl Phys 40. Neutrinos in astro, particle and nuclear physics, pp 101–112. ISSN: 0146-6410. https://doi.org/10.1016/S0146-6410(98)00014-3, https://www.sciencedirect.com/science/article/pii/S0146641098000143
25. Ianni A (2013) Solar neutrinos and the solar model. Phys Dark Univ 4. Dark TAUP 2013, pp 44–49. ISSN: 2212-6864. https://doi.org/10.1016/j.dark.2014.06.002, https://www.sciencedirect.com/science/article/pii/S2212686414000211
26. Barnett RM et al (1996) Review of particle physics. Phys Rev D 54:1–708. https://doi.org/10.1103/PhysRevD.54.1
27. Avignone FT et al (2000) ORLaND: a proposed neutrino facility at the Oak Ridge national laboratory. Phys At Nuclei 63(6):1007–1011. ISSN: 1562-692X. https://doi.org/10.1134/1.855740
28. Bernstein A et al (2020) Colloquium: neutrino detectors as tools for nuclear security. Rev Mod Phys 92:011003. https://doi.org/10.1103/RevModPhys.92.011003
29. Hellgren M, Suhonen J (2024) Charged-current neutrino-nucleus scattering off ^{127}I and ^{133}Cs. Phys Rev C 109:035802. https://doi.org/10.1103/PhysRevC.109.035802
30. Chadwick J (1932) The existence of a neutron. Proc R Soc Lond Ser A Contain Pap Math Phys Charact 136(830):692–708

31. Joliot F, Curie I (1934) Artificial production of a new kind of radio-element. Nature 133(3354):201–202. ISSN: 1476-4687. https://doi.org/10.1038/133201a0
32. Fermi E (1934) Possible production of elements of atomic number higher than 92. Nature 133(3372):898–899. ISSN: 1476-4687. https://doi.org/10.1038/133898a0
33. Hahn O, Strassmann F (1939) Über den Nachweis und das Verhalten der bei der Bestrahlung des Urans mittels Neutronen entstehenden Erdalkalimetalle. Naturwissenschaften 27(1):11–15. ISSN: 1432-1904. https://doi.org/10.1007/BF01488241
34. Meitner L, Frisch OR (1939) Products of the fission of the uranium nucleus. Nature 143(3620):471–472. ISSN: 1476-4687. https://doi.org/10.1038/143471a0
35. Haxby RO et al (1941) Photo-fission of uranium and thorium. Phys Rev 59:57–62. https://doi.org/10.1103/PhysRev.59.57
36. Onley DS, Ressler GM (1969) Electrofission of U^{238}. Phys Rev Lett 22:236–239. https://doi.org/10.1103/PhysRevLett.22.236
37. Zamani-Noor F, Onley DS (1986) Virtual photon theory in electrofission. Phys Rev C 33:1354–1366. https://doi.org/10.1103/PhysRevC.33.1354
38. Gant DHT (1939) Fission of uranium under deuteron bombardment. Nature 144(3651):707. ISSN: 1476-4687. https://doi.org/10.1038/144707a0
39. Andryushin VI, Bilen'kii SM, Gershtein SS (1971) Possible method of measuring the electromagnetic form factor of the neutrino. ZhETF Pis Red
40. Kelić A et al (2005) Cross sections and fragment distributions from neutrino-induced fission on r-process nuclei. Phys Lett B 616(1):48–58. ISSN: 0370-2693. https://doi.org/10.1016/j.physletb.2005.04.074, https://www.sciencedirect.com/science/article/pii/S0370269305006039
41. Carr R et al (2019) Neutrino-based tools for nuclear verification and diplomacy in North Korea. Sci Glob Secur 27(1):15–28. https://doi.org/10.1080/08929882.2019.1603007
42. Jaffke P, Huber P (2017) Determining reactor fuel type from continuous antineutrino monitoring. Phys Rev Appl 8:034005. https://doi.org/10.1103/PhysRevApplied.8.034005
43. Cameron AGW (1957) Nuclear reactions in stars and nucleogenesis 69(408):201. https://doi.org/10.1086/127051
44. Burbidge EM et al (1957) Synthesis of the elements in stars. Rev Mod Phys 29:547–650. https://doi.org/10.1103/RevModPhys.29.547
45. Seeger PA, Fowler WA, Clayton DD (1965) Nucleosynthesis of heavy elements by neutron capture. Astrophys J
46. Kappeler F, Beer H, Wisshak K (1989) s-process nucleosynthesis-nuclear physics and the classical model. Rep Progr Phys 52(8):945. https://doi.org/10.1088/0034-4885/52/8/002
47. Holmbeck EM, Sprouse TM, Mumpower MR (2023) Nucleosynthesis and observation of the heaviest elements. Eur Phys J A 59(2):28. ISSN: 1434-601X. https://doi.org/10.1140/epja/s10050-023-00927-7
48. Arlandini C et al (1999) Neutron capture in low-mass asymptotic giant branch stars: cross sections and abundance signatures. Astrophys J 525(2):886. https://doi.org/10.1086/307938
49. Qian YZ (2002) Neutrino-induced fission and r-process nucleosynthesis. Astrophys J 569(2):L103. https://doi.org/10.1086/340643
50. Hix WR, Thielemann FK (1999) Computational methods for nucleosynthesis and nuclear energy generation. arXiv: astro-ph/9906478 [astro-ph]
51. Janka HT et al (2007) Theory of core-collapse supernovae. Phys Rep 442(1). The Hans Bethe Centennial volume 1906–2006, pp 38–74. ISSN: 0370-1573. https://doi.org/10.1016/j.physrep.2007.02.002, https://www.sciencedirect.com/science/article/pii/S0370157307000439
52. Krane KS, Halliday D (1988) Introductory nuclear physics. Rev, Wiley, New York
53. Barbeau PS, Efremenko Yu, Scholberg K (2023) COHERENT at the spallation neutron source. Ann Rev Nucl Part Sci 73:41–68. ISSN: 1545-4134. https://doi.org/10.1146/annurev-nucl-101918-023518

54. Goldhaber M, Shrock R (2001) Subthreshold reactions involving nuclear fission. Phys Rev C 63:021303. https://doi.org/10.1103/PhysRevC.63.021303
55. Al Kharusi S et al (2021) SNEWS 2.0: a next-generation supernova early warning system for multi-messenger astronomy. New J Phys 23(3):031201. https://doi.org/10.1088/1367-2630/abde33
56. Duba CA et al (2008) HALO–the helium and lead observatory for supernova neutrinos. J Phys: Conf Ser 136(4):042077. https://doi.org/10.1088/1742-6596/136/4/042077

Chapter 2
Theoretical Background

Abstract This chapter is a foray through the theoretical framework for both neutrino interactions with matter and nuclear reactions. The discussion of neutrino interactions is partitioned into charged-current and neutral-current formulations. The nuclear theory is stilted towards concepts that impact a nucleus's journey to fission. This includes the nuclear binding energy, a macroscopic view of the nucleus found in the liquid drop model, and a microscopic view of the nucleus via the shell model. These orient the subsequent discussions of nuclear fission, its energy threshold, and various mechanisms through which the fission barrier can be lowered. Lastly, the many possible final state particles from nuclear fission are enumerated and described.

Before discussing the experimental program, a theoretical foundation is necessary to contextualize the experimental design, analysis strategy, and signal expectations. Neutrino-induced nuclear reactions involve two key components: the neutrino projectile and the target nucleus. The mechanisms through which neutrinos interact with nuclei are outlined in the first part of this chapter. The relevant assumptions and adjustments to the interaction rate calculations are also considered. Chief among the cross section corrections is the weak axial-vector coupling constant, which is of major interest to the nexus of neutrinos and nuclear physics. Following this, the nucleus's response to neutrino interactions is examined, with particular emphasis on the nuclear fission reaction channel, its threshold, and the observable final state products.

2.1 Neutrino Interaction Theory

Neutrinos interact with matter via the weak force. These interactions proceed through two possible channels: the charged-current (Fig. 2.1) and the neutral-current (Fig. 2.2). In the case of a charged-current neutrino interaction, a $W^{\pm}$-boson is exchanged, whereas in the neutral-current interaction, a Z-boson is exchanged. The details provided in this work will also only concern neutrino interactions as they occur at low energies—below 100 MeV.

T. Johnson, *The First Indication of Neutrino-Induced Nuclear Fission*,
https://doi.org/10.1007/978-3-032-01936-3_2

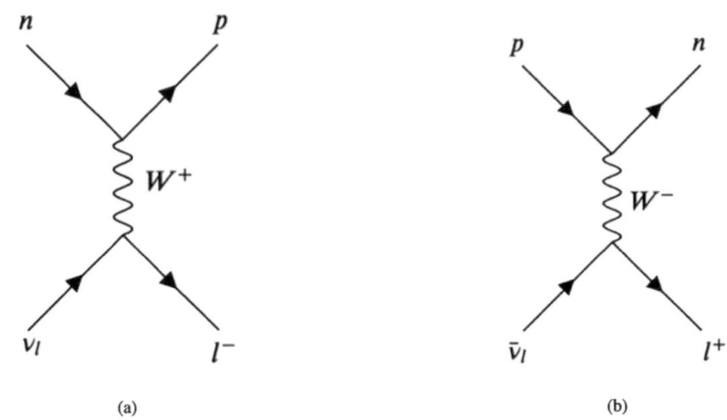

Fig. 2.1 Feynman diagrams of the charged-current interaction between a neutrino and a neutron (left) and a proton (right)

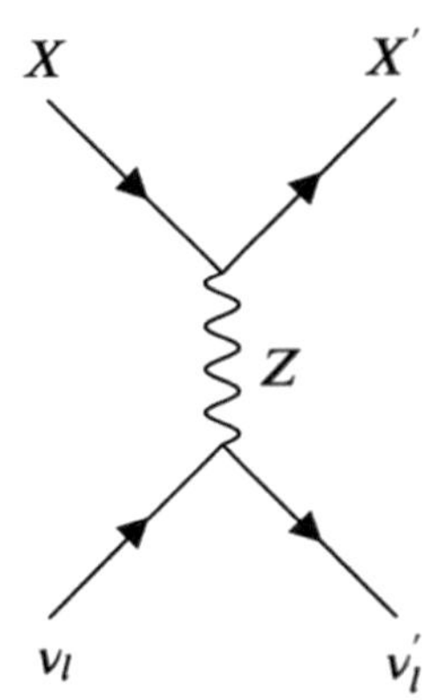

Fig. 2.2 Feynman diagram of the neutral-current neutrino-nucleus interaction via the exchange of a Z-boson

2.1.1 Charged-Current Interactions

The charged-current neutrino-nucleus interaction is mediated by the W-boson and when the neutrino's target is a nucleus, it is interchangeably referred to as "neutrino capture." Both neutrinos and anti-neutrinos may participate in this interaction but differ in whether it is a positively- or negatively-charged W-boson that is exchanged. The charged-current interaction is accompanied by the emission of a lepton in the final state in order to both conserve charge and lepton number.

Because the neutrino capture interaction requires the creation of the final state lepton, there is a kinematic threshold proportional to the lepton's rest mass. As a result, muon and tau neutrinos have a greater energetic burden to bear in order to capture on a nucleus compared the electron neutrinos.[1] Additionally, the mass

[1] The rest masses of the electron, muon and tau are 548.579909065 ± 0.000000016 keV, 105.6583755 ± 0.0000023, and 1776.93 ± 0.09, respectively [1].

difference between the initial state and final state nucleus (the Q-value) contributes to the neutrino capture threshold.

$$E_{\text{thresh}} = Q + m_l \tag{2.1}$$

where Q is the Q-value and m_l is the lepton rest mass. The neutrino capture is akin to an ordinary beta decay and shares many of the same mathematical frameworks. Thus, the underlying theory of both weak interaction processes is partially covered here. To start, there are two categories of weak transitions: allowed and forbidden.

In the allowed transition paradigm, the nucleus (or free neutron) is taken to be point-like, therefore, it cannot imbue any orbital angular momentum to the reaction products. Take the classic example of the beta decay of a free neutron. From here, there is one further distinction within the allowed transition family. If the neutrino and electron spins are coupled with a total spin of 0 (a singlet state) and the neutron shares the same spin state as the resultant proton, this is called a Fermi transition. Conversely, if the neutrino and electron are coupled with a total spin of 1 (the triplet state) and the neutron-proton spins are connected through a spin flip via the 2 × 2 Pauli matrices, $\boldsymbol{\sigma} = (\sigma_1, \sigma_2, \sigma_3)$, it is dubbed a Gamow-Teller transition. Each type of transition is depicted in Fig. 2.3. Both the Fermi and Gamow-Teller transitions have an isospin flip between the initial and final states. Either p → n or n → p.

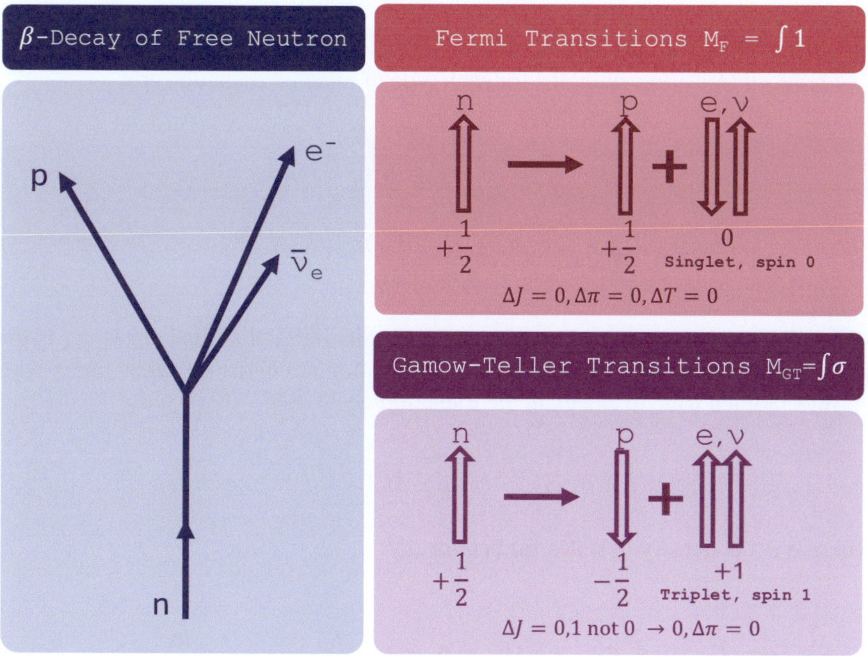

Fig. 2.3 Schematic of a β-decay of a free neutron (left) with the spin of the Fermi and Gamow-Teller transitions, adapted from [2]

Table 2.1 Breakdown of each type of allowed transition's selection rules and coupling constants [2]

Transition type	Operator	Total spin, J	Orbital angular momentum, L	Parity, π	Isospin, T	Coupling constant
Fermi	$\mathbf{T}^{\pm}$	$\Delta J = 0$	$L = 0$	$\Delta\pi = 0$	$\Delta T = 0$	G_β^2
Gamow-Teller	$\mathbf{Y}^{\pm}$	$\lvert\Delta J\rvert = 0, 1$ **not** $0 \to 0$	$L = 0$	$\Delta\pi = 0$	$\lvert\Delta T\rvert = 0, 1$	$(G_\beta g_A)^2$

The selection rules are summarized in Table 2.1.
Where the Fermi transition operator is given by

$$\mathrm{T}^{\pm} \equiv \sum_{i=1}^{A} \tau^{\pm}(i) \tag{2.2}$$

where the isospin raising and lowering operators are

$$\tau^{+} = \begin{bmatrix} 0 & 1 \\ 0 & 0 \end{bmatrix}, \tau^{-} = \begin{bmatrix} 0 & 0 \\ 1 & 0 \end{bmatrix} \tag{2.3}$$

and the Gamow-Teller operator is given by

$$\mathrm{Y}^{\pm} \equiv \sum_{i=1}^{A} \boldsymbol{\sigma}(i)\tau^{\pm}(i) \tag{2.4}$$

The reduced transition strengths for each type of allowed transition are:

$$\mathrm{B}_{\mathrm{F}}^{\pm} = \frac{|\langle N_f||\mathrm{T}^{\pm}||N_i\rangle|^2}{2\mathrm{J}_i + 1} \tag{2.5}$$

$$\mathrm{B}_{\mathrm{GT}}^{\pm} = \frac{g_A^2|\langle N_f||\sum_i \sigma(i)\tau^{\pm}(i)||N_i\rangle^2}{2\mathrm{J}_i + 1} \tag{2.6}$$

where N_i and N_f are the spin of the nucleus in its initial and final states. g_A is the weak axial-vector coupling constant. The transition strengths are taken along with Fermi's Golden Rule to calculate the transition/decay rate.

$$\frac{dW}{dtdE_e} = G_\beta^2 \frac{d\rho}{dE_e}[B_F + B_{GT}] \tag{2.7}$$

where ρ is the density of states for beta-decay

$$\rho = \frac{1}{2\pi^5} \int d^3p_f d^3p_e d^3p_{\bar{\nu}} \delta^3(\mathbf{p}_f + \mathbf{p}_e + \mathbf{p}_{\bar{\nu}})\delta(E_i - E_f - E_f^{\mathrm{Kin}} - E_e - E_{\bar{\nu}}) \tag{2.8}$$

$\mathbf{p}_\nu$, $\mathbf{p}_e$, $\mathbf{p}_f$ are the momenta of the neutrino, electron, and the final nucleus, respectively. By taking $\Delta_f = E_i - E_f$, $E_f^{\rm Kin} \sim 0$, assuming the neutrino to be of negligible mass ($p_{\bar{\nu}} = E_{\bar{\nu}}$) and integrating over $d^2p = d\phi \sin\nu d\nu p^2 dp$ to get

$$\mathrm{d}\phi = \frac{1}{2\pi^3} p_e E_e (\Delta_f - E_e)^2 dE_e \tag{2.9}$$

to get a rate of

$$\frac{dW_F}{dtdE_e} = \frac{G_\beta^2}{2\pi^3} p_e E_e (\Delta_f - E_e)^2 [B_F(f) + B_{GT}(f)] \tag{2.10}$$

g_A Quenching

A motif of the previous formulation is the weak axial-vector coupling constant, g_A. This term has several incarnations depending on the scale one is probing. The most readily accessible framework for conceptualizing g_A is when considering the beta decay of the free neutron, one expects the ratio of Fermi to Gamow-Teller strength to be 1:3. This is not true in practice and experiments instead yield a ratio of 1:3g_A^2, where g_A=1.26. This discrepancy is often attributed to the internal structure of the nucleon, however, when one employs an *ab initio* calculation from the nucleon wave functions, one arrives at a $g_A = \frac{5}{3}$ which is greater than the experimental value.

A perfunctory outline of this calculation in the Constituent Quark Model [3] ensues starting with the Gamow-Teller operator applied to the quark wave functions for the neutron to transmute it into a proton, $\langle p, m_S = 1/2 | Y^+ | n, m_S = 1/2 \rangle$. Starting with the proton's wavefunction:[2]

$$\begin{aligned} |p, m_S = 1/2\rangle = \frac{1}{\sqrt{18}} \{ & 2|d\downarrow u\uparrow u\uparrow\rangle + 2|u\uparrow d\downarrow u\uparrow\rangle + 2|u\uparrow u\uparrow d\downarrow\rangle \\ & -|u\downarrow u\uparrow d\uparrow\rangle - |u\uparrow u\downarrow d\uparrow\rangle - |u\downarrow d\uparrow u\uparrow\rangle \\ & -|u\uparrow d\uparrow u\downarrow\rangle - |d\uparrow u\downarrow u\uparrow\rangle - |d\uparrow u\uparrow u\downarrow\rangle \end{aligned} \tag{2.11}$$

Then the neutron wavefunction can be found by:

$$\begin{aligned} |n, m_S = 1/2\rangle &= \tau_Q^- |p, m_S = 1/2\rangle \\ = \frac{1}{\sqrt{18}} \{ & 2|u\downarrow d\uparrow d\uparrow\rangle + 2|d\uparrow u\downarrow d\uparrow\rangle + 2|d\uparrow d\uparrow u\downarrow\rangle \\ & -|d\downarrow d\uparrow u\uparrow\rangle - |d\uparrow d\downarrow u\uparrow\rangle - |d\downarrow u\uparrow d\uparrow\rangle \\ & -|d\uparrow u\uparrow d\downarrow\rangle - |u\uparrow d\downarrow d\uparrow\rangle - |u\uparrow d\uparrow d\downarrow\rangle \end{aligned} \tag{2.12}$$

[2] A deeper, more thorough discussion of this calculation can be found in Ref. [2] in Chap. 5. Though be aware that there is a typo in its calculation that is fixed in this work.

Lastly, the overlap is given by:

$$\begin{aligned}\langle p, m_S = {}^1\!/\!_2 | Y^+ | n, m_S = \frac{1}{2} \rangle &= \frac{1}{18} \langle p, m_S = {}^1\!/\!_2 | \\ \cdot \{ 2|u \downarrow d \uparrow d \uparrow\rangle &+ 2|d \uparrow u \downarrow d \uparrow\rangle + 2|d \uparrow d \uparrow u \downarrow\rangle \\ -|d \downarrow d \uparrow u \uparrow\rangle &- |d \uparrow d \downarrow u \uparrow\rangle - |d \downarrow u \uparrow d \uparrow\rangle \\ -|d \uparrow u \uparrow d \downarrow\rangle &- |u \uparrow d \downarrow d \uparrow\rangle - |u \uparrow d \uparrow d \downarrow\rangle \} \\ &= -\frac{1}{18}(-2 - 2 - 2 - 2 - 2 - 2 - 1 - 2 - 2 \\ &\quad -1 - 1 - 2 - 2 - 1 - 1 - 2 - 2 - 1) \\ &= {}^5\!/\!_3\end{aligned} \tag{2.13}$$

The g_A story is further complicated when the weak transitions graduate up from the quark or nucleon to the scale of a nuclear system. At this level, an even greater quenching is observed and is imposed by, as of the time of this writing, an unknown source. There are several hypotheses about the origin of the increased quenching, but they are accounted for by adopting an *effective* weak-axial vector coupling term, $g_{A_{\text{eff}}}$ [4].

$$g_A^{\text{eff}} = q g_A^{\text{free}} \tag{2.14}$$

where g_A^{free} is the measured coupling from the β-decay of the free neutron equal to 1.26 and q is the renormalization factor, which scales the effective weak-axial vector coupling constant up or down.

A common value for q is 0.77[5]. Significant theoretical work has been done to understand g_A^{eff} for light to medium mass nuclei ($0 \leq A \leq 136$), but predictive work for heavier nuclei is sparse. The trend in the lighter nuclei suggests a progressive increase in the quenching with growing nuclear sizes, but this is largely an untested hypothesis. There is a great deal of effort devoted to better understanding this coupling constant due to its importance for the search for neutrino-less double beta decay ($0\nu\beta\beta$). In the context of $0\nu\beta\beta$, the half-life is proportional to g_{eff}^4. Thus, even modest quenching of this constant results in a colossal reduction in the rate of $0\nu\beta\beta$, an already rare process with half-life limits placed on $\mathcal{O}(10^{26}$ yr). The impact of a quenched g_A on suppressing neutrino-nuclear cross sections is limited to the GT transition, leaving the Fermi transition unaffected.

Coulomb Correction

The above rate calculation is incorrect without the inclusion of the electrostatic distortion created by the interactions between the nucleus's protons and the electron (or positron). Accounting for this distortion is done with the Fermi function which has two common approaches. The first formulation is:

$$F(Z, E) = 2(1-\gamma_0)(2p_e R)^{2(\gamma_0-1)} \frac{|\Gamma(\gamma_0 - iy)^2|}{|\Gamma(2\gamma_0 - 1)^2|} e^{\pi y}$$
$$\gamma_0 = (1 - Z^2\alpha^2)^{1/2} \qquad (2.15)$$
$$y = \alpha Z E_e / p_e$$

where α is the fine structure constant, p_e is the electron momentum, E_e is the electron kinetic energy, Z is the proton number of the daughter nucleus, and R is the nuclear radius. Alternatively, one can calculate the Fermi function with the Modified Effective Momentum Approximation (MEMA) approach [6].

$$F_{\text{MEMA}} = \frac{K_{\text{eff}} E_{\text{eff}}}{KE}$$
$$K_{\text{eff}} = \sqrt{E_{\text{eff}}^2 - m_e^2} \qquad (2.16)$$
$$E_{\text{eff}} = E - V_C(0)$$
$$V_C(0) \simeq \frac{3 Z_f z_e \alpha}{2R}$$

Both formulations of the Fermi function have their separate strengths and weaknesses. The former version tends to overestimate at the higher electron energies while the latter approach tends to overestimate the lower electron energies. For this reason, some neutrino interaction event generators have taken to calculating both and accepting the lower value between the two at each electron energy—see Fig. 2.4.

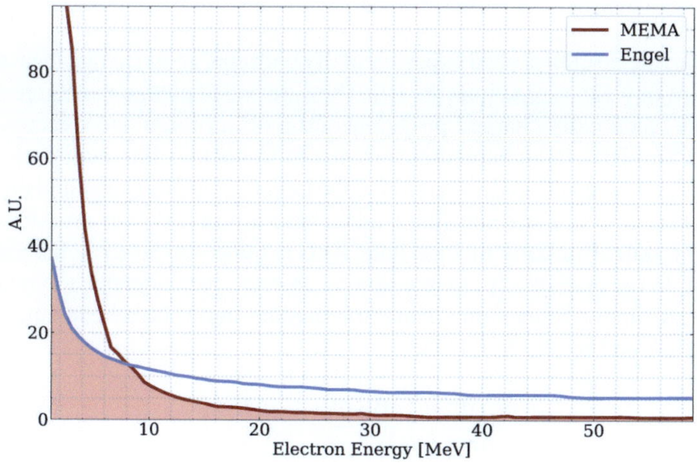

Fig. 2.4 The Coulomb correction to the charged-current cross section called the Fermi function. Two formulations are typically used. This work takes the minimum of the two functions at each electron energy value

With the Fermi function correcting the electrostatic distortion in hand, the decay rate can be written as:

$$\frac{dW}{dtdE_e} = \frac{G_\beta^2}{2\pi^2} F(Z, E_e) p_e E_e (\Delta_f - E_e)^2 [B_F + B_{GT}] \quad (2.17)$$

A more helpful form to experimentalists is the reaction cross section obtained by taking the transition probability for a particle density of 1 per interaction volume. A more comprehensive form where the coupling constants are written out more explicitly in the cross section for the kth excited state is as follows:

$$\begin{aligned}\sigma_k(E_\nu) &= \frac{G_F^2 \cos^2\theta_C}{\pi} p_e E_e F(Z, E_e)[B(F)_k + (\frac{g_A}{g_V})^2 B(GT)_k] \\ &= (1.597 \times 10^{-44}\text{cm}^2) p_e E_e F(Z, E_e)[B(F)_k + (\frac{g_A}{g_V})^2 B(GT)_k]\end{aligned} \quad (2.18)$$

where G_F is the weak coupling constant, θ_C is the Cabbibo angle, and (g_A/g_V) is typically taken to be 1.267.

Sum Rules

The total Gamow-Teller strength can be given by the Ikeda sum rule [7].

$$\begin{aligned}S_\beta^- - S_\beta^+ &= \sum_f B_{\text{GT}}^-(i \to f) - \sum_f B_{\text{GT}}^+(i \to f) \\ &= 3g_A^2(N - Z)\end{aligned} \quad (2.19)$$

This sum rule is often probed experimentally with charge-exchange measurements such as (p,n) or (^{3}He, t) reactions [8–10]. They are called charge-exchange reactions because the electrically charged projectiles drop off a proton and directly emit the remainder. Similarly to a charged-current neutrino-nucleus interaction, the charge-exchange reaction results in the creation of the target nucleus's isobar and the forward scattering angle ($\approx 0°$) in beam-line studies have a proclivity for producing a final state in a 1+ state if initially in a 0+ state. This is quite similar to a charged-current interaction. In the case of a (p, n) reaction on an even-even nucleus, the relationship between the Gamow-Teller strength and the shallow angle charge-exchange reaction could be described by Eq. 2.20 [11].

$$\frac{\sigma_{\text{GT}}(0°)g_A^2/\text{B(GT)}}{\sigma_{\text{F}}(0°)g_V^2/(N - Z)} = \left(\frac{E_p}{E_0}\right)^2 \quad (2.20)$$

where E_p is the energy of the proton, E_0 is an empirically found parameter with the value of $E_0 = 55.0 \pm 0.4$ MeV.

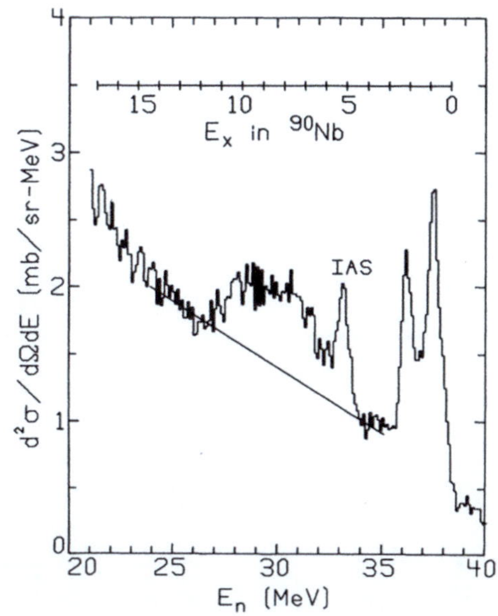

Fig. 2.5 The first experimental observation of the Giant Gamow-Teller resonance. The experiment used the ^{90}Zr(p,n)^{20}Nb charge-exchange reaction [12]

Collective Resonances

Following the charged-current neutrino-nucleus interaction, the daughter nucleus is left in an excited state. At higher energies, the density of states is so substantial that this region must be described as a continuum rather than a collection of discrete lines. Further, there are collective resonance features present in this continuous regime. For charge-exchange, allowed transitions, the primary resonances are called the Giant Gamow-Teller resonance and the Isobaric Analog State. The presence of the Giant Gamow-Teller resonance was first observed in 1975 by Doering et al. [12] (Fig. 2.5).

An example of the exploitation of the collective resonances in the isobar with a neutrino interaction is shown with ^{208}Pb nucleus in Fig. 2.6.

2.1.1.1 Inverse Beta Decay

One of the most commonly used tools for detecting neutrinos is through inverse beta decay (IBD)—Eq. 2.21. The necessary ingredient for instigating an IBD interaction is an electron anti-neutrino, thus nuclear reactor-based neutrino programs tend to gravitate towards this process.

$$\bar{\nu}_e + p \rightarrow e^+ + n \tag{2.21}$$

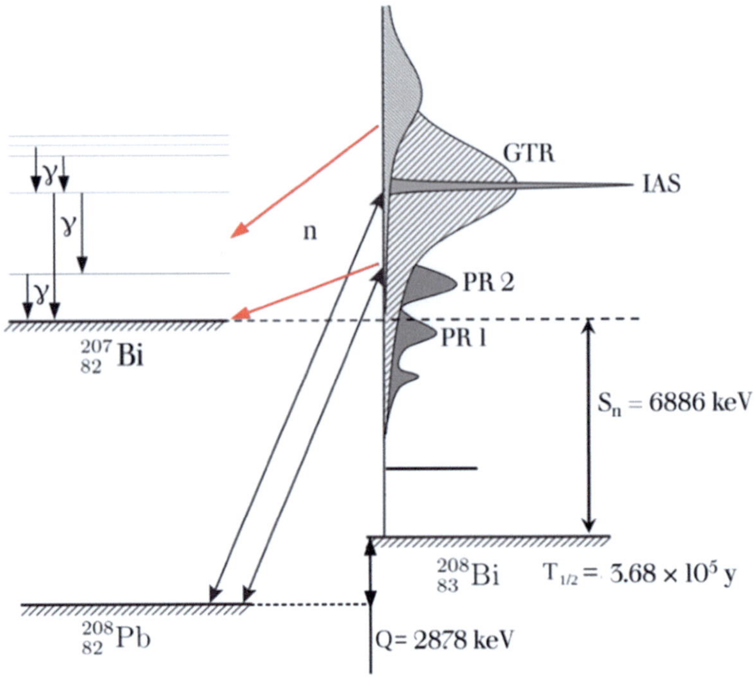

Fig. 2.6 Diagram of the nuclear transitions involved in a charged-current interaction stimulated by an electron neutrino capturing on a ^{208}Pb target nucleus. The daughter nucleus (^{208}Bi) can be at an excited state in the continuum potentially above several unbound de-excitation energy thresholds

The cross section for IBD is quite similar to its converse, but presented here for completeness.

$$\sigma_{\bar{\nu}}(E_{\bar{\nu}}) = \frac{G_\beta^2}{\pi} p_{e+} E_{e+} F(-Z, E_{e+})[B_F^+ + B_{GT}^+] \tag{2.22}$$

where the inverted charge of the final state lepton leads to a changed Fermi function where the opposite distortion between the lepton and nuclear charge occurs compared to the charged-current counterpart.

Additionally, the popularity of IBD as a detection channel stems from the neutron in the final state acting as a second observable that tends to be staggered in time from the initial detection of the positron. Often times, a hydrogen-rich detector material is utilized for its ample protons serving as neutrino targets. The hydrogen nuclei also readily moderate the emitted neutrons to thermal energies where it can be captured on the hydrogen with a resulting 2.2 MeV γ-ray. However, one can enhance the detector's capabilities for capturing neutrons by doping the detector's volume with a small amount of the neutron capture agent of their choosing such as gadolinium, lithium, boron, cadmium, etc. The adding signal from the neutron allows the experimenters to reject a great deal of background by virtue of the sought after signal having an additional feature in order to be considered a signal candidate event.

2.1.2 Neutral-Current Interactions

In addition to the charged-current interaction, there is also the prospect of neutrinos interacting via the neutral current. These cross sections tend to scale with the size of the nucleus and, with the exception of the coherent elastic neutrino-nucleus scattering interaction, have a lower cross section compared to the charged-current.

Inelastic Neutral-Current Interactions

The neutral-current cross section can be expressed as follows:

$$\sigma_{i,f}^{NC} = \frac{G_F^2 g_A^2}{\pi(2J_i + 1)}(E_\nu - E_X)^2 \times |\langle f|| \sum_k \bar{\sigma}(k) t_0(k) ||i\rangle|^2 \tag{2.23}$$

where g_A is the weak axial-vector coupling constant, E_X is the excitation energy of the final state nucleus, f, and t_0 in the reduced matrix element is the zero component of the isospin operator.

Similarly to the charged-current, there is an analogous nuclear physics process one can use to estimate the transition strengths for the neutrino process. In this case, one typically uses (p,p'), (γ,γ'), and/or (n,γn') to extract the magnetic dipole M1 transitions to intuit information on the matrix elements [13]. The basis of this exchange is the M1 transition probability takes the form of Eq. 2.24.

$$B(M1) = \frac{3}{4\pi} |\langle f|| \sum_k [\bar{l}(k) t_0(k) + (\mu_S - 0.5)\frac{\bar{\sigma}(k)}{2} + \mu_\nu \bar{\sigma}(k) t_0(k)]||j\rangle|^2 \mu_N^2 \tag{2.24}$$

Thus, one can then relate the M1 transition to the GT transition operator due to its clear structural similarities and takes the form of Eq. 2.25.

$$\begin{aligned} B(\mathrm{GT}_0) &= \frac{4\pi g_A^2}{3\mu_\nu^2} \frac{B(M1)}{\mu_N^2} \\ &= 0.308 \frac{N(M1)}{\mu_N^2} \end{aligned} \tag{2.25}$$

where μ_N is the nuclear magneton, and the isovector nuclear magnetic moment, μ_ν = 4.706.

Coherent Elastic Neutrino-Nucleus Scattering (CEνNS)

Coherent Elastic Neutrino-Nucleus Scattering (CEνNS) is a subspecies of neutral-current weak interactions in which a Z-boson is exchanged with a nuclear target where the momentum transfer is such that the nucleon structure of the nucleus is not resolved. Because the nuclear structure is unable to be resolved, the nucleus is seen as one large target rather than a collection of individual nucleons. In so doing, the nucleons can then recoil in phase with one another to fulfill "coherency." This "coherency" condition is accomplished when the de Broglie wavelength at hand is roughly equivalent to the nuclear radius, rR $\approx$ 1. Because CEvNS is a neutral-current process, it can occur with any neutrino flavor—it is "flavor blind" or "flavor democratic." The coherency condition means that neutrinos above a certain energy start to probe the finer structure of the nucleus or lose said coherency.

One benefit of the CEvNS interaction channel is that it does not have a kinematic threshold like its counterparts (e.g. inverse beta decay) have so it is able to access neutrino energies below many typical neutrino detector schemes. In the case of nuclear reactor neutrino detection or monitoring, IBD-based detectors are limited to observing neutrinos greater than 1.8 MeV. Conversely, CEvNS detectors may observe neutrinos as low energies as their specific technologies allow. Many bolometric CEvNS detectors are coming online around the world and, as of this writing, imminently close to observing CEvNS at a nuclear reactor [14, 15]. Additionally, the CEvNS cross section is the largest of any neutrino interaction channel below 100 MeV [16]—as shown in Fig. 2.7.

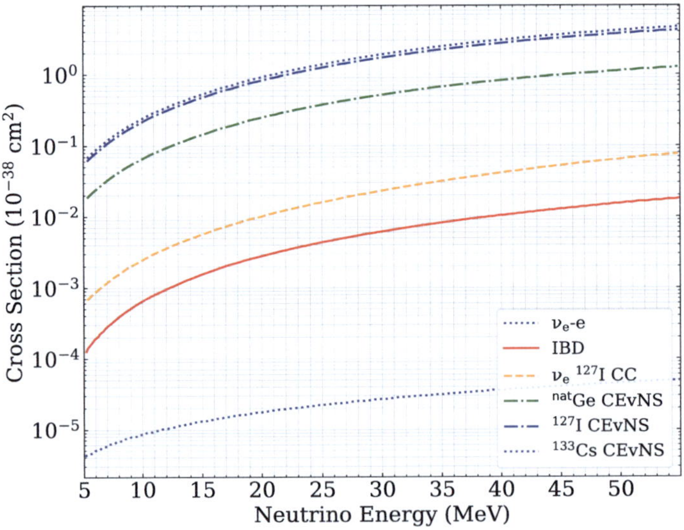

Fig. 2.7 The CEvNS cross section on ^{133}Cs, ^{127}I, and ^{nat}Ge compared to several other pertinent neutrino cross sections below 100 MeV. The cross sections were calculated with Duke CEvNS [17] and SNOwGLoBES [18]

2.2 Nuclear Theory

This work straddles the line between neutrino and nuclear physics given that it is examining a nuclear process instigated by a weak interaction-based neutrino scatter. As such, the nuclear theory side of the equation to presented here.

2.2.1 *Nuclear Structure*

One can connect protons to neutrons via the concept of isospin. Essentially, protons and neutrons can be thought of as two possible states of a nucleon but differ in their respective isospins—neutrons have an isospin of $+1/2$ while protons have an isospin of $-1/2$ in this framework.

The concept of isospin has many profound effects on the properties of the nucleus. Because both nucleons have a half-integer spin (different from their isospin), they must obey the Pauli Exclusion principle, which forbids any nucleon from inhabiting the same state as any other nucleon. However, because of the nucleon's spin, two nucleons may have the same energy as long as they have different signs of their spins—therefore, are not occupying identical states. In this way, the spin allows for degeneracy in each energy state. Isospin builds upon this principle by drawing a difference between neutrons and protons. This additional parameter makes it so that a neutron of the same energy and spin as a proton is not occupying an identical state. Therefore, the Fermi statistics of the neutron do not affect the Fermi statistics of the protons (and vice versa).

These nucleons are held in a bound state by a central potential produced by all the other nucleons in the nucleus. Additionally, these nucleons can be thought of as a gas moving about inside the nuclear volume—this is called the Fermi Gas Model. It consists of a proton gas and a neutron gas. Even in the ground state, these fermions cannot simply condense into the lowest energy state. Rather, they must fill a ladder of progressively greater energy states. These energy states are populated until all the nucleons are placed in a home. The energy of the highest rung of this energy ladder is called the Fermi energy, μ. Correspondingly, the momentum associated with this Fermi energy is the Fermi momentum, p_F [19].

The nucleons of this Fermi gas effectively move around as if they were transparent to each other. This is because a nucleon in a lower energy state can only acquire energy in discrete amounts and the energy states immediately above it are filled. A collision that would result in a nucleon being raised to an energy state that is already filled cannot occur. Therefore, a nucleon-nucleon collision must transfer enough energy to raise said nucleon to an unoccupied energy state. In other words, it must overcome the Fermi energy and raise up to the valence level. Intranuclear collisions rarely have sufficient energy to accomplish this thus rarely happen at all. This behavior is called "Pauli blocking."

There is more structure to the nucleons than the simple stacking of energy states. In fact, plotting the difference between the neutron/proton separation energies and the predicted values of the semi-empirical mass formula (Eq. 2.27) reveals stark similarities with the atomic shell model, where electrons are electrostatically bound to the nucleus and occupy orbital shells that fill at certain discrete levels to optimize stability.

However, to describe the emission of individual particles and their respective energy states, one turns to the nuclear shell model. Clearly, one typically wants to model the nucleus to accommodate the specialties of the liquid drop model (fission, α-decay, γ-ray emission) and the shell model (α-decay, neutron-emission, and γ-ray emission). This amalgamation of the two models is often referred to as the Unified model.

$$H_{\text{unified}} = H_{\text{LD}} + H_{\text{shell}} + H_{\text{interaction}} \tag{2.26}$$

2.2.2 *The Liquid Drop Model*

The liquid drop model can be roughly described as the interplay between the liquid drop's surface tension keeping it together versus the repulsive Coulombic forces between the protons in the nucleus [20]. The strength of the nucleus's surface tension is well described by the surface term in the Bethe-Weizsäcker mass formula shown below [21]:

$$\mathrm{B}(A, Z) = a_V A - a_S A^{2/3} - a_C \frac{Z^2}{A^{1/3}} - a_a \frac{(N - Z)^2}{A} + \delta(A) \tag{2.27}$$

where the coefficients

$$a_V = 15.753 \text{ MeV} \tag{2.28}$$

$$a_S = 17.804 \text{ MeV} \tag{2.29}$$

$$a_C = 0.7105 \text{ MeV} \tag{2.30}$$

$$a_a = 23.69 \text{ MeV} \tag{2.31}$$

$$\delta(A) = \begin{cases} 33.6A^{-3/4} & \text{if N and Z are even} \\ -33.6A^{-3/4} & \text{if B and Z are off} \\ 0, & \text{A = N + A is odd} \end{cases} \tag{2.32}$$

a_S is the surface term. Nucleons on the inside of the nucleus experience a pull in all directions thereby canceling out any effect. Conversely, nucleons on the surface are exclusively pulled from the inside. Therefore, this collective pull on the surface nucleons by the inner nucleons is proportional to the surface area of the drop, $4\pi r^2 \sim A^{2/3}$.

The Coulombic term is informed by the uniformly distributed protons throughout the nuclear volume directly combatting the surface term. Typically, the surface term overpowers the Coulomb repulsion (hence how nuclei are able to exist), but the electrostatic force increases with greater proton number, Z ($Q^2/R \sim Z^2/A^{1/3}$). At higher Z, the Coulomb force can win out even as the nucleus deforms which increases the surface area.

The liquid drop model does not perfectly characterize the binding energy of the nucleus and its divergence can be, in part, described by the shell structure of the nucleus much like the electrons in the atom's electron cloud.

2.2.3 The Shell Model

The nucleons appear to fill shells evidenced by sudden, dramatic increases in binding energy where shell closures occur [22]. The appearance of these shells works to enhance the stability of the nucleus compared to simply considering the terms involved in the Liquid Drop model. The occupancy numbers corresponding to the maximized stability are called "magic numbers" and occur at 8, 20, 28, 50, 82, and 126.

2.2.4 The Theory of Nuclear Fission

Nuclear fission can be conceptualized as a tug-of-war between the surface tension of the liquid drop and the electrostatic repulsion of the constituent charged protons in the nucleus. The dynamics of this enduring nuclear struggle is governed largely by the distortions experienced by the liquid drop as these affect the strengths of both the surface and Coulomb terms. The nucleus can be described as a sphere with distortions characterized by Legendre polynomials:

$$R(\theta) = \frac{R_0}{\lambda}\left(1 + \sum_{n=1} \alpha_n P_n(\cos(\theta))\right) \tag{2.33}$$

where R_0 is the radius of the undistorted sphere, P_n are the Legendre polynomials of order n, and λ is simply a normalization factor to retain a constant nuclear volume. The primary parameter describing the deformation of the sphere are the quadrupole and hexadecapole coefficients, α_2 and α_4. A scan of spherical distortions at various α_2 and α_4 values is shown in Fig. 2.8. One can then represent the potential energy across a mapping in α_2 and α_4 to outline the evolution of a nucleus at varying degrees of distortion—this is called the potential energy mapping. Adding a greater number of these Legendre coefficients provides an increasingly vivid image of a specific nucleus's affinity upon agitation.

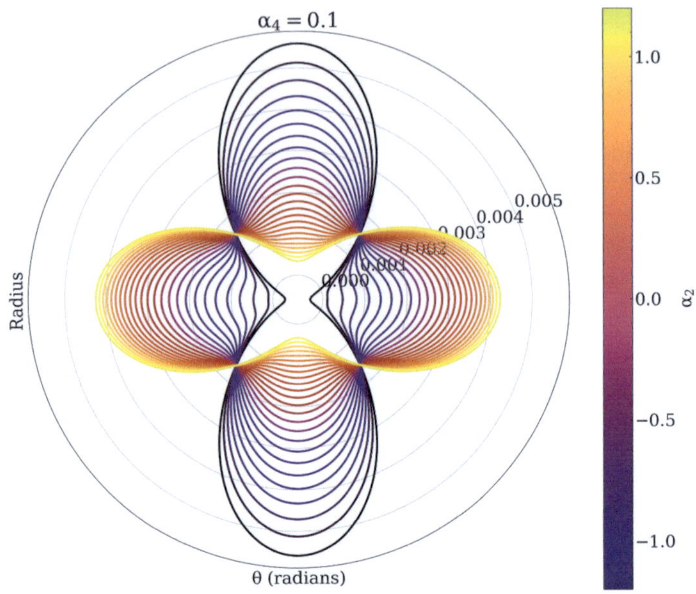

Fig. 2.8 The distortion of a sphere via the varying of the α_2 and α_4 Legendre coefficients. This is the mathematical framework for describing distortions of the nucleus

The point in the potential energy mapping at which the repulsive, Coulomb term becomes dominant is called the saddle point. The energy at the saddle point is called the fission barrier.

2.2.5 *Fission Barriers*

In the potential energy mapping, there exists a most efficient path from the valley associated with the undistorted sphere and the low-lying region where the nucleus splits into two fragments. The inflection point after which there is a descent into the two-fragment valley. An illustration of an arbitrary potential energy mapping with the aforementioned features is shown in Fig. 2.9.

2.2.6 *Barrier Lowering Processes*

Angular Momemtum

When considering nuclear fission induced by thermal neutrons or through spontaneous fission, it is justified to ignore the effects of angular momentum, However, when fission is induced by higher energy particles, it is no longer proper to neglect

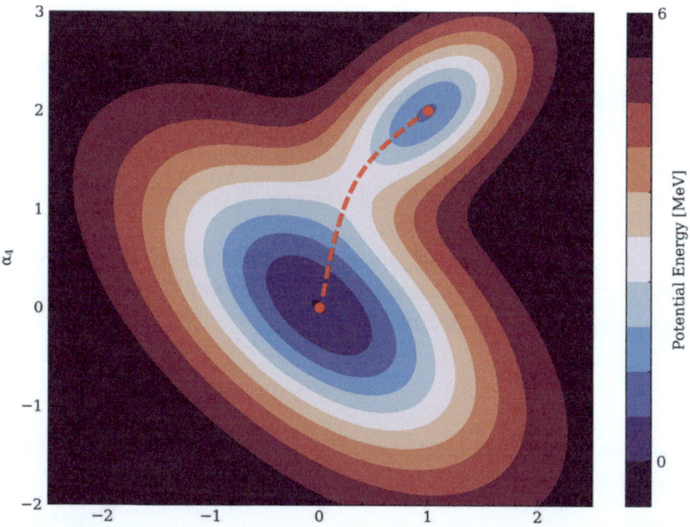

Fig. 2.9 Illustration of a hypothetical potential energy mapping across the two Legendre polynomial coefficients describing distortion and elongation. Overlaid is the most efficient path of distortions from the spherical ground state to the saddle point and then to the valley indicative of the two-fragment fission

the role of angular momentum of the daughter nucleus. When a projectile imparts angular momentum to the target nucleus, the fission barrier is reduced because an additional term must be included in the formula expressing the total change in energy between parent and daughter nuclei.

$$\Delta E_{\text{tot}} = \Delta E_S + \Delta E_C + \Delta E_{\text{rot}} \tag{2.34}$$

where E_S and E_C are the familiar surface and Coulomb terms and E_{rot} is the newly initiated rotational energy term accounting for the fact that the nucleus is deformed by the rotation. This addition effectively lowers the fission barrier. Calculations performed by Kelic et al. [23] show a reduction of the fission barrier by approximately 4 MeV for neutron-rich nuclei excited to roughly 20–30 MeV.

A more widely applicable tool for judging the effect of angular momentum on the fission probability is Eq. 2.35.

$$\frac{\Gamma_F}{\Gamma_N} = 5.25A^{-2/3}T^{-1}\exp\frac{(B_N - B_F - (E_R^I - E_S^I))}{T} \tag{2.35}$$

where E_S^I is the rotational energy at the saddle point, and EI_R is the rotational energy of the residual nucleus after neutron emission, I is the total angular momentum, and T is the nuclear temperature. The addition of angular momentum decreases the fission barrier by (E_S^I - E_R^I).

Shell Wash-out

It is well-established that at higher excitation energies, the fission probability of the compound nucleus is enhanced due to shell effects. According to the Strutinsky approach [24], the fission barrier is:

$$B_\mathrm{f} = B_\mathrm{LD} + B_\mathrm{shell} + B_\mathrm{pair} \tag{2.36}$$

By assuming an even-even nucleus, the pairing term can be neglected. The liquid drop component has a weak dependence on the excitation energy. Conversely, the shell term goes like the ansatz in Eq. 2.37 [25].

$$B_\mathrm{shell}(\mathrm{E}_X) = B_\mathrm{shell}(0)\exp(-\gamma_D \mathrm{E}_X) \tag{2.37}$$

where γ_D is the shell-correction damping parameter. This term has been taken to be $\gamma = 0.064$ MeV and $\gamma = 0.05$ in MeV depending on the context [26].

Fermi Motion

A consequence of treating the nucleus as a Fermi gas with freely moving nucleons is that the movement of the nucleons can be factored into the kinematics of a neutrino scatter. The motion of a nucleon inside the nuclear volume, defined by the Fermi gas model, is called Fermi motion. Rather than considering the collision as a neutrino impacting a stationary object, one must instead acknowledge the chance that the motion of the nucleon can be anti-parallel to the projectile, thereby changing the effective energy of the neutrino according to the expression [19]:

$$\begin{aligned} E_\nu^\mathrm{eff} &= \frac{(s^\mathrm{eff} - M_n^2)}{2M_n} \\ s^\mathrm{eff} &= M_n^2 + 2E_\nu(E_n - k_n\cos(\theta)) \\ E_n &= \sqrt{k_n^2 + M_n^2} \end{aligned} \tag{2.38}$$

where k_n is the neutron momentum and M_n is the neutron mass. As an example, the Fermi motion is applied to the neutrino energy spectrum for the electron neutrinos emitted from a core-collapse supernova is shown in Fig. 2.10. This is a minute correction but would prove interesting for experiments dealing with sub-threshold neutrino-nucleus reactions. In the case of a sub-threshold reaction, a small fraction of neutrinos that are just shy of reaching the energy threshold will be able to successfully initiate the process.

2.2.7 *Final State Particles*

Nuclear fission is marked by an enormous range of products it can create—there are hundreds of possible combinations. A brief summary of the principal categories of

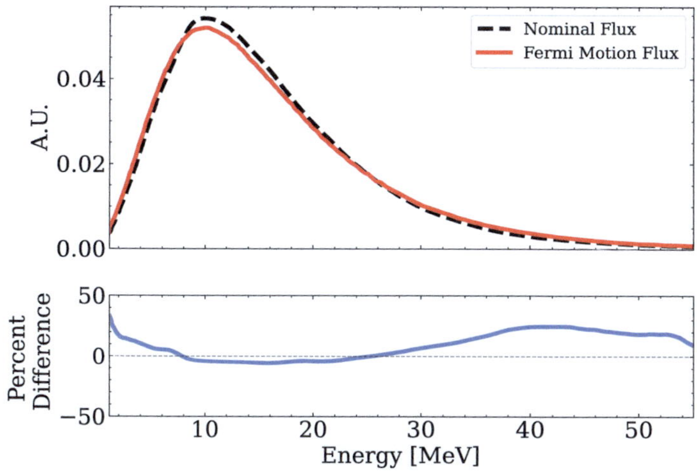

Fig. 2.10 Example of the effect of Fermi motion on the electron neutrino energy spectrum from a core-collapse supernova. There is a slight blurring of the original distribution that enhances the flux of neutrinos are both higher and lower energies. This may have the effect of vaulting a small fraction of neutrinos above an energy threshold that otherwise wouldn't be achieved

fission products is enumerated here. Throughout this exploration of the many fission products, a subset of commonly studied elements recur to indicate the general features of each class of products. These common elements are ^{232}Th, ^{235}U, ^{238}U, and ^{252}Cf.[3].

Fission Fragments

After nuclear fission, the large nucleus explodes into a small number of larger chunks. These chunks are called fission fragments. Typically, there are two fission fragments with a handful of neutrons, but a smaller portion of the time, a third fission fragment is produced in a process called ternary fission. The most common form of ternary fission is the emission of alpha particles along with the large fission fragments. However, for rare-events such as this, the small number of nuclear fission events can be assumed to be a typical binary fission.

The fission fragments each possess roughly half of the protons in the nucleus and as separate entities, are abruptly jettisoned by the immense electrostatic force between the two like-charges. A large sum of kinetic energy is achieved by these fission fragments. A heuristic for the average total kinetic energy for the fission fragments provided in Fig. 2.11. This yields the following expression for fission fragment kinetic energy:

$$\bar{E}_K = 0.1065\left(\frac{Z^2}{A^{1/3}} + 20.1\right) \tag{2.39}$$

[3] These isotopes are commonly of interest for nuclear power, nuclear weapons, and nuclear physics calibrations thus enjoy a large volume of study in many experimental setups.

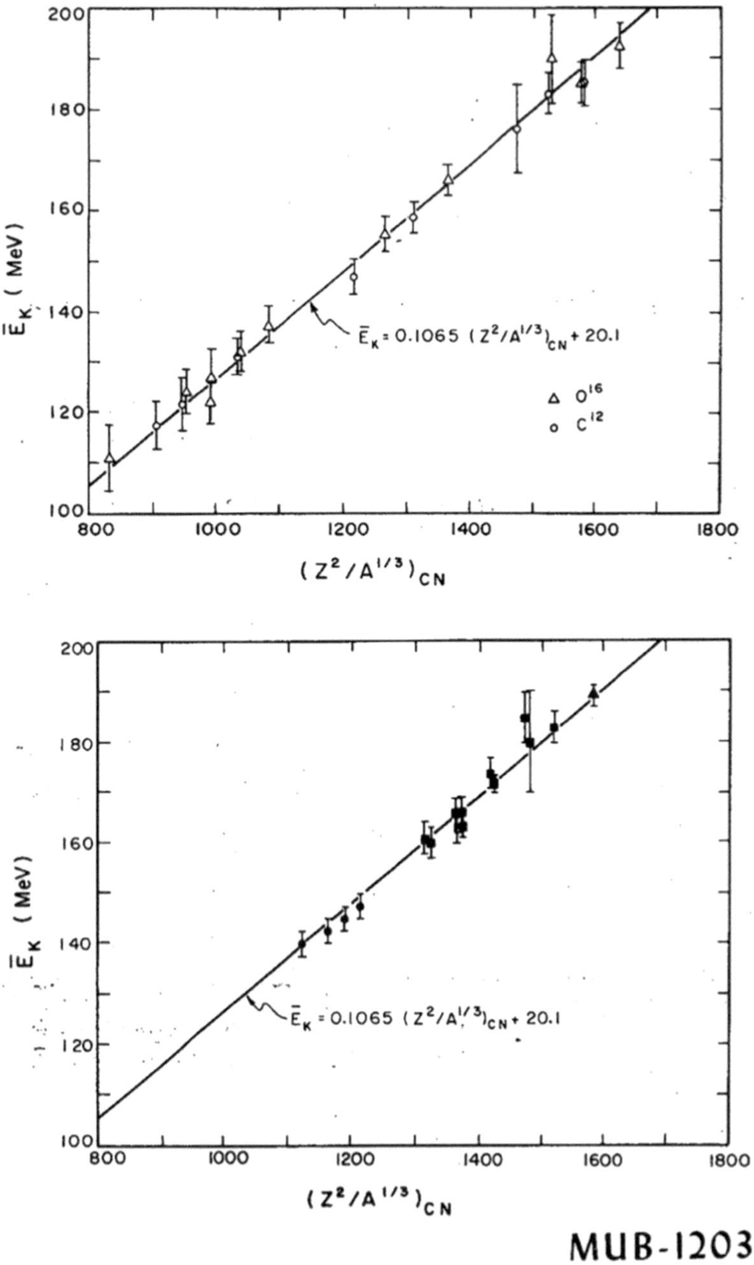

Fig. 2.11 The average total kinetic energy of the fission fragments as a function of the $Z^2/A^{1/3}$ before neutron emission, from Ref. [27]. The upper panel is composed of a combination of 25 MeV ^{3}He-induced fission reactions and spontaneous fission. The lower is composed of fission reactions induced by ^{12}C and ^{16}O

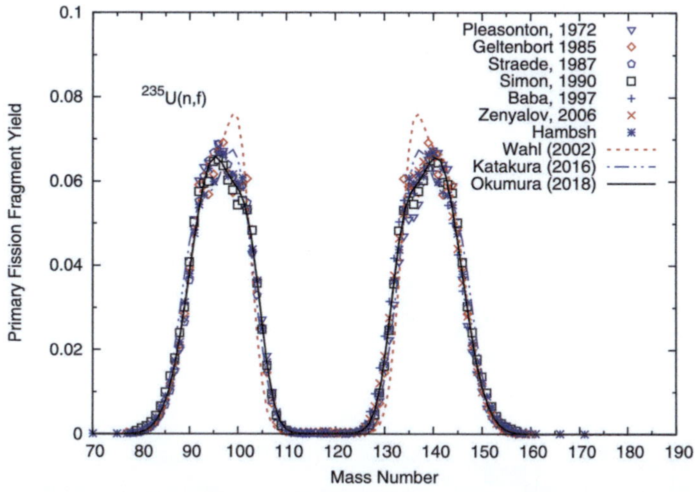

Fig. 2.12 The fission fragment mass distribution exhibited from thermal neutron-induced nuclear fission of ^{235}U, from Ref. [31]. There is a large variety of possible radioisotopes that are produced making it difficult to predict

where $\bar{E}_K$ is the average total kinetic energy of the fission fragments. Additionally, the fragments significantly disrupt the electron cloud of their parent nucleus and carry a high net charge along with them. The net charge induces a myriad of electronic interactions with the atoms they propagate past leading to energy loss through ionization and rapid termination.

When the fission fragments' excitation energy and angular momentum are exhausted leaving them in their ground state, they are then referred to as "fission products" rather than fragments (Fig. 2.12). These fragments are still quite unstable and will more than likely radioactively decay to stability via beta-decay several times [28]. The exact identities of these fission product yields are the subject of immense study for many reasons. A notable motivation for examining the various fission products is that their distributions following the fission of ^{235}U, ^{238}U, ^{239}Pu, and ^{241}Pu determine the expected neutrino flux from nuclear reactors. Discrepancies between modeled reactogenic neutrino spectra and measured values have proliferated several "reactor anomalies" which continue to command attention from experimenters [29].[4]

Prompt Neutrons

After the parent nucleus shatters into fragments, these fragments are likely in a highly excited state. The primary mechanism for exhausting this excitation energy is through the emission of neutrons. As such, the number of prompt neutrons, or

[4] The tension between the modeled nuclear reactor neutrino flux and measured values have been attributed to flaws in the nuclear data or, more provocatively, the existence of a hypothetical fourth-generation "sterile" neutrino that does not couple to matter in the same way as the known three "active" neutrino flavors [30].

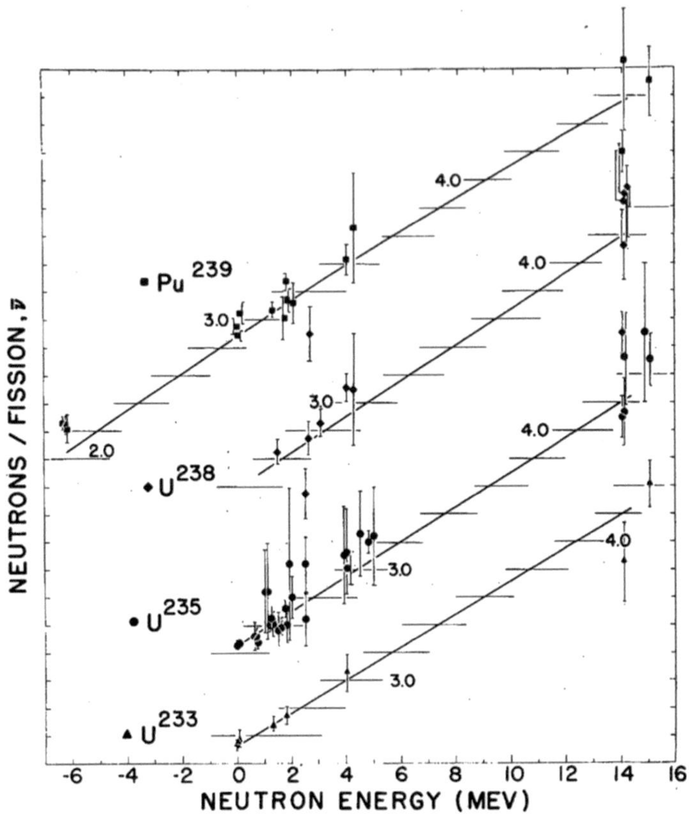

Fig. 2.13 The dependence of the average neutron multiplicity from fission on the energy imparted into the nucleus via the impinging projectile—in this case, a neutron

the multiplicity, is proportional to the extent to which the nucleus is excited—see Fig. 2.13.

Prompt neutron emission terminates when the excitation energy decreases to below the neutron separation energy, S_n.

Prompt γ-rays

Prompt γ-rays exhaust the fission fragments' angular momentum, and to a lesser extent, their kinetic energy. Using ^{235}U as an example, the time difference between the initial nuclear fission and detection of the prompt γ-rays is 2×10^{-8} seconds. The understanding amongst the nuclear physics community is that the majority of the neutron emission has completed by the time the prompt γ-ray emission commences [27]. There is, however, an intermediate time frame in which both neutron and prompt γ-ray emission happen simultaneously and actually compete statistically, $\Gamma_n(E^*)/\Gamma_\gamma(E^*)$. The average total γ-ray energy was measured as 7.2 ± 0.8 MeV per fission and the average number of photons between 0.3 and 10 MeV was

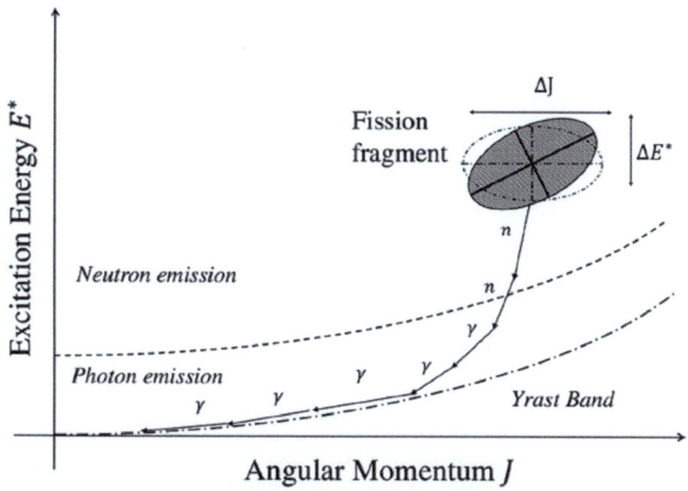

Fig. 2.14 Schematic of how the fission fragments exhaust excitation energy and angular momentum via both neutron and prompt γ-ray emission though with different contexts, from Ref. [33]

7.4 ± 0.8 per fission. Figure 2.14 illustrates how the fission fragments exhaust their excitation energy and angular momentum. This schematic is quite instructive in how it depicts the manner in which prompt γ-ray emission takes over as the main thrust in draining the fission fragments once the Yrast[5] line is reached [33].

The energy spectra from the prompt γ-rays was measured in Ref. [34] emitted from the spontaneous fission of ^{252}Cf and neutron-induced fission of ^{235}U—see Fig. 2.15.

Delayed Products

As previously established in this section, after the neutron-rich fission fragments are depleted of their energy and angular momentum, they are now fission products and highly unstable. Stability can then be reached through a series of beta decays—three per fission product on average. The beta-decays will largely transition to a low-lying excited state or even the ground state in the daughter nucleus. However, a minority of the time, the resultant excited state will lie above the unbound threshold for emitted a neutron—see Fig. 2.16. This kind of neutron emission is dubbed beta-delayed neutron emission because the half-lives of the fission products vary and depend on the decay lengths associated longer timescales of the weak interaction which takes noticeably longer than the other decay avenues present in nuclear fission i.e. the strong

[5] From footnote 1 in Ref. [32]: 'By yrast' level of a given nucleus, at a given angular momentum, is meant the level with least energy at the angular momentum. The English language seems not to have a graceful superlative form for adjectives expressing rotation. Professor F. Ruplin (of the Germanic Languages Department of the State University of New York, Stony Brook) suggested the use of the Swedish adjective *yr* for designating these special levels. This word derives from the same Old Nourse verb *hvirfla* (to whirl) as the English verb *whirl*, and forms the natural superlative, *yrast*. It can thus be understood to mean 'whirlingest,' although literatally translated from Swedish it means 'dizziest' or 'most bewildered"'.

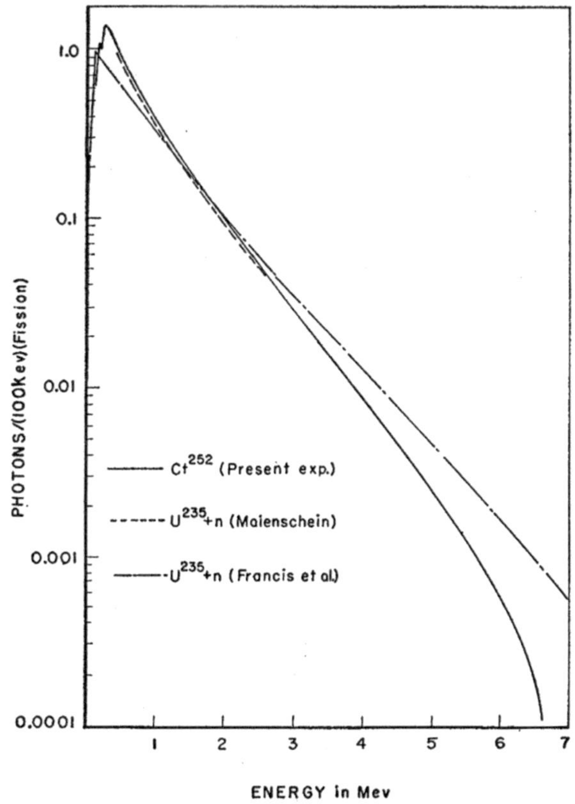

Fig. 2.15 The energy spectrum of prompt fission γ-rays, from Ref. [27]

interaction $\mathcal{O}(10^{-20}$ sec) and the electromagnetic $\mathcal{O}(10^{-10}$sec). Delayed neutrons are quite rare, sitting at approximately 1% of all fission-related neutrons.

Delayed neutrons have an enormous range of progenitor nuclei, also known as neutron precursors. There are too many progenitor nuclei to conveniently treat individually, so instead are grouped into six camps according to their half-lives. The results of the measurements performed in Ref. [36] for ^{238}U and ^{232}Th are shown in Table 2.2. The decay group with the highest probability have half-lives of 1.93 $\pm$ 0.07 and 2.16 $\pm$ 0.08 for ^{238}U and ^{232}Th, respectively. This naturally poses a challenge for experiments that irradiate samples of either isotope at intervals smaller than these half-lives—it would be difficult to ascertain the start time of the reaction if multiple cycles of irradiation have occurred before the delayed product is emitted and detected.

Delayed γ-rays abide by nearly the same underlying mechanism of decay from excited states in the precursor nuclei. Like prompt γ-rays, there are on average seven delayed γ-rays emitted per fission event [37].

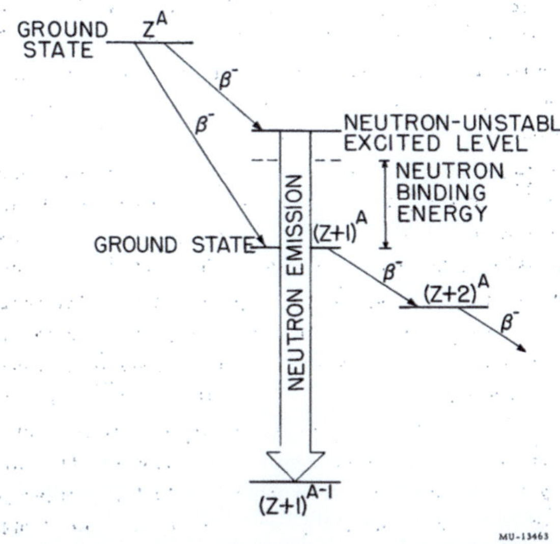

Fig. 2.16 Beta-delayed neutron schematic drawing illustrating the context through which neutrons can be emitted through beta-decay if transitioning to an excited state above the neutron separation energy, from Ref. [35]

Table 2.2 The six half-life groups associated with beta-delayed neutron emission for ^{238}U and ^{232}Th, from Ref. [36]

Group #	Half-life [sec]	Relative abundance [%]
^{238}U		
1	52.38 ± 1.29	0.013 ± 0.001
2	21.58 ± 0.39	0.137 ± 0.002
3	5.00 ± 0.19	0.162 ± 0.020
4	1.93 ± 0.07	0.388 ± 0.012
5	0.490 ± 0.023	0.225 ± 0.013
6	0.172 ± 0.009	0.075 ± 0.005
^{232}Th		
1	56.03 ± 0.95	0.034 ± 0.002
2	20.75 ± 0.66	0.150 ± 0.005
3	5.74 ± 0.24	0.155 ± 0.021
4	2.16 ± 0.08	0.446 ± 0.015
5	0.571 ± 0.0402	0.172 ± 0.013
6	0.211 ± 0.019	0.943 ± 0.006

References

1. Navas S et al (2024) Review of particle physics. Phys Rev D 110:030001. https://doi.org/10.1103/PhysRevD.110.030001
2. Grotz K, Klapdor HV (1990) Weak interaction in nuclear particle and astrophysics. IOP Publishing Ltd
3. Entem DR et al (2025) The constituent quark model. arXiv: 2504.07897 [hep-ph]
4. Engel J, Menéndez J (2017) Status and future of nuclear matrix elements for neutrinoless double-beta decay: a review. Rep Prog Phys 80(4):046301. https://doi.org/10.1088/1361-6633/aa5bc5
5. Martínez-Pinedo G et al (1996) Effective g_A in the pf shell. Phys Rev C 53:R2602–R2605. https://doi.org/10.1103/PhysRevC.53.R2602
6. Engel J (1998) Approximate treatment of lepton distortion in charged-current neutrino scattering from nuclei. Phys Rev C 57:2004–2009. https://doi.org/10.1103/PhysRevC.57.2004
7. Krmpotić F et al (1997) Ikeda sum rule, self-consistency and double-beta decay in the renormalized quasiparticle random phase approximation. Nucl Phys A 612(2):223–238. ISSN: 0375-9474. https://doi.org/10.1016/S0375-9474(97)80010-3, https://www.sciencedirect.com/science/article/pii/S0375947497800103
8. Scholten O, Bertsch GF, Toki H (1983) Gamow-teller strength in the continuum. Phys Rev C 27:2975–2976. https://doi.org/10.1103/PhysRevC.27.2975
9. Anderson BD et al (1990) Gamow-teller strength in the ^{54}Fe(p, n)^{54}Co reaction at 135 MeV. Phys Rev C 41:1474–1485. https://doi.org/10.1103/PhysRevC.41.1474
10. Flanders BS et al (1989) Gamow-teller strength in the ^{208}Pb(p, n)^{208}Bi reaction at 134.3 MeV. Phys Rev C 40:1985–1992. https://doi.org/10.1103/PhysRevC.40.1985
11. Taddeucci TN et al (1987) The (p, n) reaction as a probe of beta decay strength. Nucl Phys A 469(1):125–172. ISSN: 0375-9474. https://doi.org/10.1016/0375-9474(87)90089-3, https://www.sciencedirect.com/science/article/pii/0375947487900893
12. Doering RR et al (1975) Observation of giant gamow-teller strength in (p, n) reactions. Phys Rev Lett 35:1691–1693. https://doi.org/10.1103/PhysRevLett.35.1691
13. Semenov SV (2017) Inelastic scattering of neutrinos on nuclei. Bull Russ Acad Sci Phys 81(6):735–738. https://doi.org/10.3103/S1062873817060193
14. Agnolet G et al (2017) Background studies for the MINER coherent neutrino scattering reactor experiment. Nucl Instrum Methods Phys Res Sect A: Accel Spectrom Detect Assoc Equip 853:53–60. ISSN: 0168-9002. https://doi.org/10.1016/j.nima.2017.02.024, https://www.sciencedirect.com/science/article/pii/S0168900217302085
15. Strauss R et al (2017) The nucleus experiment: a gram-scale fiducial-volume cryogenic detector for the first detection of coherent neutrino–nucleus scattering. Eur Phys J C 77(8):506. ISSN: 1434-6052. https://doi.org/10.1140/epjc/s10052-017-5068-2
16. Akimov D et al (2017) Observation of coherent elastic neutrino-nucleus scattering. Science 357(6356):1123–1126. https://doi.org/10.1126/science.aao0990
17. Scholberg K (2017) Dukecevns: CEvNS-related code repository. https://github.com/schol/dukecevns. Accessed 1 Sept 2024
18. Gaba R et al (2024) Supernova neutrino signal using SNOwGLoBES. In: Jena S et al (ed) Proceedings of the XXV DAE-BRNS high energy physics (HEP) symposium, 12–16 Dec 2022, Mohali, India. Springer Nature Singapore, Singapore, pp 1162–1164
19. Saraswat K et al (2018) Charged current quasi elastic scattering of muon neutrino with nuclei. Indian J Phys 92(2):249–257. ISSN: 0974-9845. https://doi.org/10.1007/s12648-017-1093-0
20. Bethe HA, Bacher RF (1936) Nuclear physics A stationary states of nuclei. Rev Mod Phys 8:82–229. https://doi.org/10.1103/RevModPhys.8.82
21. Weizsäcker CFV (1935) Zur Theorie der Kernmassen. Zeitschrift f ü r Physik 96(7):431–458. ISSN: 0044-3328. https://doi.org/10.1007/BF01337700
22. Heyde KLG (1994) The nuclear shell model. In: The nuclear shell model: study edition. Springer, Berlin, Heidelberg, pp 58–154. ISBN: 978-3-642-79052-2. https://doi.org/10.1007/978-3-642-79052-2_4

23. Kolbe E, Langanke K, Fuller GM (2004) Neutrino-induced fission of neutron-rich nuclei. Phys Rev Lett 92:111101. https://doi.org/10.1103/PhysRevLett.92.111101
24. Strutinsky VM (1967) Shell effects in nuclear masses and deformation energies. Nucl Phys A 95(2):420–442. ISSN: 0375-9474. https://doi.org/10.1016/0375-9474(67)90510-6, https://www.sciencedirect.com/science/article/pii/0375947467905106
25. Denisov VY, Sedykh IY (2018) Calculation of the fission width of an excited nucleus with the fission barrier dependent on excitation energy. Phys Rev C 98:024601. https://doi.org/10.1103/PhysRevC.98.024601
26. Denisov VY, Hofmann S (2000) Formation of superheavy elements in cold fusion reactions. Phys Rev C 61:034606. https://doi.org/10.1103/PhysRevC.61.034606
27. Hyde E (1964) The nuclear properties of the heavy elements. Prentice-Hall Inc
28. Bhatia C et al (2015) Exploratory study of fission product yields of neutron-induced fission of ^{235}U,^{238} U, and ^{239}Pu at 8.9 MeV. Phys Rev C 91:064604. https://doi.org/10.1103/PhysRevC.91.064604
29. Mueller TA et al (2011) Improved predictions of reactor antineutrino spectra. Phys Rev C 83:054615. https://doi.org/10.1103/PhysRevC.83.054615
30. Dasgupta B, Kopp J (2021) Sterile neutrinos. Phys Rep 928:1–63. ISSN: 0370-1573. https://doi.org/10.1016/j.physrep.2021.06.002, https://www.sciencedirect.com/science/article/pii/S0370157321002696
31. Kawano T et al (2023) Fission fragments and fission products. In: Talou P, Vogt R (eds) Nuclear fission: theories, experiments and applications. Springer International Publishing, Cham, pp 141–262. 978-3-031-14545-2. https://doi.org/10.1007/978-3-031-14545-2_2
32. Grover JR (1967) Shell-model calculations of the lowest-energy nuclear excited states of very high angular momentum. Phys Rev 157:832–847. https://doi.org/10.1103/PhysRev.157.832
33. Marin S et al (2021) Structure in the event-by-event energy-dependent neutron-γ multiplicity correlations in ^{252}Cf(sf). Phys Rev C 104:024602. https://doi.org/10.1103/PhysRevC.104.024602
34. Smith AB, Fields PR, Friedman AM (1956) Prompt gamma rays accompanying the spontaneous fission of Cf^{252}. Phys Rev 104:699–702. https://doi.org/10.1103/PhysRev.104.699
35. Ledoux X et al (2015) Delayed neutron measurements of 232Th neutron-induced fission. Ann Nucl Energy 76:514–520. ISSN: 0306-4549. https://doi.org/10.1016/j.anucene.2014.10.012, https://www.sciencedirect.com/science/article/pii/S0306454914005519
36. Keepin GR, Wimett TF, Zeigler RK (1957) Delayed neutrons from fissionable isotopes of uranium, plutonium, and thorium. Phys Rev 107:1044–1049. https://doi.org/10.1103/PhysRev.107.1044
37. Cárdenas ES et al (2015) Comparison of fission signatures from delayed -ray and neutron emissions. Nucl Instrum Methods Phys Res Sect A Accel Spectrom Detect Assoc Equip 792:67–73. ISSN: 0168-9002. https://doi.org/10.1016/j.nima.2015.04.039, https://www.sciencedirect.com/science/article/pii/S0168900215005276

Chapter 3
Neutrinos at the SNS

Abstract This chapter explores the experimental site where the neutrino-induced nuclear fission experiment is deployed. The experiment takes place at Oak Ridge National Laboratory's Spallation Neutron Source (SNS) as a part of the COHERENT collaboration. We start with a brief history of the COHERENT collaboration beginning with the discovery of the collaboration's namesake process of coherent elastic neutrino-nucleus scattering (CEvNS). A brief review follows of the COHERENT detector systems that are directly relevant to the neutrino-induced nuclear fission measurement. These detectors either share the same core technologies as the neutrino-induced nuclear fission instrument or help to constrain shared systematics that impact all neutrino measurements taking place in the SNS facility.

The Spallation Neutron Source (SNS) at Oak Ridge National Laboratory (ORNL) is a premier facility for conducting neutron-related science and is the most intense neutron source in the world, but serendipitously, it is simultaneously an invaluable asset for neutrino physics. A byproduct of the spallation process are neutrinos of several flavors isotropically emitted from the target. The intensity of the SNS beam lends itself nicely to producing a burst of neutrinos. Similar facilities are coming online around the world with burgeoning neutrino programs to pursue, in some cases, very complementary neutrino physics searches [1–3].

3.1 The COHERENT Experiment

The COHERENT Neutrino Program precipitated from the initial mission of discovering coherent elastic neutrino-nuclear scattering (CEvNS) using the SNS neutrino flux [4]. This multi-pronged initiative takes on the grand challenge of detecting CEvNS by deploying a suite of detector sub-systems geared towards parallel and symbiotic studies. The divisions of the COHERENT program are permeable and interconnected but categorized into CEvNS detectors, inelastic neutrino-nucleus detectors, and background monitoring detectors (Figs. 3.1, 3.2 and 3.3).

T. Johnson, *The First Indication of Neutrino-Induced Nuclear Fission*,
https://doi.org/10.1007/978-3-032-01936-3_3

Fig. 3.1 The Spallation Neutron Source (SNS) facility at Oak Ridge National Laboratory (ORNL). The COHERENT collaboration is established in the basement corridor

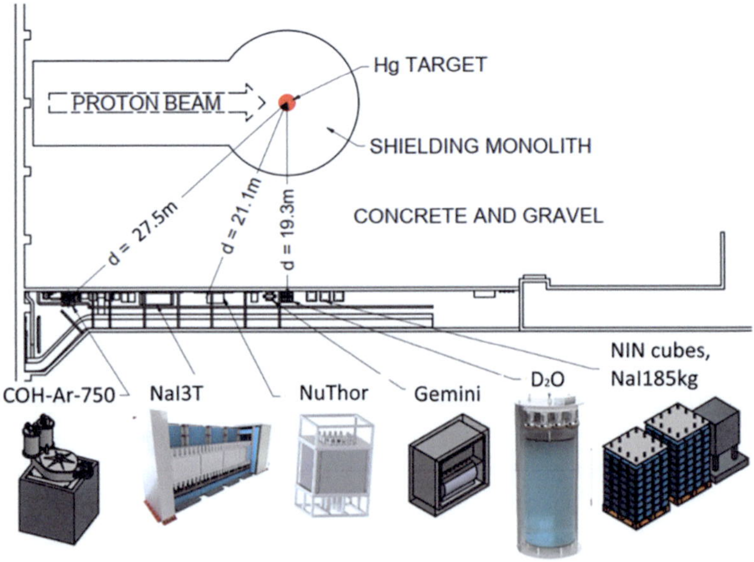

Fig. 3.2 Schematic of Neutrino Alley in the basement corridor of the Spallation Neutron Source. Several of the deployed neutrino detectors in the COHERENT detector suite are indicated in their positions in the space. Also shown in the Figure is the direction of the proton beam parallel to the length of Neutrino Alley. The liquid mercury target is as little as 19.3 m away from the walls outlining the access corridor. From the COHERENT Collaboration [5]

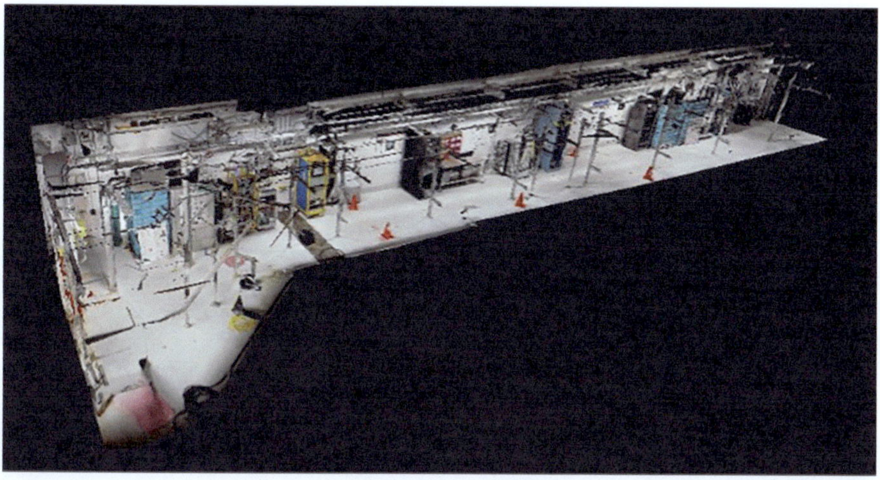

Fig. 3.3 A 3D map of the Neutrino Alley in Oak Ridge National Laboratory's Spallation Neutron Source. This is a previous configuration of Neutrino Alley. An updated 3D scan would depict many new detector sub-systems. From the COHERENT Collaboration

3.2 Neutrino Alley

The foundation of the COHERENT neutrino program is the facility operated in the basement corridor of the SNS called "Neutrino Alley" [6]. The importance of the Neutrino Alley stems from the relatively low neutron background. The large quantity of concrete and gravel backfill between Neutrino Alley and the SNS target efficiently attenuates the neutrons to a manageable level for a low background, rare-event search. Despite Neutrino Alley's neutron quietness, there is still a non-zero leakage of high-energy neutrons originating from the SNS target that are referred to as Beam-Related Neutrons (BRNs).

A deeper discussion of BRNs will ensue in a later section, but relevant to the critical importance of Neutrino Alley is that the timing of the BRNs in this highly shielded space emulates the timing structure of the protons striking the target. At less optimal spaces in the SNS, not only is the total number of BRNs greatly increased, but the timing of the neutrons also bleeds into the interstitial space between beam spills which directly and adversely affects the quality of neutrino-related searches that depend on knowing the timing structure of the neutrinos.

There is a modest 8 meters-water-equivalent overburden to help reduce the background imposed by the cosmic muon flux. Furthermore, Neutrino Alley greatly benefits from a high neutrino flux because it is located a mere 20 meters away from the SNS target.

Since the Neutrino Alley is a repurposed service corridor, the space usage is severely constrained to fit within one meter from the wall to not restrict movement in case of emergency. So all neutrino detectors must fit within this small footprint. This is particularly ironic because neutrino detectors tend to be large mass and even fill kiloton-sized caverns [7–9].

3.2.1 The CEvNS Program

The CEvNS detectors in the COHERENT suite are specially selected to value low energy thresholds to reliably measure the minuscule observable for a CEvNS interaction, a nuclear recoil. The region-of-interest in energy is often in the few keV range. It is an ongoing effort to lower the energy thresholds and increase the energy resolutions of these technologies [10–12].

Studying CEvNS presents a singular opportunity to probe the Standard Model with high precision, because the CEvNS cross section boasts such a low theoretical uncertainty. An experiment with well-understood and small uncertainties would be able to explore possible deviations from the Standard Model predictions and imply certain emergent concepts like new weak interaction mediators besides the typical Z-boson or an anomalous neutrino magnetic moment [13, 14].

CsI CEvNS Discovery

The inaugural CEvNS detector in the COHERENT program was the 14.6 kg CsI[Na] scintillator deployed in 2017 with an energy threshold of 6.5 keVnr [15].[1] After its initial campaign, the data was unblinded and analyzed, finding a rejection of the null-hypothesis with 6.8 sigma—thus becoming the world's first observation of CEvNS and kicking off a worldwide field around this simple, but powerful low-energy neutrino process.

3.2.2 Continuing the CEvNS Legacy

The CEvNS interaction cross section can be well described with the following expression:

$$\frac{d\sigma}{d\cos\theta} = \frac{G^2}{8\pi}[Z(4\sin^2\theta_W - 1) + N]^2 E^2(1 + \cos\theta) \tag{3.1}$$

Where Z is the proton number, N is the neutron number, θ_W is the weak mixing angle, G is the weak coupling constant, E is the neutrino energy, and θ is the scattering angle. It just so happens that $4\sin^2\theta_W$ is approximately equal to 1 so the Z

[1] keVnr refers to energy of the nuclear recoil. This distinction is made because radiation detectors often have a "quenching factor" in which a nuclear recoil will produce a diminished observable signature compared to an electronic recoil of the same energy.

term essentially cancels out leaving behind just the neutron number term. Because the CEvNS cross section is essentially proportional to the number of neutrons in the target nucleus squared, detecting CEvNS using a range of nuclear targets allows for experimenters to view the breadth of the Standard Model prediction to examine possible deviations from theory that might indicate new physics. This partially motivates the COHERENT experiment deploying multiple neutrino detectors of varying target materials in parallel (See Fig. 3.4).

Following the discovery, COHERENT successfully measured CEvNS on both liquid argon [17] and germanium [18]. A synopsis of the COHERENT CEvNS efforts is shown in the following Table 3.1.

The CEvNS detector in the COHERENT suite that is most related to the work reported in this thesis is the NaI effort, so a brief description of its physics goals and instrument are detailed in the following section.

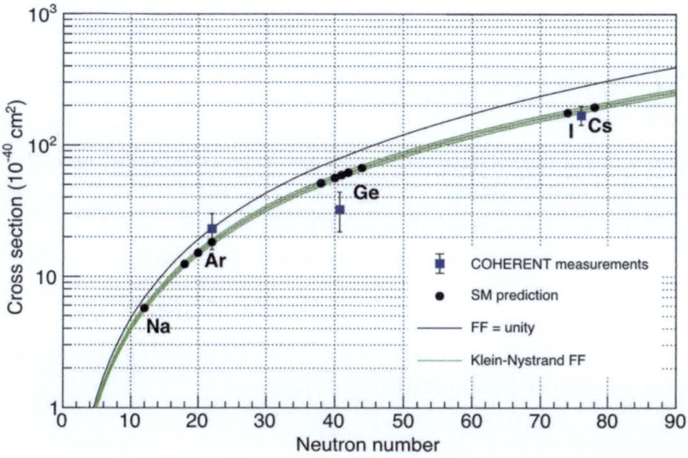

Fig. 3.4 The N^2 dependence of the CEvNS cross section with the current three CEvNS measurements overlaid to indicate adherence to the predicted Standard Model cross sections [16]

Table 3.1 COHERENT's currently deployed or previous CEvNS measurement campaigns along with their respective technologies and the mass of their active detector material

Detector sub-system name	Technology	Mass (kg)
CsI	CsI[Na] Scintillator	14.6
Ar	Single-Phase Liquid Argon	24
NaI	NaI[Tl] Scintillator	2,425
Ge	P-type Point Contact	16

3.2.3 NaI

The lightest nucleus studied by a COHERENT CEvNS detector is sodium with the NaI ton-scale experiment or NaIvETe (depicted in Fig. 3.5) [19]. NaIvETe is composed of a large set of NaI[Tl] scintillator detectors repurposed for a CEvNS search. These crystals were not manufactured for a neutrino rare-event search like other NaI-based experiments like NEON [20] or DAMA [21]. As such, there is a sizable radioactive presence which inhibits discovery potential until a large target mass is accumulated. The commissioning, deployment and analysis is currently underway at the time of this writing.

Because the sodium nucleus is an odd-even nucleus with an unpaired proton, the CEvNS rate is influenced by the weak-axial vector coupling constant, g_A. Thus, a precision measurement of the ^{23}Na CEvNS rate provides information on this coupling constant that also comes into play for inelastic neutrino-nucleus interactions.

3.2.4 INCOHERENT Program

A burgeoning wing of the COHERENT collaboration is the set of inelastic neutrino-nucleus scattering measurements conducted over a range of detector sub-systems deployed in Neutrino Alley. Several of these detectors are quite informative towards the goals of this work, and are therefore enumerated here to summarize their respective purposes, results and implications to the neutrino-induced nuclear fission mission.

A brief synopsis of the significance of each INCOHERENT detector towards the neutrino-induced nuclear fission initiative ensues:

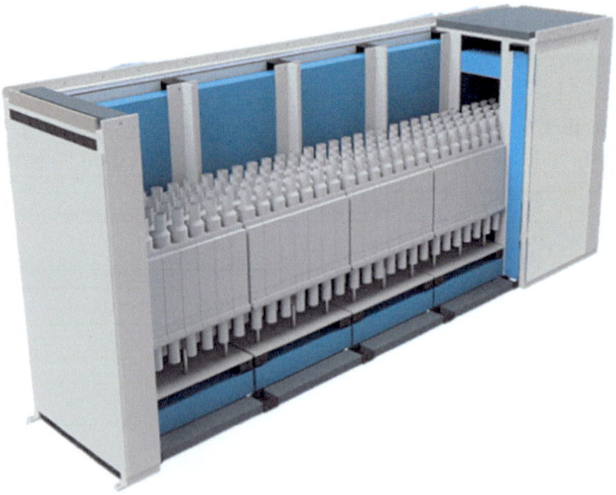

Fig. 3.5 The COHERENT NaIvETe detector CAD design of 5 modules totaling approximately 2425 kg of instrumented NaI scintillating crystals. Courtesy of Dr. Sam Hedges from the COHERENT Collaboration

Table 3.2 COHERENT's currently deployed or previous Inelastic Neutrino-Nucleus campaigns along with their respective target nucleus and the mass of the target nucleus in the apparatus. The best fit amplitude of the NINs population was notably smaller than the prediction value

Detector sub-system name	Target nucleus	Mass (kg)
Lead neutrino cube	Lead	900
Iron neutrino cube	Iron	700
NaIvE-185	Sodium & Iodine	185
D_2O	Deuterium & Oxygen	600
NuThor	Thorium	52

- The Lead Neutrino Cube (and the Eljen cell detector) made strides towards measuring the cross section of neutrino-induced neutron (NIN) emission [22] on natural lead. Ultimately, the cross section was smaller than expectations leading to a limit on the cross section and invaluable information about the predictions yielded by MARLEY [23].
- NaIvE-185 furthered the campaign of measuring neutrino-nucleus cross sections on larger nuclear systems. Although not as severe as lead, the measured cross section of ^{127}I returned lower than theoretical predictions leading to further speculation of a atomic mass dependent quenching of the cross section.
- The D_2O detector greatly benefits all detectors in the COHERENT detector suite by reducing the uncertainty in the neutrino flux as it continues to operate and accumulate statistical precision (Table 3.2).

3.2.5 *Eljen Cell*

Prior to the construction and deployment of the Neutrino Cubes[2], two EJ-301 liquid scintillator cells dubbed the "Eljen cells" were deployed in the shielding package of the CsI[Na] detector that were used to make the first CEvNS measurement. The purpose of the Eljen cell detector was to measure the in-situ background induced by beam-related neutrons (BRNs) and neutrino-induced neutrons (NINs). The CsI[Na] shielding features 2.2 tons of lead that are all targets for electron neutrinos originating from the SNS liquid mercury target to produce NIN events. Because the Eljen cell is a liquid scintillator detector, the neutrons observed by it are identified using pulse shape discrimination to distinguish between the elastic scatting from the neutrons and the electromagnetic scattering from photons which leverages the differences in the integrated charge at different portions of the recorded waveforms read out from the scintillators.

[2] Dedicated neutrino-induced neutron emission detectors that will be discussed in the subsequent section.

The parameters used to describe the power law model for the beam-related neutrons interacting with the volume inside of the CsI[Na] shielding package was bench marked using the data extracted from the Eljen cells and demonstrated good agreement between the model and data. The Eljen Cell observed a modest NIN population. A non-zero lead NIN observation is favored at 2.9σ.

3.2.6 Neutrino Cubes

The Neutrino Cubes (Nubes) were dedicated neutrino-induced neutron (NIN) detectors intended to assess the impact the neutrons may have for CEvNS searches. As such, the Neutrino cubes were composed of common shielding materials: 900 kgs of lead and 700 kgs of iron. 3D rendering of the lead neutrino cube is shown in Fig. 3.6. Both are sources of NIN backgrounds with the potential to mimic the nuclear recoil signature of CEvNS events. The NINs may do so by propagating to the CEvNS detector and elastically scattering in the medium. Benchmarking the magnitude of this background is especially important because this background is coincident in time with the neutrino beam pulse and cannot be treated easily like the steady-state background (Fig. 3.7).

The Lead Neutrino Cube detector collected 62.3 GWHr of beam data at the SNS before unblinding the dataset. Using a combination of charge-exchange (p,n) scattering measurements [24] and the MARLEY neutrino event generator tool, 346 NIN events were predicted to be observed in the dataset. Despite this, the best fit value for the number of NINs was 37^{+72}_{-37} NIN counts [25].

When combined with the result from the Eljen Cell detector, the extent that the theoretical calculations over predict the measured NIN rate are depicted in Fig. 3.8. The joint fit of the two datasets results in a cross section scaling factor of $0.29^{+0.17}_{0.16}$

Fig. 3.6 A CAD model of the Neutrino Cube geometry. (Left) the façade of the detector where the outer water shielding is visible (Middle) the muon veto panels inside the water shielding (Right) the cast lead with openings to house the four liquid scintillator detectors

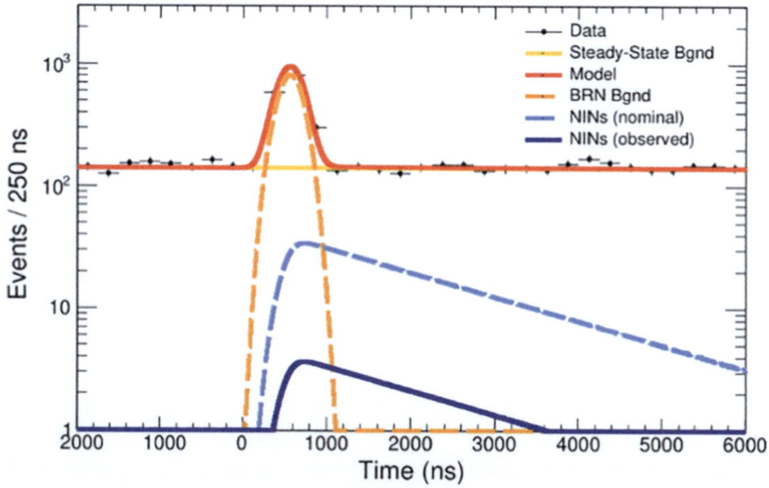

Fig. 3.7 The maximum likelihood fit in time for the Pb Neutrino Cube's unblinded dataset. The best fit value for the NIN counts was 37^{+72}_{-37} which is consistent with no NINs. From the COHERENT Collaboration [25]

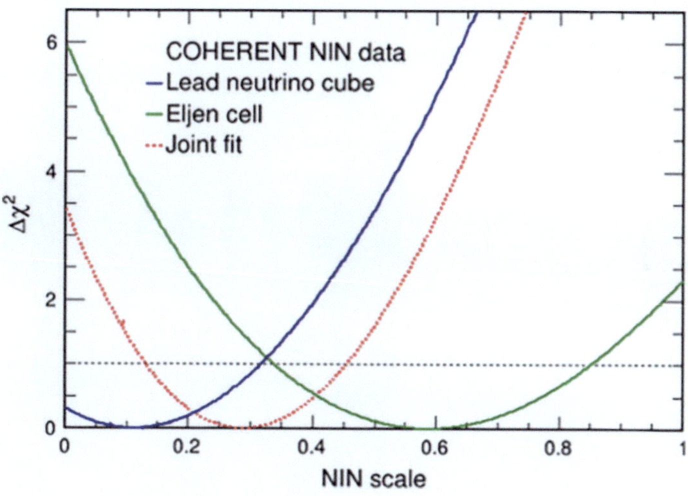

Fig. 3.8 The likelihood profiles for the Lead neutrino cube, Eljen cell and the joint fit of the two detectors. One corresponds to the predicted NIN rate calculated with MARLEY. The joint analysis yields a best fit scaling factor of $0.29^{+0.17}_{0.16}$. From the COHERENT Collaboration [25]

compared to the predicted interaction rate. This result will be leveraged for making lead NIN background predictions in this work. Prior to this work, lead was the heaviest nucleus used in a neutrino measurement campaign in this energy regime and this result may suggest a systematic reduction in the neutrino-nucleus cross sections for increasingly large nuclei.

3.2.7 NaIvE-185

The NaIvE-185 detector is a charged-current detector composed of 24 7.7 kg NaI[Tl] scintillating crystals. These crystals were inherited from the Department of Homeland Security's Advanced Spectroscopic Portal Program. Immediately surrounding the NaI[Tl] scintillators is a $^1/_2$ inch thick steel shield designed to reduce the backgrounds posed by primarily the hot-off gas (HOG) pipe carrying radioactive gas from the SNS target through Neutrino Alley emitting a perpetual stream of 511 keV gamma-rays during beam operation. A muon veto was constructed from plastic scintillator panels and was instrumented with a number of photomultiplier tubes for readout in order to reduce the contamination of the measured signal by cosmic radiation (Fig. 3.9).

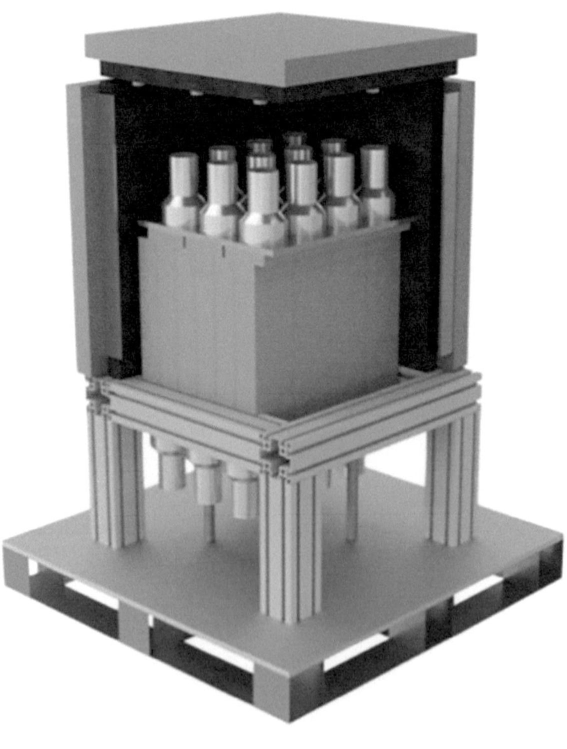

Fig. 3.9 A 3D rendering of the NaIvE-185 detector showing the configuration of the 24 7.7 kg NaI[Tl] scintillators [26]

The NaIvE-185 detector operated 18.7 meters from the SNS target and accumulated an integrate beam exposure of 22.7 GWhrs. After a maximum likelihood fit was performed on the processed data, the best fit value for the number of observed ^{127}I charged-current interactions was determined to be 546^{+121}_{-108} [26]. This represents a 5.6σ rejection of the null hypothesis and an affirmative observation of this neutrino-nucleus interaction. The reported cross section is $9.2^{+2.1}_{-1.8} \times 10^{-40}$cm^2 which amounts to approximately 40.9% of the theoretical prediction of the cross section computed with MARLEY. Of particular note here is that the predicted number of events was approximately 1,300 and was the product of inputs from data-driven matrix elements derived from (p,n) charge-exchange measurements performed at particle accelerators [27] and the event generation software MARLEY.

The tension between the predicted and observed cross sections on ^{127}I is still subject to a great deal of discussion at the time of this work, though there might be deeper implications for all charged-current neutrino-nucleus scattering experiments at this energy regime. In reaction to these measurements, study has commenced to diagnose the discrepancy between experiment and theory [28]. That phenomenological work suggests that a quenched value of weak axial-vector coupling constant, g_A, could at least partially account for observations. A quenched g_A could then be implemented into the cross section formulation by charged-current form factor, F_A^{CC}. The NaIvE-185 and Lead Neutrino Cube results indicating lower measured cross sections than predicted may indicate that the neutrino-induced nuclear fission cross section on thorium could be suppressed as well (Fig. 3.10).

3.2.8 D_2O

The D2O detector is a highly anticipated detector subsystem for the entirety of the COHERENT detector suite because it will be instrumental for bettering the precision of our neutrino flux estimates [29]. A 3D rending of the D2O detector is shown in Fig. 3.11. At present, the uncertainty associated with the neutrino flux normalization is the largest shared uncertainty for all neutrino detectors deployed to the SNS. This represents as major barrier to conducting precision physics with both the CEvNS detectors and, to a lesser extent, the inelastic detectors. At present, the systematic uncertainty on the neutrino flux normalization is $\sim$10% and stems from the estimate of the fraction of protons colliding with the liquid mercury target that yields a pion (0.09 pions per proton) at the high energies of the SNS beam ($\sim$1 GeV), and is based on the simulation work performed by the COHERENT collaboration for the nominal beam configuration. However, because the neutrino capture cross section of deuterium has a minuscule 2–3% theoretical uncertainty, measuring charged-current events in heavy-water will allow the uncertainty of the neutrino flux normalization to be diminished to a comparably low 2–3% with enough beam exposure [30]. Figure 3.12 illustrates the progression of the statistical precision with SNS beam exposure.

The improved systematic uncertainty on the neutrino flux from the SNS will be helpful in diminishing the overall uncertainty budget for all neutrino detectors.

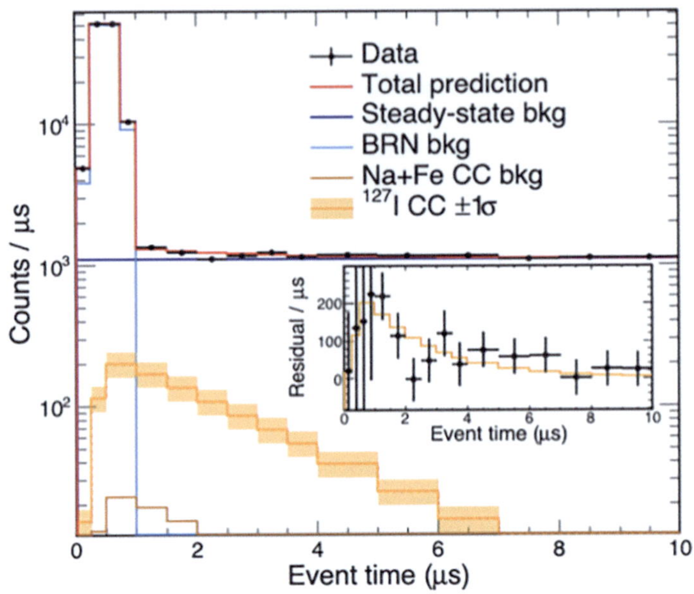

Fig. 3.10 The maximum likelihood fit in time from the data analyzed from the NaIvE-185 detector. The major components in this time fit at the beam-related neutrons (BRNs) in the prompt region from 0-1 microseconds, the steady-state backgrounds primarily from the hot off gas pipe and cosmic muons that were unsuccessfully tagged by the muon veto system, and the events originating from the charged-current interactions of ^{127}I (as well as much smaller components from Na and Fe) [26]

Although this will have a lesser effect on the inelastic neutrino detectors (like a neutrino-induced nuclear fission detector) because their primary uncertainties remain to be the cross section uncertainty stemming from the complexity of the nuclear physics.

3.3 Neutrino Production at the SNS

In addition to being the world's most intense pulsed source of neutrons, the Spallation Neutron Source is an invaluable tool for neutrino physics because it simultaneously produces a pulsed, isotropic supply of neutrinos. A series of four conventional and then superconducting linear accelerators brings packets of protons to approximately 1 GeV. Subsequently, these protons are sent into a storage ring where they will be circulated numerous times in which the roughly 1 millisecond long proton packets are constricted into 695 nanosecond packets for a sharp, high-power collision with the downstream target. The SNS target is made of liquid mercury and as a liquid metal is able to be its own coolant which contributes significantly to a long-lived and stable system (Figs. 3.13, 3.14 and 3.15).

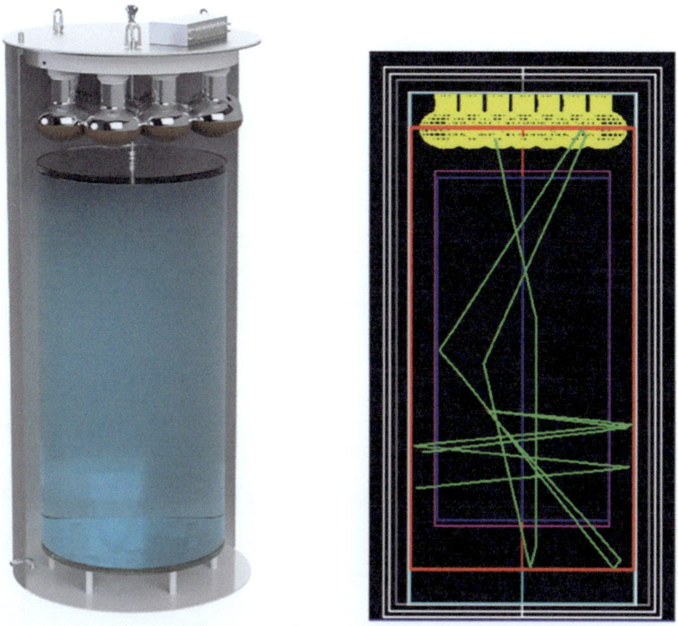

Fig. 3.11 (left) a 3D rendering of the D2O detector illustrating the inner volume filled with 600 kgs of heavy-water (right) A Geant4 event display depicting the scattering of light through the detector volume originating in the heavy-water inner volume and eventually reaching a photomultiplier tube at the top of the apparatus [29]

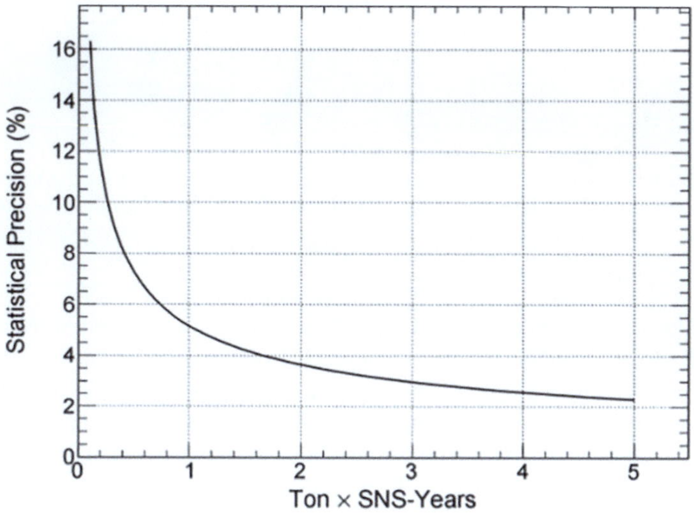

Fig. 3.12 The statistical precision of the SNS neutrino flux normalization as a function of the beam-exposure of the D2O detector. The nominal uncertainty goal of 2% is projected to be achieved in approximately 5 Ton x SNS Years [29]

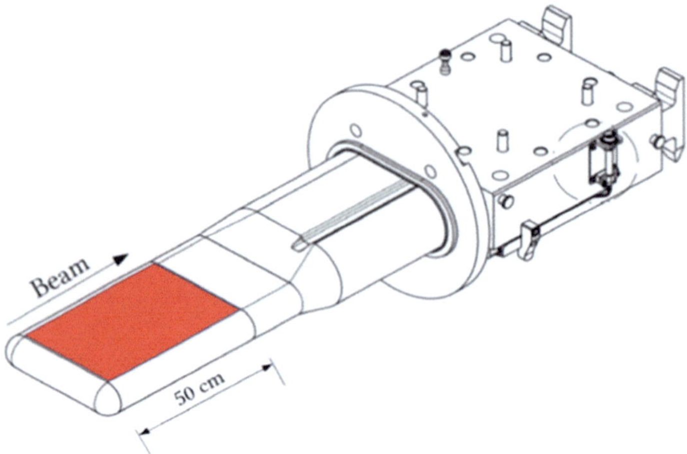

Fig. 3.13 Engineering drawing of the SNS liquid mercury target [31]. The liquid mercury acts as both the target and the coolant

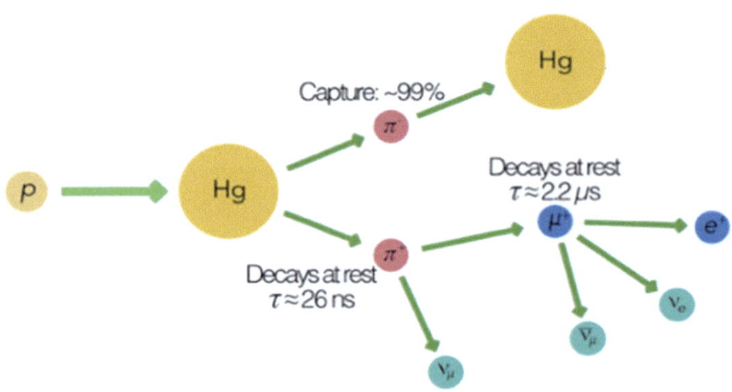

Fig. 3.14 The neutrino production process for the Spallation Neutron Source (SNS), a pion decay-at-rest (π Dar) source [32]. The different half-lives of the pion and muon result in the characteristic timing profile of the SNS neutrino pulse

The neutrinos are produced as a by-product of the pions emitted in the aftermath of the high energy protons striking the mercury. Both positive and negative pions are produced but the vast majority of the negative pions are captured by the ubiquitous nuclei. Conversely, the positive pions repeatedly scatter and eventually come to rest in the mercury where they will then decay, with a 26-nanosecond half-life, into both positive muons and muon neutrinos. These muon neutrinos are dubbed the "prompt" neutrinos. The positive muons will then decay with a half-life of 2.2 microseconds into positrons, electron neutrinos and muon anti-neutrinos. These two latter neutrino flavors are called the "delayed" neutrinos. A pictorial schematic is shown in the figure below.

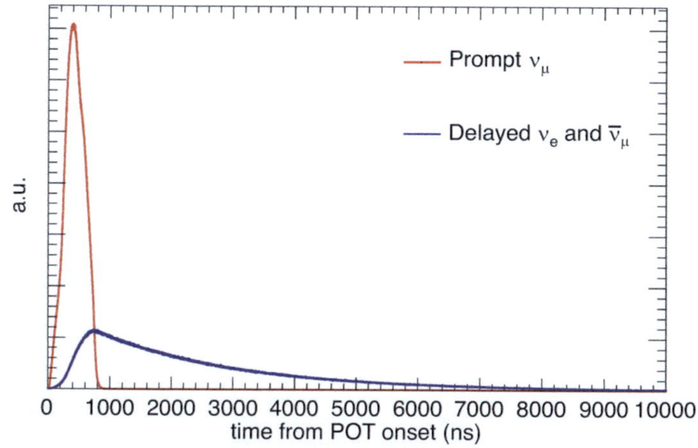

Fig. 3.15 Timing distribution of the prompt and delayed neutrino pulses at the SNS. The prompt pulse consists of ν_μ's from the two-body decay of the pion. The short half-life leads to the timing distribution of the prompt pulse to closely mirror the timing distribution of the protons-on-target. The delayed neutrino pulse is comprised of equal parts ν_e's and $\bar{\nu_\mu}$'s from the three-body decay of the muon. The comparatively long half life of 2.2 microseconds leads to an elongated timing profile. From the COHERENT Collaboration [33]

The neutrinos emanating from the SNS can be categorized into two camps based on their respective timing structure. The exponential decays associated with the decay of the pion (muon) for the prompt (delayed) neutrinos is then convolved with the timing distribution of the protons-of-target (POT) trace. The products of the convolution of the exponential decays and average POT trace are shown below.

In addition to having differing timing profiles, the neutrino flavor contributions also differ in their energy distributions based on the radioactive decays that begat them. Because the muon neutrinos in the prompt pulse are produced via the two-body decay of the pion, their resultant energy is a mono-energetic peak centered at 29.79 MeV. This is easily calculated with [34]:

$$\mathrm{P}(E_\nu|\nu_\mu) = \delta\left(E_\nu - \frac{m_\pi^2 - m_\mu^2}{2m_\pi}\right) = \delta(E_\nu - 29.79\mathrm{MeV}) \tag{3.2}$$

Here, E_ν is the neutrino energy, m_π is the mass of the pion, and m_μ is the mass of the muon. However, the delayed neutrinos (electron- and muon anti-neutrinos) are produced in a three-body decay. Their respective energies can be computed with the following expression:

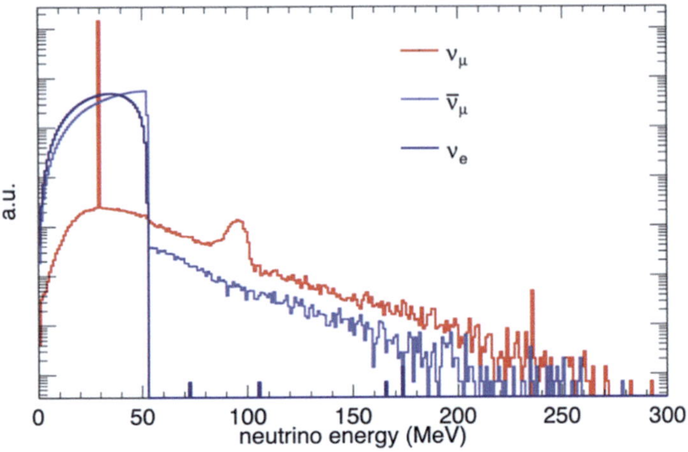

Fig. 3.16 The energy distributions of the three neutrino flavors emitted from the SNS target as simulated in GEANT4 to propagate the production through the SNS target and target assembly. From the COHERENT Collaboration [31]

$$\mathrm{P}(E_\nu|\nu_e) = \frac{12}{W^4}E^2(W - E_{nu}) \tag{3.3}$$

$$\mathrm{P}(E_\nu|\overline{\nu}_\mu) = \frac{6}{W^4}E_\nu^2\left(W - \frac{2}{3}E_\nu\right) \tag{3.4}$$

where W = 52.83 MeV. These three energy distributions generated from a Geant4 simulation [35] (as well as the decay-in-flight component) are shown in Fig. 3.16.

Because the basis of the work is firmly situated on the electron neutrinos, it is worth viewing this energy distribution—which is also called the Michel spectrum—in isolation. The endpoint of this energy distribution is approximately 52 MeV (Fig. 3.17).

3.3.1 *Beam Power and Energy*

For a sustained period of time prior to this work, the SNS proton beam was typically operated around a beam power of 1.4 MW and beam energy of 1 GeV which according GEANT4 simulations, yields 2.36 x 10^{15} neutrinos per second [31]. The relationship between the beam operation conditions and the expected neutrino flux is well encapsulated by the parametric formula:

$$\frac{\nu}{t} = P\left(p_3E^2 + p_2E + p_1 + \frac{p_0}{E}\right) \tag{3.5}$$

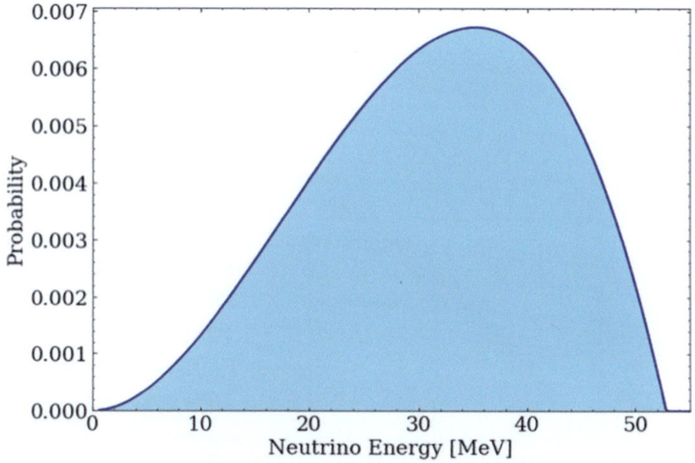

Fig. 3.17 The nominal shape of the Michel electron neutrino energy spectrum with an end point of 52.8 MeV

Table 3.3 SNS target fit parameters to model the neutrino flux as a function of the beam energy and power. These parameters were extracted when the target's beam window was simulated as aluminum

Fit parameter	Aluminum PBW	Inconel PBW
$p_3(\text{GeV}^{-3})$	0.28(2)	0.27(2)
$p_2(\text{GeV}^{-2})$	−1.12(6)	−1.09(6)
$p_1(\text{GeV}^{-1})$	1.79(6)	1.75(5)
p_0	−0.68(2)	−0.67(2)
b	99.99(1)	100.04(1)
$m(\text{GeV}^{-1})$	−0.48(1)	−0.53(1)

where E is the beam energy, P is the beam power, and the p_x's are parameters organized in Table 3.3. Note that the parameters shown here are produced when simulating the proton beam window (PBW) is made of aluminum rather than Inconel.

Additionally, there was a relatively long beam outage between the summer of 2023 to the summer of 2024 for the Proton Power Upgrade (PPU) in which the SNS beam was prepared to reach even greater beam powers which bring with it even greater neutrino luminosities. The beam current is planned to increase from 26 to 38 mA where the neutrino flux should increase linearly with this improvement. Figure 3.18 succinctly illustrates the relationship between beam operating values and the neutrino flux emitted.

This parametric approach for estimating the neutrino flux is adopted for the signal predictions presented later in this work.

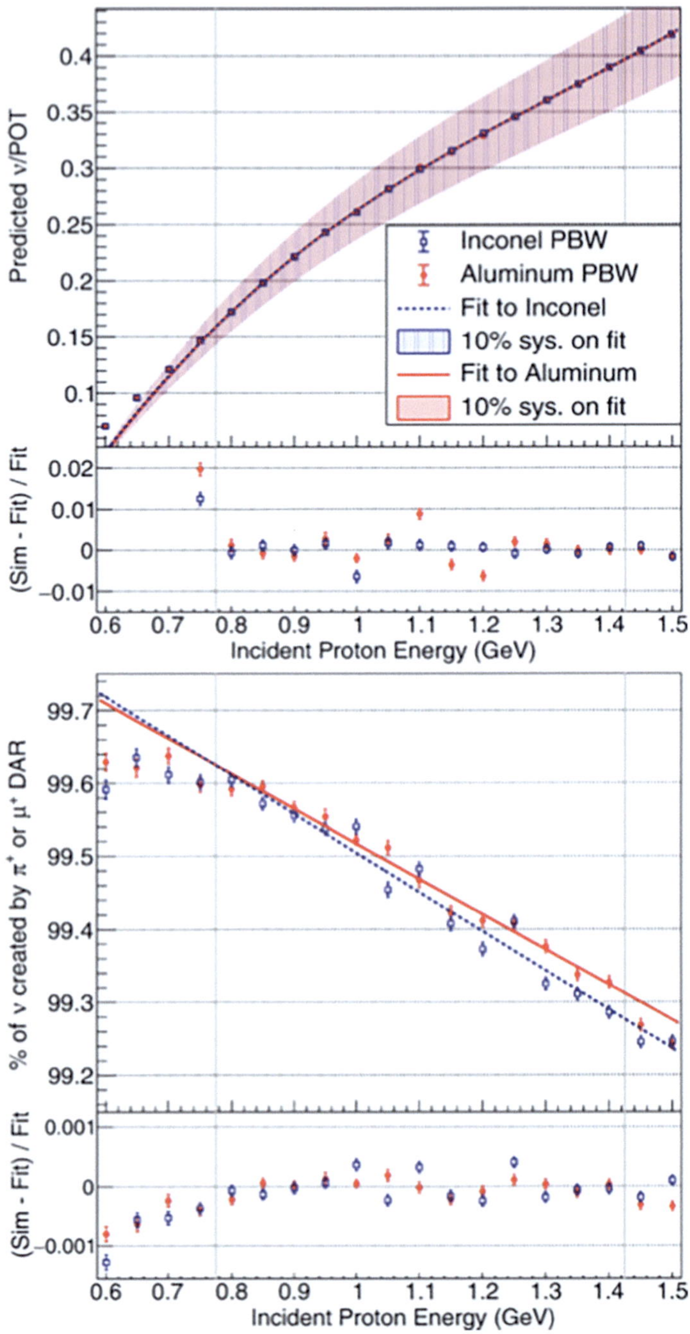

Fig. 3.18 Relationship between SNS proton beam energy and the neutrino flux emitted from the target [31]

3.3.2 *Neutrino Oscillations in the Standard Model*

It can be shown that a negligible fraction of the neutrino flux emanating from the SNS target undergoes flavor oscillations away from the electron flavor (the focus of this work). An estimate is performed here for good measure.

$$P(\nu_e \rightarrow \nu_e) \approx 1 - \sin^2(2\theta_{13})\sin^2(1.27\Delta m^2_{23}\frac{L}{E}) \tag{3.6}$$

Where neutrino alley is approximately 20 meter away from the SNS target and the neutrino energies range from 0 to 53 MeV for the Michel electron neutrino spectrum. A negligibly small fraction of the originating electron neutrinos from the SNS are predicted to oscillate out of the electron flavor at the baselines of the COHERENT detector suite. Notably, only the inelastic interaction-based neutrino detectors would be sensitive to such oscillations (Fig. 3.19).

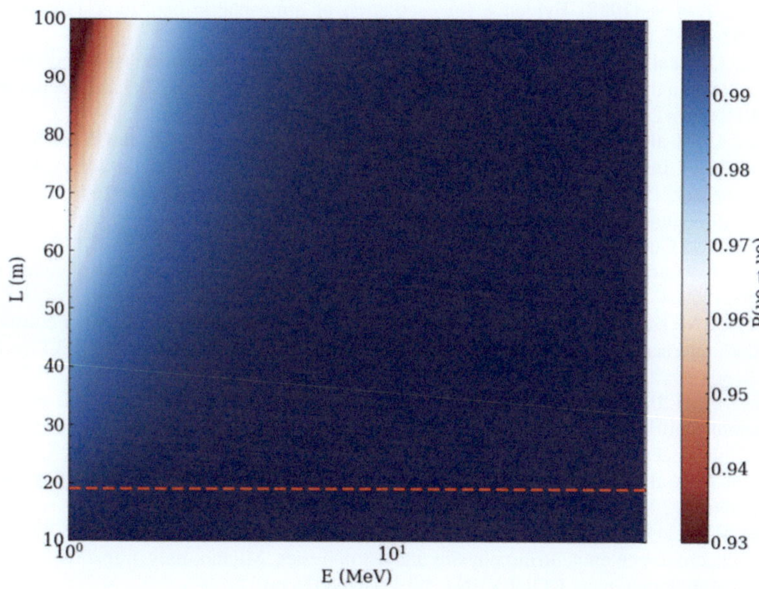

Fig. 3.19 Electron neutrino survival probability at energies and baselines applicable to the SNS electron neutrino flux and Neutrino Alley. At approximately 20 m, the survival probability is nearly 100 %

References

1. Huang MY, Guo XH, Young BL (2016) Detection of supernova neutrinos at spallation neutron sources. Chin Phys C 40(7):073102. https://doi.org/10.1088/1674-1137/40/7/073102
2. Ajimura S et al (2021) The JSNS2 detector. Nucl Instrum Methods Phys Res Sect A: Accel Spectrom Detect Assoc Equip 1014, 165742 (2021). ISSN: 0168-9002. https://doi.org/10.1016/j.nima.2021.165742; https://www.sciencedirect.com/science/article/pii/S0168900221007270
3. Abele H et al (2023) Particle physics at the European spallation source. Phys Rep 1023:1–84. ISSN: 0370-1573. https://doi.org/10.1016/j.physrep.2023.06.001, https://www.sciencedirect.com/science/article/pii/S0370157323001898
4. Efremenko Y, Hix WR (2009) Opportunities for neutrino physics at the spallation neutron source (SNS). J Phys: Conf Ser 173(1):012006. https://doi.org/10.1088/1742-6596/173/1/012006
5. Akimov D et al (2022) The COHERENT experimental program. arXiv: 2204.04575 [hep-ex]
6. The COHERENT collaboration et al (2022) Monitoring the SNS basement neutron background with the MARS detector. J Instrum 17(03):P03021. https://doi.org/10.1088/1748-0221/17/03/P03021
7. DUNE Collaboration et al (2022) Snowmass neutrino frontier: DUNE physics summary. arXiv: 2203.06100 [hep-ex]
8. Fukuda Y et al (1998) Evidence for oscillation of atmospheric neutrinos. Phys Rev Lett 81:1562–1567. https://doi.org/10.1103/PhysRevLett.81.1562
9. Ahmad QR et al (2002) Direct evidence for neutrino flavor transformation from neutral-current interactions in the Sudbury neutrino observatory. Phys Rev Lett 89:011301. https://doi.org/10.1103/PhysRevLett.89.011301
10. Augier C et al (2023) Results from a prototype TES detector for the Ricochet experiment. Nucl Instrum Methods Phys Res Sect A: Accel Spectrom Detect Assoc Equip 1057:168765. ISSN: 0168-9002. https://doi.org/10.1016/j.nima.2023.168765, https://www.sciencedirect.com/science/article/pii/S0168900223007568
11. The CONUS Collaboration et al (2024) The CONUS+ experiment. arXiv: 2407.11912 [hep-ex]
12. Aguilar-Arevalo AA et al (2024) Searches for CE NS and physics beyond the standard model using skipper-CCDs at CONNIE. arXiv: 2403.15976 [hep-ex]
13. Khan AN, McKay DW, Rodejohann W (2021) CP-violating and charged current neutrino non-standard interactions in CEνNS. Phys Rev D 104:015019. https://doi.org/10.1103/PhysRevD.104.015019
14. Dutta B et al (2016) Sensitivity to Z-prime and nonstandard neutrino interactions from ultralow threshold neutrino-nucleus coherent scattering. Phys Rev D 93:013015. https://doi.org/10.1103/PhysRevD.93.013015
15. Akimov D et al (2017) Observation of coherent elastic neutrino-nucleus scattering. Science 357(6356):1123–1126. eprint: https://doi.org/10.1126/science.aao0990
16. Green M (2024) COHERENT: CEvNS and more at the SNS. In: Presented at the XXXI international conference on neutrino physics and astrophysics, Milano, Italy. https://neutrino2024.org/presentations/Green_COHERENT_NEUTRINO24.pdf
17. Akimov D et al (2021) (2021) First measurement of coherent elastic neutrino-nucleus scattering on argon. Phys Rev Lett 126:012002. https://doi.org/10.1103/PhysRevLett.126.012002
18. S. Adamski et al., First detection of coherent elastic neutrino-nucleus scattering on germanium (2024). arXiv: 2406.13806 [hep-ex]
19. Hedges S (2024) Measurement of quenching factors in NaI[Tl]. In: Low energy neutrino-nucleus interactions at the spallation neutron source. Springer Nature Switzerland, Cham, pp 45–65. ISBN: 978-3-031-68110-3. https://doi.org/10.1007/978-3-031-68110-3_4
20. Choi JJ et al (2025) First direct search for light dark matter using the NEON experiment at a nuclear reactor. Phys Rev Lett 134:021802. https://doi.org/10.1103/PhysRevLett.134.021802
21. Bernabei R et al (2000) Search for WIMP annual modulation signature: results from DAMA/NaI-3 and DAMA/NaI-4 and the global combined analysis. Phys Lett B 480(1), 23–31. ISSN: 0370-2693. https://doi.org/10.1016/S0370-2693(00)00405-6

22. Domogatskii GV, Nadezhin DK (1978) Neutrino production of bypassed isotopes, and the possible role of neutrinos in nucleosynthesis. Soviet Astron 22. Provided by the SAO/NASA Astrophysics Data System, pp 297–305. https://ui.adsabs.harvard.edu/abs/1978SvA....22..297D
23. Gardiner S (2021) Simulating low-energy neutrino interactions with Marley. Comput Phys Commun 269:108123. ISSN: 0010-4655. https://doi.org/10.1016/j.cpc.2021.108123, https://www.sciencedirect.com/science/article/pii/S0010465521002356
24. Wakasa T et al (2012) Complete sets of polarization transfer observables for the ^{208}Pb(**p**,**n**) reaction at 296 MeV and Gamow-Teller and spin-dipole strengths for ^{208}Pb. Phys Rev C 85:064606. https://doi.org/10.1103/PhysRevC.85.064606
25. An P et al (2023) Measurement of $^{nat}\mathrm{Pb}(\nu_e, \mathrm{X}n)$ production with a stopped-pion neutrino source. Phys Rev D 108:072001. https://doi.org/10.1103/PhysRevD.108.072001
26. An P et al (2023) Measurement of electron-neutrino charged-current cross sections on ^{127}I with the COHERENT NaIνE detector. Phys Rev Lett 131:221801. https://doi.org/10.1103/PhysRevLett.131.221801
27. Fujita Y et al (2002) $M1$ and Gamow-teller transitions in $T = 1/2$ nuclei ^{23}Na and ^{23}Mg. Phys Rev C 66:044313. https://doi.org/10.1103/PhysRevC.66.044313
28. Hellgren M, Suhonen J (2024) Charged-current neutrino-nucleus scattering off ^{127}I and ^{133}Cs. Phys Rev C 109:035802. https://doi.org/10.1103/PhysRevC.109.035802
29. COHERENT collaboration et al (2021) A D2O detector for flux normalization of a pion decay-at-rest neutrino source. J Instrum 16(08):P08048. https://doi.org/10.1088/1748-0221/16/08/P08048
30. Mosconi B et al (2007) Model dependence of the neutrino-deuteron disintegration cross sections at low energies. Phys Rev C 75:044610. https://doi.org/10.1103/PhysRevC.75.044610
31. Akimov D et al (2022) Simulating the neutrino flux from the Spallation neutron source for the COHERENT experiment. Phys Rev D 106:032003. https://doi.org/10.1103/PhysRevD.106.032003
32. Scholz BJ (2017) First observation of coherent elastic neutrino-nucleus scattering. PhD thesis. Chicago U. https://doi.org/10.1007/978-3-319-99747-6. arXiv: 1904.01155 [nucl-ex]
33. Akimov D et al (2018) COHERENT 2018 at the Spallation neutron source. arXiv: 1803.09183 [physics.ins-det]
34. Avignone FT (2002) Potential new neutrino physics at stopped-pion sources. Prog Part Nucl Phys 48(1):213–222. ISSN: 0146-6410. https://doi.org/10.1016/S0146-6410(02)00127-8
35. Agostinelli S et al (2003) Geant4—a simulation toolkit. Nucl Instrum Methods Phys Res Sect A: Accel Spectrom Detect Assoc Equip 506(3):250–303. ISSN: 0168-9002. https://doi.org/10.1016/S0168-9002(03)01368-8

Chapter 4
Neutrino-Induced Fission on Thorium

Abstract We establish the predictions for neutrino-induced nuclear fission and what target material is most ideal for this search. After determining that target material is thorium, we set out to make estimates of the charged-current neutrino interaction cross section, the expected branching ratios from the decaying daughter nucleus, and then define the signal for neutrino-induced nuclear fission.

Before endeavoring to measure neutrino-induced nuclear fission, it is crucial to establish a basis for estimating the interaction rate, identifying the resulting observables, and selecting the most suitable atom to split. Building on the understanding of the SNS as a neutrino source introduced in Sect. 3.3, along with the theoretical framework of both charged-current (Sect. 2.1.1) and neutral-current (Sect. 2.1.2) neutrino-nucleus interactions, the interaction cross section can be computed. This then enables modeling the thorium target nucleus's likelihood of decaying through nuclear fission after stimulation by SNS neutrinos.

4.1 Target Nucleus Selection

It is worth enumerating why thorium is an ideal target nucleus for this search. To start, thorium is a "fissionable" nucleus. A handy rule of thumb for fissionability is $\frac{Z^2}{A}$ [1]. A greater fissionability equates to a greater tendency to undergo nuclear fission upon stimulation by a projectile. This reflects the destabilizing influence of accumulating protons, which increasingly repel each other in this liquid drop-like nucleus. This criterion guides the experimenter to the heavier end of the periodic table toward the actinide series and its nearest neighbors. However, this fissionability heuristic also forebodes the dangers of spontaneous fission. A nucleus which is too fissionable can undergo spontaneous fission without external provocation. This is particularly pernicious when hunting for a rare nuclear fission process. Therefore, the fission barrier should be low enough for the neutrinos to vault the daughter nucleus above it, but not so low that there is a heavy background of spontaneous fissions. As one can see from Fig. 4.1, thorium isotopes are quite stable to spontaneous fission compared to the other fissionable nuclei.

T. Johnson, *The First Indication of Neutrino-Induced Nuclear Fission*,
https://doi.org/10.1007/978-3-032-01936-3_4

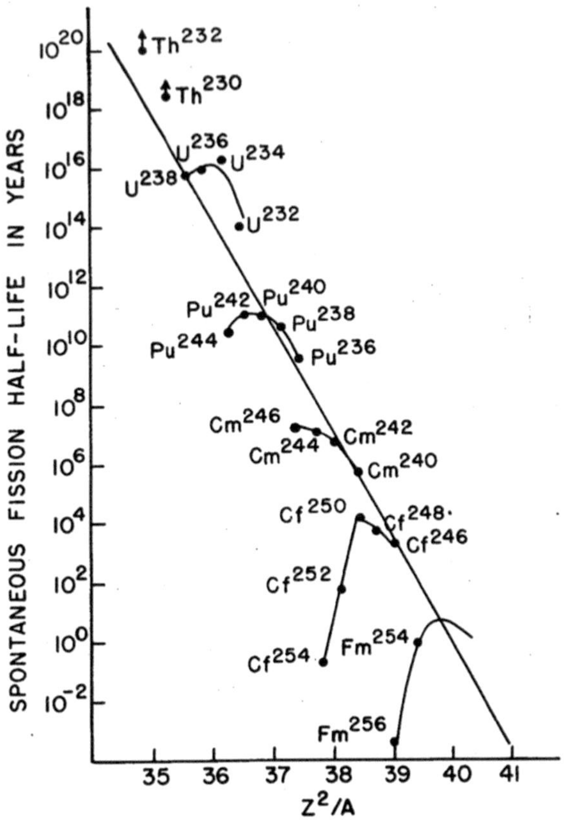

Fig. 4.1 The spontaneous fission half-lives of several isotopes in the actinide series of the periodic table. ^{232}Th has the most favorable half-life of the nuclei [2]

Next, thorium exists almost entirely as the ^{232}Th. The remaining naturally occurring isotopes of thorium (shown in Table 4.1) are present in trace (extremely small) amounts. Thorium exclusively existing as a single isotope greatly simplifies the prediction work required for this search. Also, ^{232}Th is an even-even nucleus which

Table 4.1 The natural abundance of the thorium isotopes. Essentially, all naturally occurring thorium exists in the A = 232 isotope [4]

Isotope	Natural abundance [%]
227	Trace
228	Trace
229	Trace
230	0.02
231	Trace
232	100.0
233	Trace
234	Trace

is widely considered to be a simpler nuclear system for performing calculations thanks to its ground state having spin and parity of 0+ and excited states following predictable pattern across nuclear sizes [3].

4.2 Charged-Current Neutrino-Nucleus Interaction on Thorium

The main channel for inducing neutrino-induced nuclear fission on the target nuclei is through the charged-current neutrino capture on ^{232}Th nucleus, leaving it in a highly excited state thanks to the high-lying Gamow-Teller resonance (GGTR) and the Isobaric Analog State (IAS). A schematic of the neutrino-induced nuclear fission via a charged-current interaction is shown in Fig. 4.2. These resonances are characterized by a high degree of collective motion among many, if not all, nucleons in the nucleus. The transition probability between the ground state of the target nucleus (^{232}Th) and the continuum state in the isobar (^{232}Pa) is quantified by the β-decay strength function, S_β(E). The intensity of the strength function is greatest at the nuclear excitation energy values corresponding to the centroids of the nuclear resonances, which, in turn, have a significantly greater probability of leaving the new nucleus with an excitation energy of the resonant centroids. Because the centroids of these two resonances are degenerate for neutron-rich nuclei and lie above the fission barrier for ^{232}Pa, the augmented capture cross section readily leads to fission (Fig. 4.3).

$$\nu_e +^{232}_{90} \mathrm{Th}_{142} \rightarrow e^- +^{232}_{91} \mathrm{Pa}^*_{141} \tag{4.1}$$

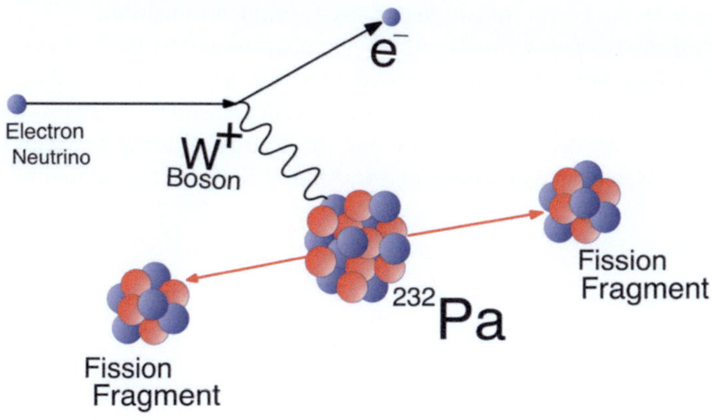

Fig. 4.2 An illustration of the neutrino-induced nuclear fission process for an electron neutrino capturing on a thorium nucleus creating an electron and a protactinium nucleus that subsequently undergoes nuclear fission

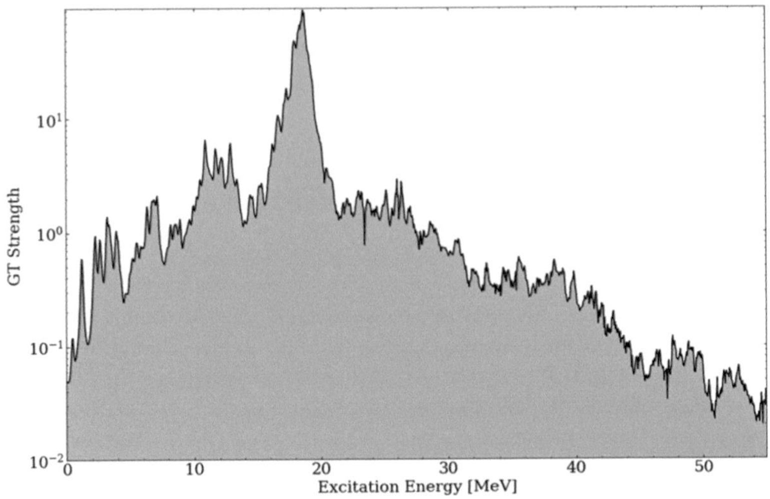

Fig. 4.3 The beta strength function for a neutrino capture on thorium computed by Dr. Evan Ney and Professor Jonathan Engel

Because the β-decay strength function has yet to be measured experimentally for ^{232}Th into ^{232}Pa, a theoretical calculation must suffice. This calculation was performed by collaborators Professor Jonathan Engel and Dr. Evan Ney at the University of North Carolina at Chapel Hill. The calculation was performed using the code PYNFAM [5].

The result of the calculation is shown in Figs. 4.3 and 4.4 below. It is worth noting that this β-strength function only includes the allowed Gamow-Teller transitions and does not include the Fermi transitions or any forbidden transitions.

The Fermi transition is almost completely encapsulated by the Isobaric Analogue State (IAS) which has a centroid that is degenerate with the centroid of the easily identifiable resonance called the Giant Gamow-Teller Resonance. QRPA calculations of the Gamow-Teller Resonance show that the energy splitting between the Giant Gamow-Teller Resonance and the Isobaric Analog State is represented by [6]:

$$\Delta_{\text{GGTR - IAS}} = E_{\text{GGTR}} - E_{\text{IAS}} \tag{4.2}$$

$$= 7.0 - 57.8T_0/A \text{ MeV} \tag{4.3}$$

$$T_0 = (N - Z)/2 \tag{4.4}$$

Here, $\Delta_{\text{GGTR - IAS}}$ is the energy between the centroids of the Giant Gamow-Teller Resonance, E_{GGTR}, and the Isobaric Analog State, E_{IAS}. T_0 is the parent nucleus's ground state isospin. Using Eq. 4.3, the energy splitting between the predominant resonances in ^{232}Th is a mere 0.522 MeV.

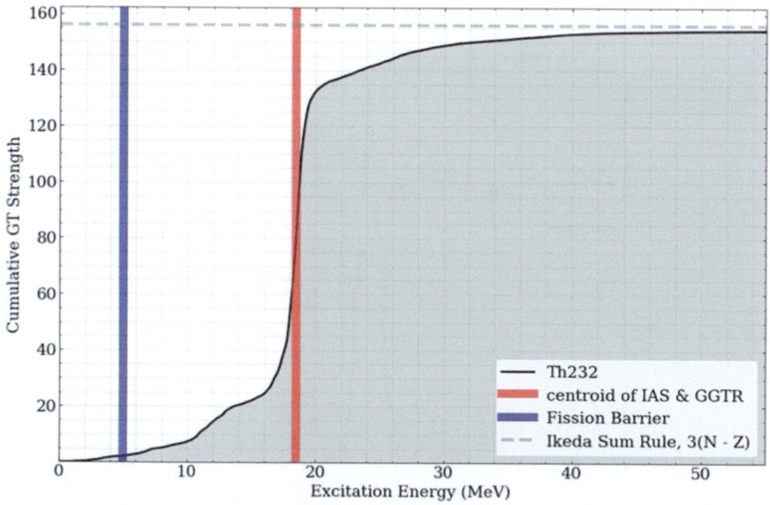

Fig. 4.4 The cumulative strength derived from Fig. 4.3. The final strength is described by the Ikeda sum rule ($\approx$ 3(N–Z)). Overlaid on the cumulative strength are the locations of ^{232}Pa's fission barrier and the centroid of the IAS and GGTR

Predictive work is not solely limited to theoretical predictions, because there are several experimental studies that have identified the excitation energy of the IAS for the charge-exchange reaction Eq. 4.5 at $E_X = 18.48 \pm 0.02$ MeV.

$$^{232}\mathrm{Th}(^{3}\mathrm{He}, \mathrm{tf})^{232}\mathrm{Pa} \tag{4.5}$$

As an aside, this same study determined that the probability of fission after excitation of the IAS is $P_f = 0.065$ with an estimated accuracy of 20%. The IAS of this charge-exchange reaction is visualized in Fig. 4.5.

To acquire a more realistic view of this charged-current neutrino capture on ^{232}Th, the β-strength function values are adjusted with the weak axial vector coupling constant, g_A^2. Where g_A is taken to be 1.26. This correction accounts for the deviation from experimental and theoretical values of the Gamow-Teller transitions strengths.

Next, the Fermi strength is encapsulated by placing a narrow, delta function-like resonance at 18.48 MeV with a strength of (N–Z) [8]. The Gamow-Teller strength and the Fermi strength are combined and utilized in the neutrino-nucleus event generator MARLEY to calculate the nuclear excitation energies for a large number of neutrino interactions, Fig. 4.6. The largest contribution to the expected neutrino interactions results in nuclear excitations at the combination of the IAS and GGTR.

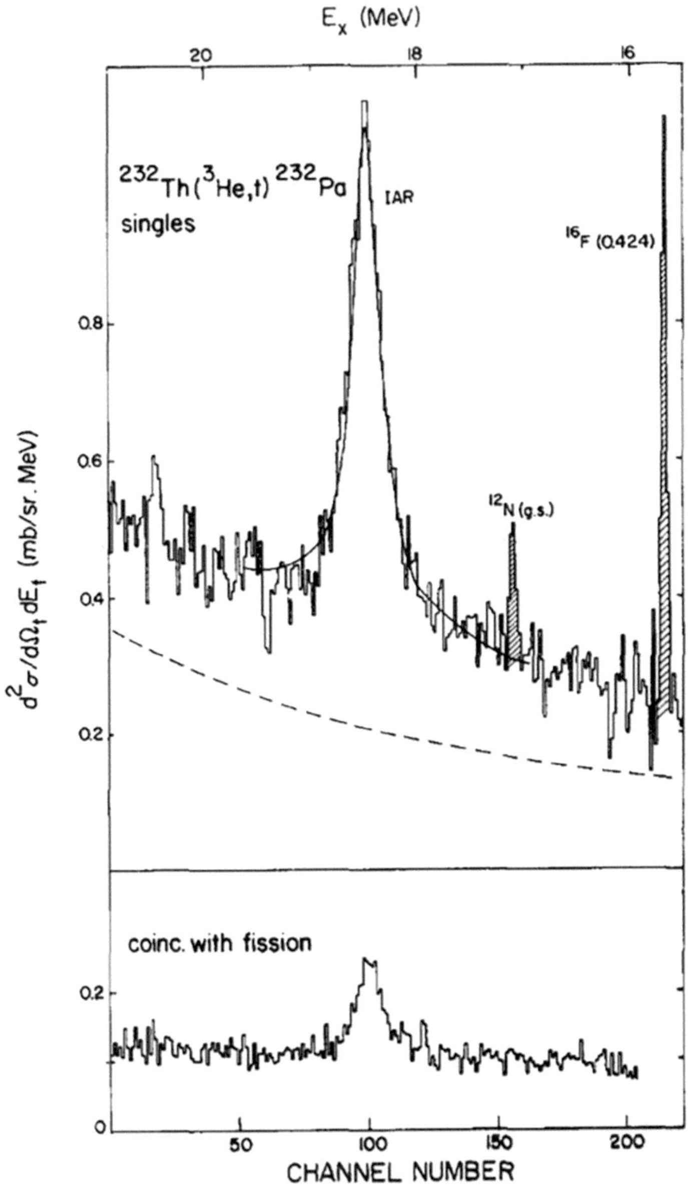

Fig. 4.5 The identification of the ^{232}Th(^{3}He, tf)^{232}Pa charge-exchange IAS [7]. The center of the IAS was found to at 18.48 ± 0.02 MeV

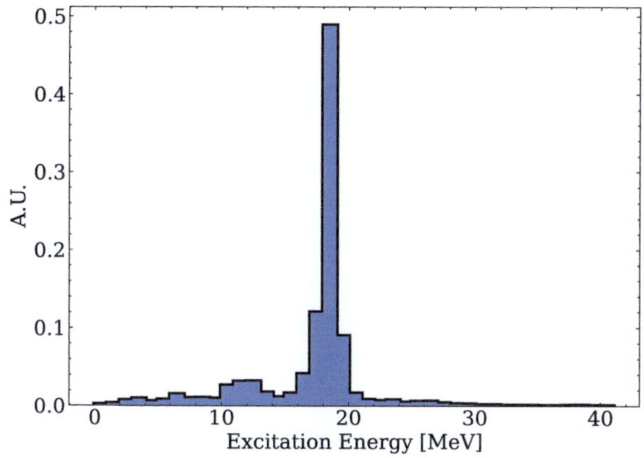

Fig. 4.6 Distribution of nuclear excitation energies precipitated from 10,000 SNS Decay-at-Rest spectrum electron neutrinos produced in MARLEY. This mirrors the distribution of the beta-strength function used as a seed for this calculation. Note the large contribution centered at 18.7 MeV is the Giant Gamow-Teller Resonance and the Isobaric Analog State

Armed with the β-strength function, incorporating both Fermi and Gamow-Teller transitions, it becomes possible to compute the neutrino capture cross section on ^{232}Th as a function of neutrino energy with Eq. 4.6.

$$\sigma = \frac{1}{\pi}\int_{\epsilon_{min}}^{\epsilon_{max}} G_F^2 \cos\theta_C \mathrm{p}_e \mathrm{E}_e \mathrm{F}(\mathrm{Z},\mathrm{E}_e) S(E) dE \tag{4.6}$$

where G_F is the weak coupling constant, θ_C is the Cabbibo angle, p_r is the momentum of the outgoing lepton, E_e is the energy of the outgoing lepton, F(E, E_e) is the Fermi function (the degree to which the electron wavefunction is distorted by the nuclear charge) and S(E) is the β-strength function. The kinematic threshold for this reaction mostly depends on the mass difference between the parent and daughter nuclei and the rest mass of the lepton in the final state. Note that this particular expression is for a collective structure of excited states.

One can alternatively express the cross section for a series of discrete excited states. The cross section for a neutrino capturing on a nucleus to produce the kth excited state in the daughter nucleus is given in Eq. 4.8

$$\sigma_k = \frac{G_F^2 \cos_C}{\pi} p_e E_e F(Z, E_e) \left(B(F)_k + \left(\frac{g_A}{g_V}\right)^2 B(GT)_k \right) \tag{4.7}$$

$$= (1.597 \times 10^{-44} \mathrm{cm}^2) p_e E_e F(Z, E_e) \left(B(F)_k + \left(\frac{g_A}{g_V}\right)^2 B(GT)_k \right) \tag{4.8}$$

As indicated in the Fig. 4.7, the cross sections rapidly increase with increasing neutrino energy. They further increase after the neutrino passes the centroid of the two major nuclear resonances that aid the charged-current interaction process. As such, the majority of the charged-current neutrino interactions on ^{232}Th at the SNS are induced by higher energy neutrinos in the Michel spectrum as shown in Fig. 4.8.

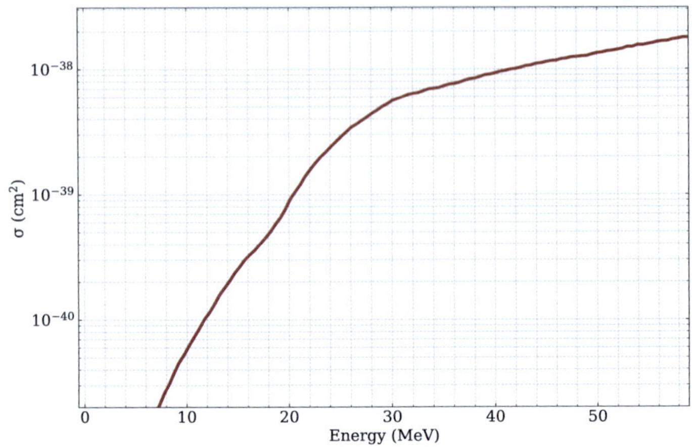

Fig. 4.7 The charged-current cross section of neutrino capture on thorium as a function of the neutrino energy computed through MARLEY's cross section executable. The inputs to this calculation were the theoretically produced Gamow-Teller strengths and an Isobaric Analogue State placed at the experimentally determined 18.48 MeV with a strength of (N–Z). To compute the Fermi Function term in the cross section, MARLEY takes the minimum of the two possible of formulations (MEMA or Fermi version)

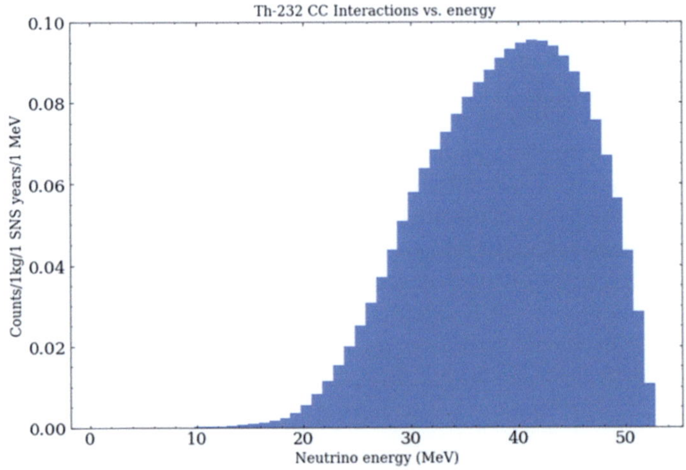

Fig. 4.8 The distribution of expected charged-current events based on the probability of each energy bin inducing an interaction. The neutrino energy spectrum used is the SNS's Michel electron neutrino spectrum

4.2.1 Final State Lepton Energy Distribution

Because the charged-current interaction is instigated by the capture of an electron neutrino, an electron is emitted to conserve charge and lepton number. This electron's kinetic energy is the remainder of the energy not allocated to the nuclear excitation and the energy needed to both span the mass difference between parent and daughter nuclei and the rest mass of the electron itself. For thorium, the average kinetic energy is approximately 22 MeV though spans a wide distribution. Predictions for this kinetic energy distribution shown here were performed using MARLEY. The presence of this relatively high-energy lepton in the final state has previously been used as the observable signal for neutrino interaction events. However, the practicality of searching for this lepton should be judged on a case-by-case basis. For thorium, the target mass is radioactive and this will likely be prohibitive to successfully measuring the final-state lepton (Fig. 4.9).

4.2.2 Cross Check with Fermi-Dirac Supernova Flux-Averaged Cross Section

There are several methods for predicting the cross section of charged-current neutrino-nucleus scattering processes–a handful of authors have endeavored to perform this calculation for neutrinos capturing on ^{232}Th. However, they performed and subsequently reported this calculation as the flux-averaged cross section in the context of a core-collapse supernova. Although the energies and flavors of the SNS and

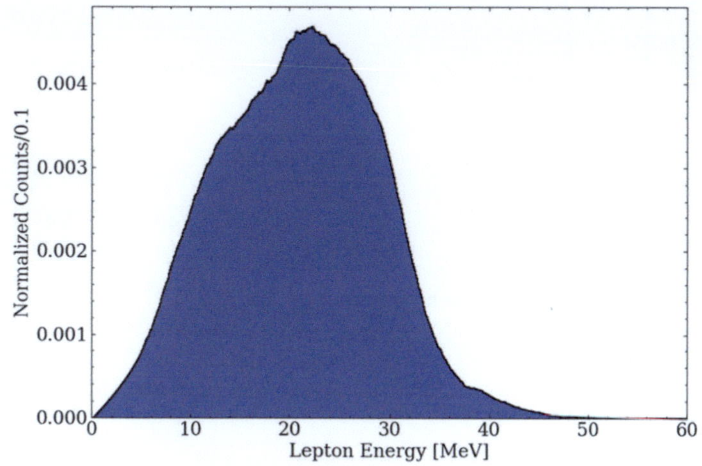

Fig. 4.9 The kinetic energy distribution of the charged-current lepton—in this case an electron. This electron final state prediction was made using MARLEY

core-collapse supernova have a non-zero overlap, this flux-averaged cross section is not directly applicable to the purposes here. As a cross-check, the cross section as a function of neutrino energy shown in the previous section is convolved with a simple Fermi-Dirac supernova neutrino energy spectrum to match the context of the authors in [9]. The expression for the Fermi-Dirac spectrum is given by Eq. 4.9

$$F(E) \approx \frac{E^2}{\exp(E/T - \mu) + 1} \tag{4.9}$$

where E is the neutrino energy, T is the temperature of the distribution and μ is the chemical potential. For the cross-check performed here, we take the T = 4 MeV and $\mu = 0$ as shown in Fig. 4.10.

The above neutrino energy distribution is combined with the neutrino cross section shown in Eq. 4.6 to calculate the flux-averaged cross section for a core-collapse supernova. This result is then compared to the calculation performed by Kolbe et al. and displayed in Table 4.2 [9]. The MARLEY calculation is 26% different from the Kolbe et al. calculation. The origin of the discrepancy between calculation methods is unknown at this time.

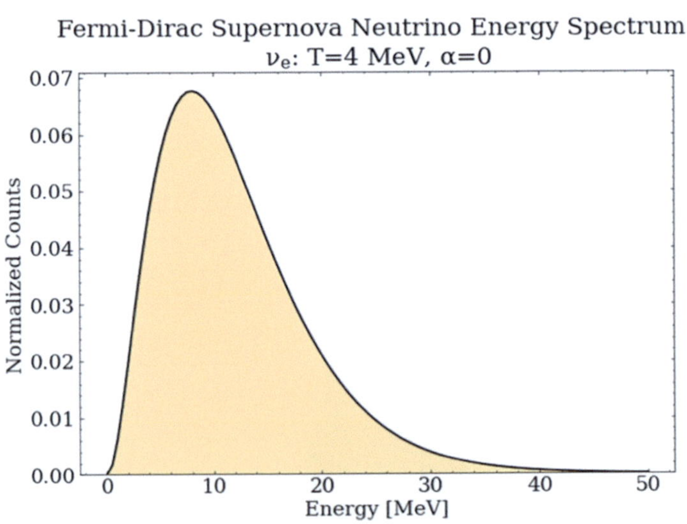

Fig. 4.10 The Fermi-Dirac supernova neutrino energy spectrum for T = 4 MeV and chemical potential = 0

Table 4.2 A comparison of the Fermi-Dirac Supernova neutrino spectrum (4 MeV, $\mu = 0$) averaged cross section for ^{232}Th using MARLEY versus a calculation performed by Kolbe et al., from Ref. [9]

MARLEY	Kolbe et al.
$5.66 \times 10^{-40}\text{cm}^2$	$4.49 \times 10^{-40}\text{cm}^2$

4.2.3 *Cross Check with Uranium-238 Cross Section*

At the time of this work, there is a severe dearth of predictive work conducted for neutrino-induced nuclear fission, especially for ^{232}Th. Further, much of the cross section calculations are reported as flux-averaged cross sections, which are only directly comparable for other computations for the exact neutrino spectrum used by the authors. It is much more helpful to make comparisons with the cross sections reported as a function of the neutrino energy. There are a few calculations performed on adjacent nuclei in the periodic table with similar atomic masses and, more importantly, neutron excesses. These computations are reported as a cross section as a function of the neutrino energy. One such comparison is shown in Fig. 4.11. The ^{238}U cross section [10] was calculated with a Random Phase Approximation framework. The two computations are well aligned at higher neutrino energies with a notable deviation at lower energies. This may promulgate from the difference in the isotopes' respective charge-current thresholds derived from the mass differences between the target and daughter nuclei.

4.3 De-Excitation Via Nuclear Fission

After the neutrino capture occurs on the thorium target nucleus, there is a great deal of excitation energy that the nucleus must remove via any available channel. Due to the relatively low threshold provided by protactinium's fission barrier, nuclear fission can readily occur. The fission probability can be imperfectly expressed using Eq. 4.10. The alternative decay channel in this expression is through neutron emission [11].

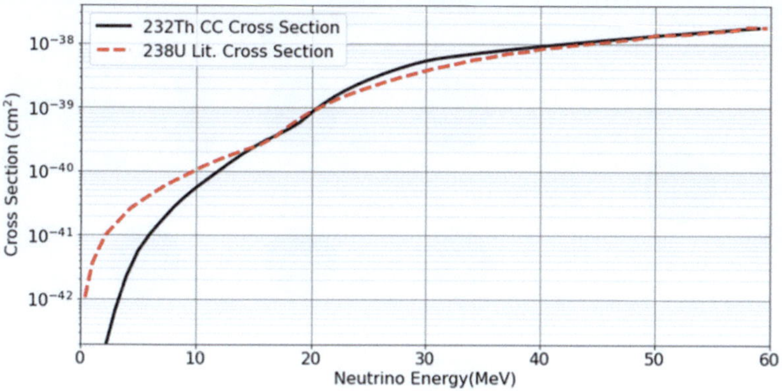

Fig. 4.11 Comparison of the electron neutrino charged-current cross section on ^{232}Th (black) and ^{238}U (red) [10]

$$P_f = \frac{1}{1 + 4(m_n/\hbar^2)R^2 T \exp((B_f - S_n)/T)} \quad (4.10)$$

where R is the nuclear radius in femptometers given by R = $1.2A^{1/3}$ fm, A is the atomic mass number of the nucleus, B_f is the fission barrier, S_n is the neutron separation energy, m_n is the neutron mass, $\hbar$ is the reduced Planck's constant, and T is the nuclear temperature. As a brief aside, the nuclear temperature is quite similar to the excitation energy of the nucleus but differs slightly and can be characterized by Eq. 4.11 [12].

$$\frac{1}{T} = \left(\frac{a}{E^*}\right)^{1/2} - \gamma/E^* \quad (4.11)$$

where a is:

$$a \approx \frac{A}{8}\text{MeV}^{-1} \quad (4.12)$$

The second term becomes negligibly small with increasing number of nucleons in the nucleus as well as excitation energy.

However, this formulation is only applied to first-chance fission where the excitation energy given to the nucleus is immediately exhausted through nuclear fission. Conversely, there can also be second-, third- (and so on) chance fission, where rather than terminating the energy exhaustion process with an immediate fission, some of the excitation energy is partially siphoned off by a neutron emission–then there may be enough leftover energy to undergo fission (See Fig. 4.12). Further, with higher exci-

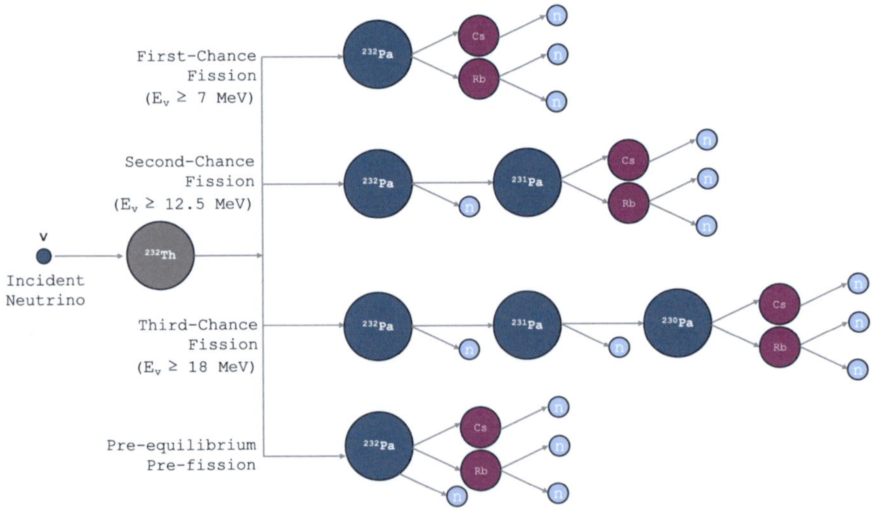

Fig. 4.12 Schematic of the possible fission ordering after excitation by neutrinos on a thorium nucleus

tation energies, there is a systematic washing out of the shell effects of the nucleus, which directly depresses the fission barrier relative to the neutron separation energy of the nucleus, expressed here as $\Delta U = B_F - S_n$.

In order to accommodate the probability of second and third chance nuclear fission as well as shell wash out, this approach emulates the work in [10] using the statistical decay software package called ABLA07, which must be coupled with a particle transport code (in this case INCL). The work here can be replicated using the Liege Intranuclear cascade and ablation code INCL/ABLA [13]. Neutron-rich nuclei like ^{232}Th and ^{232}Pa dominantly de-excite through nuclear fission and neutron emission depending on the input fission barrier and neutron separation energy. According to the Reference Input Parameter Library (RIPL-3) and ENSDF-VII, ^{232}Pa has a neutron separation energy of 5.549 MeV and a fission barrier of 6 MeV. Additionally, ^{232}Th has a neutron separation energy of 6.44 MeV and a fission barrier of 5.8 MeV.

The multiple-chance de-excitations are computed with ABLA07 at a range of compound nucleus excitation energies with user-defined restrictions placed on the compound nucleus's angular momentum to emulate neutrino capture final state nuclei. The output fission probabilities as a function of the excitation energy from ABLA07 are shown in Fig. 4.13. The onset of the probability distribution around 6 MeV corresponds to the location of the fission barrier. Subsequent enhancements of the fission barrier indicate the onset of the second-, third-, fourth-, and fifth-chance fission thresholds. Additionally, the notable dip following the onset of the first- and second-chance fission thresholds corresponds to the onset of the 1n and 2n separation energy thresholds, where increased statistical competition occurs.

The ABLA07-generated fission probabilities are compared to data measured from ^{232}Th(^{3}He,t)^{232}Pa transfer reactions [14]. Conveniently, this comparison can be made because the compound nucleus has no memory of how it was produced, just a current

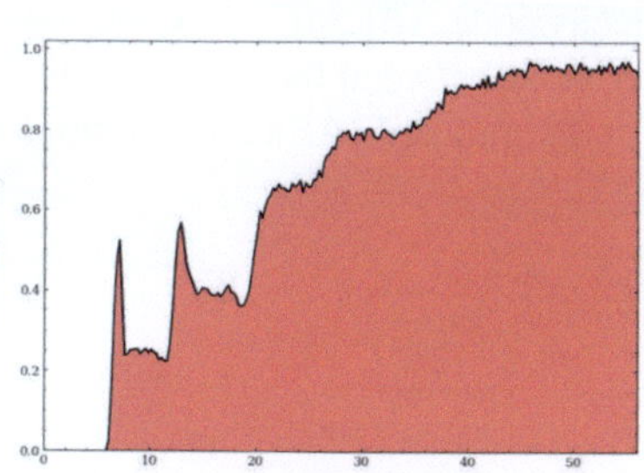

(a) The probability of nuclear fission of Pa-232 at a range of excitation energies. The computation was performed using ABLA07.

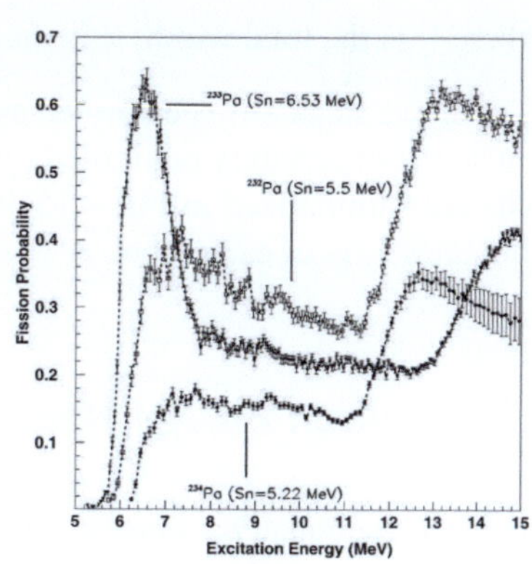

(b) Fission Probability of ^{233}Pa at a range of excitation energies [14].

Fig. 4.13 The probability of decaying via nuclear fission or neutron emission at a range of excitation energies knowledge of how much energy is possessed (and must expel). The information concerning the formation of the compound nucleus is destroyed by the numerous collisions between the nucleons in the intra-nuclear cascade in which the energy injected by the projectile is equilibrated equally to all of the constituent nucleons (Fig. 4.14).

4.3.1 Isobaric Analog State Decay Cross Check

As alluded to in Fig. 4.5, there is data available from the charge-exchange reaction of ^{232}Th(ν_e,e^-)^{232}Pa to help benchmark the ABLA07 simulations. As such, ABLA07 was executed to emulate the conditions of this reaction in data for direct comparison. The procedure for this test was as follows:

- Events were simulated in equal proportions with nuclear excitation energies in increments of 0.05 MeV beginning at 18.0 MeV and ending at 18.75 MeV. This covers the central value of the resonance with room on both ends.
- Every simulated event is evaluated for whether it experienced a nuclear fission as well as how many neutrons were emitted in the fission or neutron emission.
- The individual contribution of each excitation energy simulation is weighted by the shape of the Breit-Wigner structure of the IAS. Where according to [7], it is centered at 18.48 ± 0.02 MeV with a total width $\Gamma = 306 \pm 20$ keV and the Breit-Wigner distribution expressed in Eq. 4.13.

$$f(E) = M^2 \frac{\Gamma}{(E - \omega)^2 + \Gamma^2} \tag{4.13}$$

where Γ is the total width, ω is the centroid, and M^2 is the amplitude.

The above steps were performed using the decay code ABLA07 and Gemini++ to estimate the uncertainty quantified by the differences between the two models. The results are summarized in Fig. 4.15. The data and simulated IAS de-excitations are within error bars of each other.

4.3.2 Neutron Energy Modeling

In order to adequately simulate neutrino-induced nuclear fission and neutrino-induced neutron (NIN) emission events in a detector geometry, the energy distributions of these emitted neutrons must be modeled and then imported into a particle transport simulation package such as MCNP [15]. The energies for both neutrino-induced nuclear fission neutrons and NINs are modeled with the Watt fission spectrum

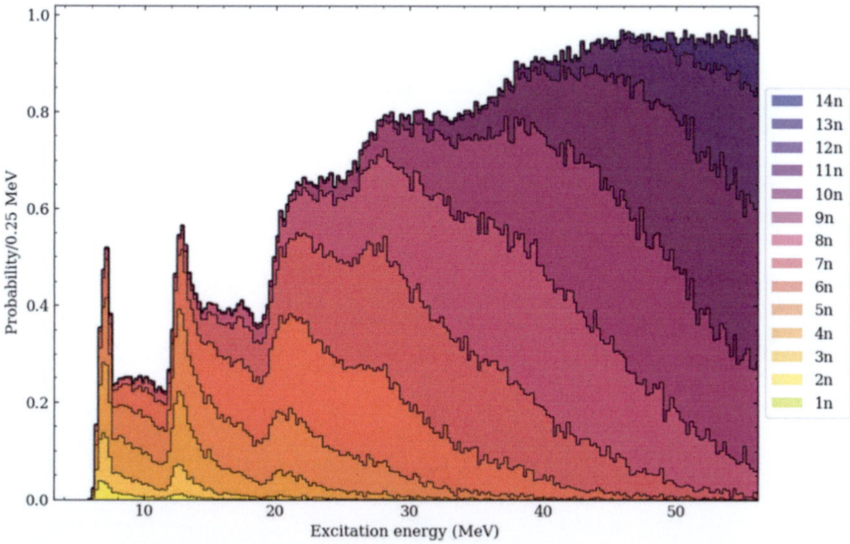

(a) The fission probability at a range of excitation energies. The distribution is broken down by the neutron multiplicities for the underlying nuclear fissions.

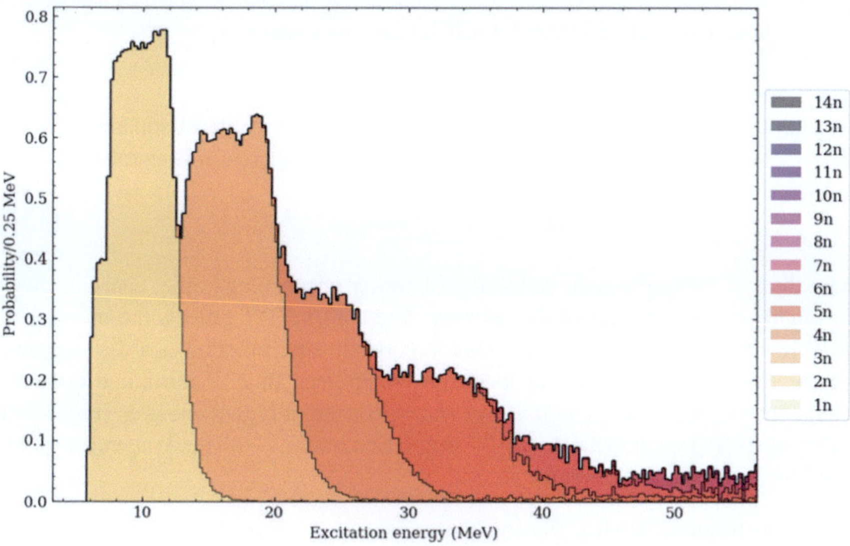

(b) The neutron emission probability at a range of excitation energies. The distribution is broken down by the neutron multiplicities for the underlying neutron emission events.

Fig. 4.14 The nuclear modeling of output. The modeling was conducted with the INCL/ABLA software for the de-excitation of ^{232}Pa at a range of excitation energies

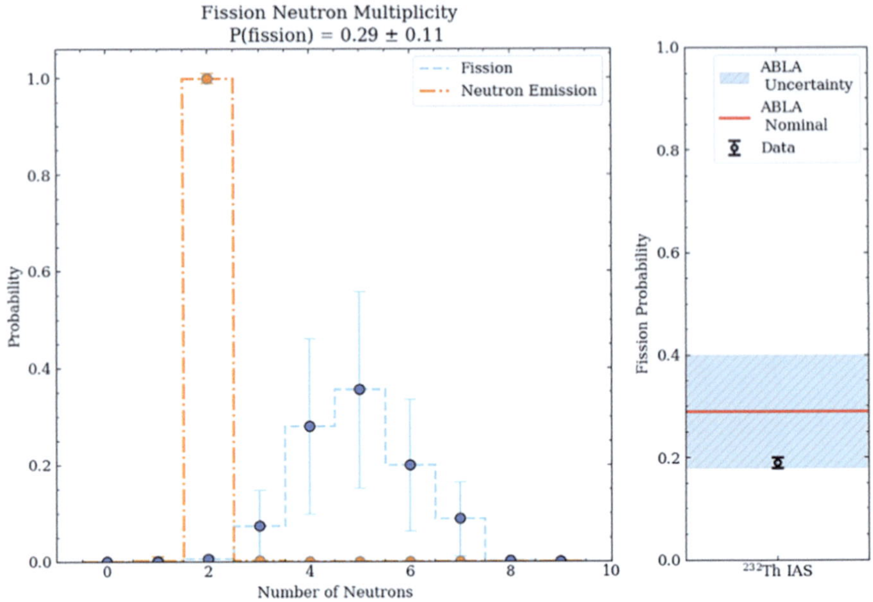

Fig. 4.15 (left) Results of an INCL/ABLA simulation of the IAS of ^{232}Pa. (right) a comparison of the fission probability calculated by ABLA and from data measured in [7]

[16] with the empirical parameters, *a* and *b*, resulting from a least-squares fit to the ABLA07 simulation data. The expression for the Watt spectrum is as follows:

$$f(E) = Ce^{-E/a}\sinh(bE)^{1/2} \tag{4.14}$$

where the $e^{-E/a}$ expresses the reduced probability of reaching higher energies given the roughly thermal nature of the nucleus. The $\sinh(bE)^{1/2}$ reflects the increase in available states at higher energies. This hyperbolic sine function is a development over the more basic Maxwellian spectrum which uses an $E^{1/2}$ term, and does not match experimental data as well as the Watt spectrum at higher energies (Fig. 4.16).

The least-squares fit to the ABLA07 simulation outputs with the Watt curve yields the following best fit parameters:

- Neutrino-Induced Nuclear Fission Neutrons:
  a = 1.00152 MeV, b = 0.00284 MeV^{-1}
- Neutrino-Induced Neutron Emission:
  a = 0.59255 MeV, b = 0.01501 MeV^{-1}.

The neutrino-induced nuclear fission and NIN kinetic energy distributions used going forward are shown in Fig. 4.17. These parameters are then easily integratable into the source input in MCNP simulations.

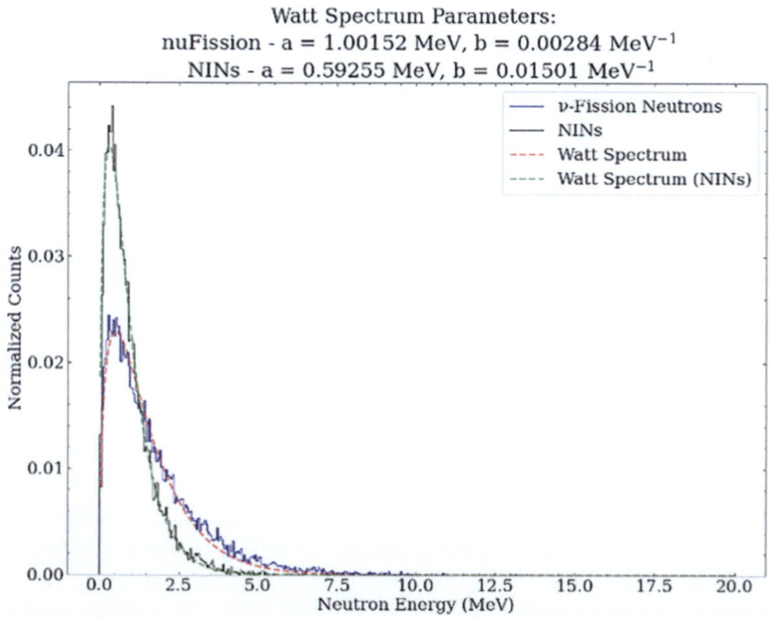

Fig. 4.16 The neutron kinetic energies from both neutron emission and nuclear fission in ABLA07. Each population is individually fit with the Watt curve described in Eq. 4.14. The best fit parameters are listed above the plot

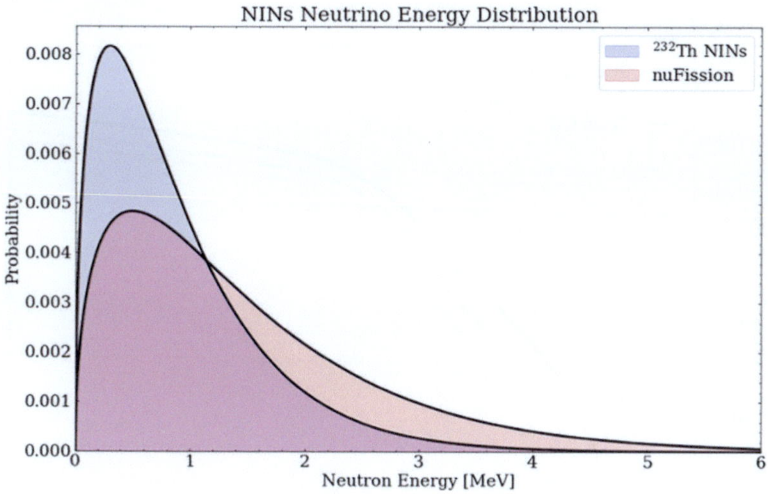

Fig. 4.17 The neutron kinetic energy distributions yielded from INCL/ABLA nuclear simulations fit with Watt curves for neutrino-induced nuclear fission (red) and neutrino-induced neutrons (blue)

4.4 Neutrino-Induced Nuclear Fission Cross Section

The branching ratios yielded by ABLA07 can then be combined with the charged-current cross section to compute a neutrino-induced nuclear fission partial cross section as a function of the neutrino energy (Fig. 4.18).

The partial cross sections can be further broken down and examined by demarcating the population of events by the number of neutrons emitted in their respective final states. These partial cross sections segregated by both neutrino-induced nuclear fission/NINs and neutron multiplicity are shown in Fig. 4.19.

With this foundation established, it is now possible to incorporate knowledge of the SNS neutrinos. As previously established, the SNS emits a population of electron neutrinos according the Michel energy spectrum with an average energy of 30 MeV and a maximum energy of 52 MeV.

This information on the neutron multiplicity of thorium neutrino-induced nuclear fission events with SNS electron neutrinos can be concisely represented in the following transfer matrix in Fig. 4.20. The transfer matrix greatly diminishes the computational burden of studies requiring alterations to the electron neutrino energy distribution (e.g. a sterile neutrino disappearance oscillation study).

One can additionally project the smearing matrix into the neutron multiplicity basis to better read the expected distribution of neutron multiplicities with SNS electron neutrinos. Along with the neutrino-induced nuclear fission events, Fig. 4.21 illustrates the distribution of neutron emission events expected at the SNS.

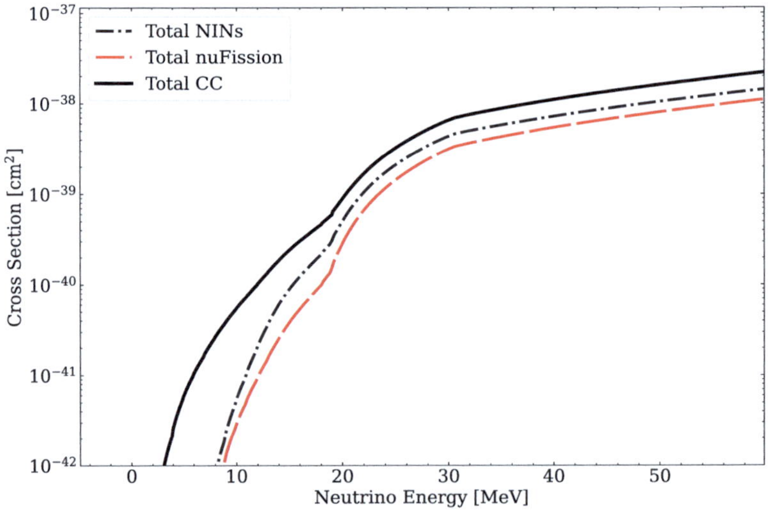

Fig. 4.18 The total charged-current cross section (solid black line), the neutrino-induced neutron emission partial cross section (blue dot-dashed line), and the neutrino-induced nuclear fission partial cross section (dashed red line)

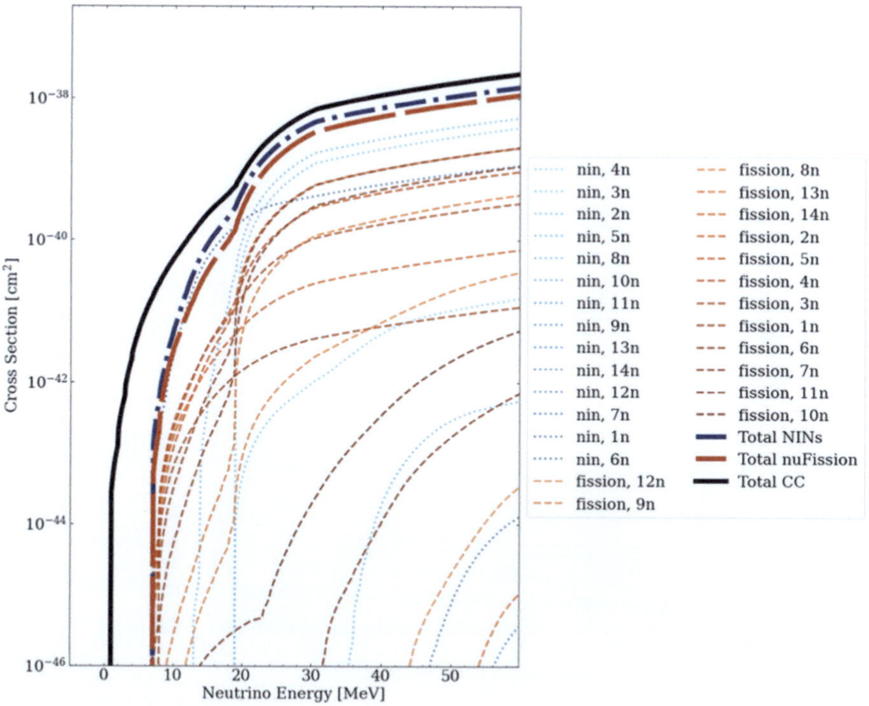

Fig. 4.19 The partial cross sections as a function of neutrino energy broken out into the individual neutron multiplicities for both neutrino-induced neutrons and neutrino-induced nuclear fission events

4.5 Neutral-Current Neutrino-Nucleus Interactions on Thorium

Even though the neutral-current interaction is not thought to be the primary catalyst for neutrino-induced nuclear fission, a fraction of the total neutral-current interactions can still successfully excite the ^{232}Th nuclei above the fission barrier. Neutral-current interactions are flavor-blind, so the decreased cross section can be, at least in part, made up for by the increase in neutrino flux from the other flavors (roughly a factor of three increase in flux at the SNS). The neutral-current interaction's capacity to highly excite nuclei is hampered by the fact that the incoming neutrino is also present in the final state, therefore does not deposit the entirety of its energy into the nucleus. However, the neutral-current opens the possibility of deploying this detector to a nuclear reactor which outputs electron anti-neutrinos at a higher flux than the sum of all neutrino flavors at the SNS (albeit with a substantially lower average energy).

After the neutral-current neutrino-nucleus interaction, the ^{232}Th nucleus is left in an excited state. The degree of that excitation depends on the neutral-current transition operator, the zero-component of the Gamow-Teller operators [17]. Because specific

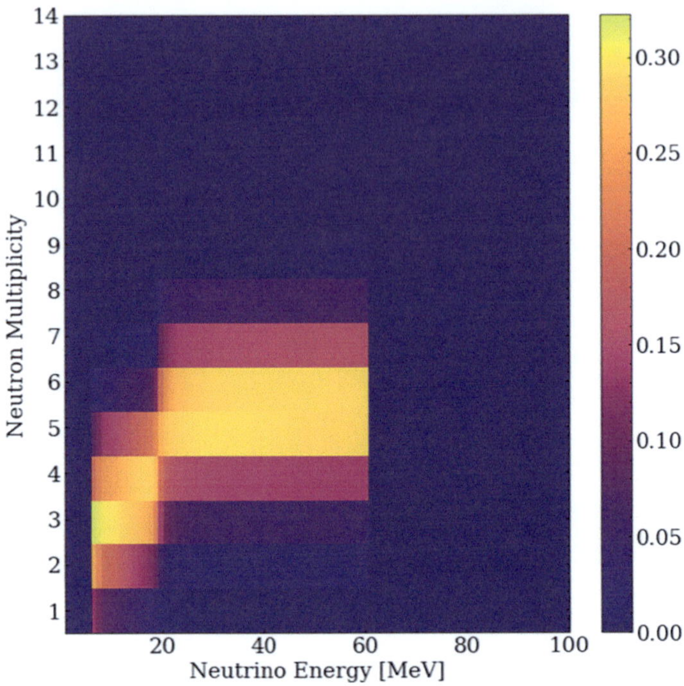

Fig. 4.20 Smearing matrix to take in a neutrino energy spectrum and output the distribution of true neutron multiplicities per neutrino-induced nuclear fission event. The columns are normalized to one to correctly normalize the contributions to the smeared vector

data about inelastic neutral-current neutrino-nucleus scattering does not exist for most nuclei, one must extrapolate the Gamow-Teller strength from either theoretical or experimental M1 transition strength data using the formula 4.15

$$B(\mathrm{GT}_0) = \frac{4\pi g_A^2 B(M1)}{1\mu_\nu^2 \mu_N^2} = 0.308\frac{B(M1)}{\mu_N^2} \tag{4.15}$$

The neutral, zero-component of the Gamow-Teller operator is valid under the conditions:

- $\mathrm{J}_f - J_i = 0, 1$ (but not for $\mathrm{J}_f = \mathrm{J}_i = 0$)
- There is no change in the isospin.

The theoretically calculated M1 strengths are gathered and displayed in Fig. 4.22.

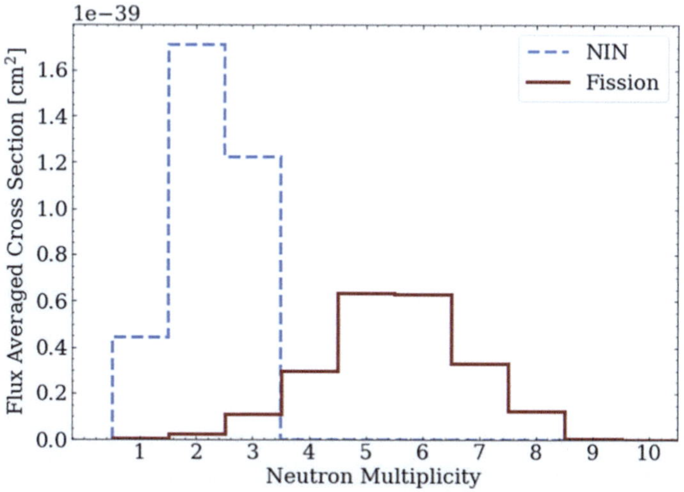

Fig. 4.21 The neutron multiplicity distributions of neutrino-induced nuclear fission (dark red) and neutrino-induced neutron emission (blue). There's a characteristically greater average neutron multiplicity for fission which allows for strong discriminatory power against neutron emission backgrounds

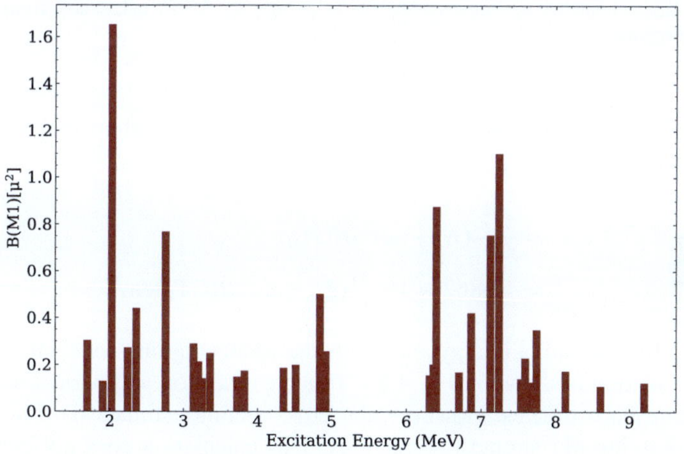

Fig. 4.22 The theoretical M1 magnetic dipole transition strength values for Thorium [18]. The M1 strengths are foundation for making neutral-current neutrino-nucleus interaction cross section predictions

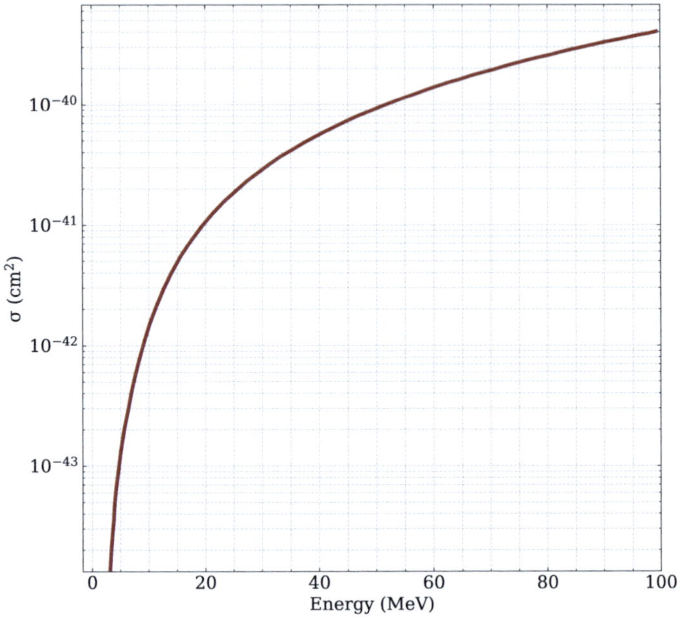

Fig. 4.23 Neutral-current neutrino-nuclear cross section on ^{232}Th calculated from B(M1) to $B(GT_0)$ conversion

The $B(GT_0)$ strengths can then be used to compute the neutral-current cross section with the expression

$$\begin{aligned} \sigma &= \frac{G_F^2}{\pi}(E_\nu - \omega)^2 B(\mathrm{GT}_0) \\ &= 1.6862 \times 10^{-44}(E_\nu - \omega)^2 B(\mathrm{GT}_0)\mathrm{cm}^2 \end{aligned} \tag{4.16}$$

where ω is the excitation energy and G_F is the weak coupling constant. This result of this calculation is shown in Fig. 4.23. This cross section will be instrumental for reactor neutrino estimates. However, the cross-section estimate is only as strong as the underlying M1 strength predictions. This calculation does not consider the contribution from forbidden transitions which make up a significant portion of other heavy, neutron-rich nuclei such as Pb and Bi.

Additionally, using the above M1 strengths converted to GT strength, one can input directly into the MARLEY Monte Carlo event generator to precipitate the following flux-averaged, neutral-current cross sections for each neutrino flavor comprising the SNS flux:

$$\sigma_{\bar{\nu}_\mu} = 6.52 \times 10^{-41} \text{cm}^2 \tag{4.17}$$

$$\sigma_{\nu_\mu} = 1.88 \times 10^{-40} \text{cm}^2 \tag{4.18}$$

$$\sigma_{\nu_e} = 4.69 \times 10^{-41} \text{cm}^2 \tag{4.19}$$

The neutral-current neutrino-nucleus interactions expected from these three flux-averaged cross sections results in 0.13 interactions per kg ^{232}Th per SNS year.

4.5.1 Event Rate at Nuclear Reactors

As mentioned in previous sections of this work, there is potential to apply neutrino-induced nuclear fission to nuclear reactor monitoring initiatives. However, nuclear reactors exclusively radiate electron anti-neutrinos that are primarily low energy (See Fig. 4.24). Because of this, the only charged-current interactions possible are anti-neutrino captures. For neutron-rich nuclei, like ^{232}Th, this reaction channel is "Pauli suppressed." Nuclei with significant neutron excesses do not want to take on additional neutrons. One can think of this restriction in terms of the Fermi Gas model in which the protons and neutrons in the nucleus can be thought of as a two separate, non-interacting gases constrained within the volume of the nucleus. Both the neutron and proton gases construct energy levels like a ladder. An anti-neutrino capture would transmute a proton into a neutron, but there are no open neutron shells that can be reached through allowed transitions. This means that producing a neutron

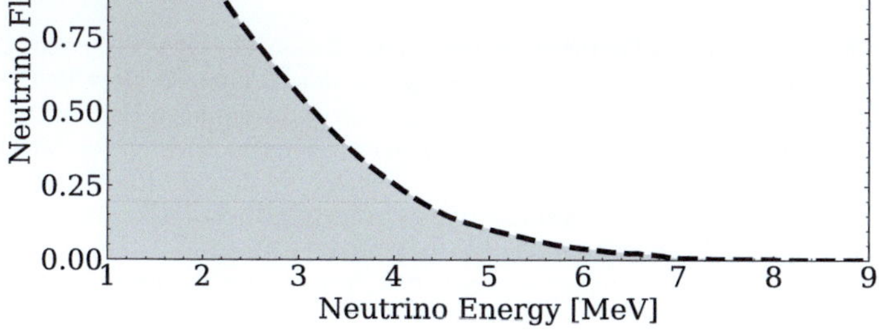

Fig. 4.24 The energy spectrum of electron anti-neutrinos emitted from nuclear reactors from Ref. [19]. There are number of different calculations for the reactor neutrino energy spectrum but this one is depicted here to illustrate the non-zero component above 5–6 MeV, where the energy thresholds for unbound nuclear de-excitations typically lie

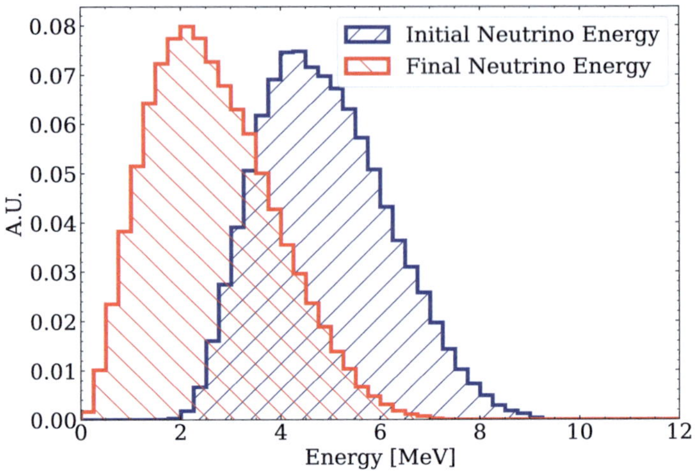

Fig. 4.25 (red) The initial kinetic energy of the reactor neutrinos impinging upon the thorium target mass and successfully interacted via the neutral-current. (blue) the kinetic energy of the scattered reactor neutrino after undergoing a neutral-current interaction with the thorium nucleus

in the final state requires that the transition be through a forbidden transition which are suppressed by a factor of $\mathcal{O}(10^4)$.

The combination of the lower average neutrino energy at nuclear reactors and the neutral-current's inefficiency at exciting nuclei may prove detrimental to this endeavor. Figure 4.25 illustrates the dilemma with a computation done in MARLEY. Further study is needed to apply neutrino-induced nuclear fission to nuclear reactors for monitoring. Particular attention should be paid to the forbidden transitions and also improving the neutral-current predictions in MARLEY overall.

References

1. Bohr N, Wheeler JA (1939) The mechanism of nuclear fission. Phys Rev 56(5):426–450. https://doi.org/10.1103/PhysRev.56.426. https://link.aps.org/doi/10.1103/PhysRev.56.426
2. Hyde E (1964) The nuclear properties of the heavy elements. Prentice-Hall, Inc.
3. Scharff-Goldhaber G, Weneser J (1955) System of even-even nuclei. Phys Rev 98(1):212–214. https://doi.org/10.1103/PhysRev.98.212. https://link.aps.org/doi/10.1103/PhysRev.98.212
4. Meija J et al (2016) Isotopic compositions of the elements 2013 (IUPAC technical report). Pure Appl Chem 88(3):293–306. https://doi.org/10.1515/pac-2015-0503
5. Ney EM (2021) Towards a complete theory of beta decay in density functional theory. https://doi.org/10.17615/ceap-7e60
6. Nakayama K, Pio Galeão A, Krmpotić F (1982) On the energetics of the Gamow-Teller resonances. Phys Lett B 114(4):217–221. ISSN: 0370-2693. https://doi.org/10.1016/0370-2693(82)90480-4. https://www.sciencedirect.com/science/article/pii/0370269382904804

7. van der Werf SY et al (1981) Fission probability of the IAR in the actinide region studied via the (3He, tf) reaction. Phys Lett B 105(2):111–115. ISSN: 0370-2693. https://doi.org/10.1016/0370-2693(81)91001-7. https://www.sciencedirect.com/science/article/pii/0370269381910017
8. Osterfeld F (1992) Nuclear spin and isospin excitations. Rev Mod Phys 64(2):491–557. https://doi.org/10.1103/RevModPhys.64.491. https://link.aps.org/doi/10.1103/RevModPhys.64.491
9. Kolbe E, Langanke K, Fuller GM (2004) Neutrino-induced fission of neutron-rich nuclei. Phys Rev Lett 92(11):111101. https://doi.org/10.1103/PhysRevLett.92.111101. https://link.aps.org/doi/10.1103/PhysRevLett.92.111101
10. Kelić A et al (2005) Cross sections and fragment distributions from neutrino-induced fission on r-process nuclei. Phys Lett B 616(1):48–58. ISSN: 0370-2693. https://doi.org/10.1016/j.physletb.2005.04.074. https://www.sciencedirect.com/science/article/pii/S0370269305006039
11. Siemens PJ, Jensen AS (1987) Elements of nuclei: many-body physics with the strong interaction, 1st edn. CRC Press. https://doi.org/10.1201/9780429493904
12. Morrissey DJ, Benenson W, Friedman WA (1994) Measurement of temperature in nuclear reactions. Annu Rev Nucl Part Sci 44:27–63. ISSN: 1545-4134. https://doi.org/10.1146/annurev.ns.44.120194.000331. https://www.annualreviews.org/content/journals/10.1146/annurev.ns.44.120194.000331
13. Heikkinen A, Kaitaniemi P, Boudard A (2008) Implementation of INCL cascade and ABLA evaporation codes in Geant4. J Phys Conf Ser 119(3):032024. https://doi.org/10.1088/1742-6596/119/3/032024
14. Petit M et al (2004) Determination of the 233Pa (n, f) reaction cross section from 0.5 to 10 MeV neutron energy using the transfer reaction 232Th (3He, p) 234Pa. Nucl Phys A 735(3):345–371. ISSN: 0375-9474. https://doi.org/10.1016/j.nuclphysa.2004.02.017. https://www.sciencedirect.com/science/article/pii/S0375947404002064
15. Pozzi SA, Padovani E, Marseguerra M (2003) MCNP-PoliMi: a Monte-Carlo code for correlation measurements. Nucl Instrum Methods Phys Res Sect A Accel Spectrometers Detect Assoc Equip 513(3):550–558. ISSN: 0168-9002. https://doi.org/10.1016/j.nima.2003.06.012. https://www.sciencedirect.com/science/article/pii/S0168900203023027
16. Watt BE (1952) Energy spectrum of neutrons from thermal fission of U^{235}. Phys Rev 87(6):1037–1041. https://doi.org/10.1103/PhysRev.87.1037. https://link.aps.org/doi/10.1103/PhysRev.87.1037
17. Lee HC (1978) Neutral current and the nuclear scattering of reactor antineutrinos. Nucl Phys A 294(3):473–491. ISSN: 0375-9474. https://doi.org/10.1016/0375-9474(78)90232-4. https://www.sciencedirect.com/science/article/pii/0375947478902324
18. Nojarov R et al (1993) Orbital and spin M1 excitations in actinide nuclei. Nucl Phys A 563(3):349–386. ISSN: 0375-9474
19. Avignone FT, Greenwood ZD (1980) Calculated spectra of antineutrinos from the fission products of ^{235}U, ^{238}U, and ^{239}Pu, and antineutrino-induced reactions. Phys Rev C 22(2):594–605. https://doi.org/10.1103/PhysRevC.22.594. https://link.aps.org/doi/10.1103/PhysRevC.22.594

Chapter 5
The NuThor Detector

Abstract This chapter is devoted to outlining the entire design, commissioning and operation of the neutrino-induced nuclear fission detector–named NuThor. We start with the characterization of the thorium metal plates sourced from Oak Ridge National Lab that act as the neutrino target in NuThor. Then, the design of the NuThor instrument is described. Once successfully deployed, NuThor's phases of operation are detailed and Science Run 2 is determined to be the most ideal subject for the analysis conducted throughout the remainder of this thesis. Lastly, the expected signal and backgrounds are summarized with an exhaustive explanation of why each signal candidate and potential background source is addressed in the measurement.

The NuThor detector is the first dedicated neutrino-induced nuclear fission experiment. NuThor was deployed to Oak Ridge National Laboratory's Spallation Neutron Source as a part of the COHERENT Experiment introduced in Sect. 3.1. The centerpiece of the NuThor detector is the thorium target mass, chosen based on the criteria established in Sect. 4.1. The thorium is in the form of metal cylinders and was sourced from Oak Ridge National Laboratory–managed by The Advanced Radiation Detection, Imaging, Data Science, and Applications (ARDIDSA) group and the Nuclear Material Control and Accountability (NMC&A) program. Of the neutrino-induced nuclear fission products detailed in Sect. 2.2.7, prompt fission neutrons were the signal that the instrument was designed to detect, but several possible signal candidates are discussed. The thorium is hermetically sealed in a Neutron Multiplicity Meter (NMM) to observe the neutrons that are emitted from the nuclear fission events. This section outlines the approaches taken to design NuThor and the final product after many rounds of design, simulation, optimization and construction.

T. Johnson, *The First Indication of Neutrino-Induced Nuclear Fission*,
https://doi.org/10.1007/978-3-032-01936-3_5

5.1 The Thorium

A total of 52.0 kg of thorium metal was loaned to the NuThor detector for this initiative. This 52 kg target consists of fifteen 6-inch diameter plates (depicted in Fig. 5.1). Due to the unprecedented nature of bringing thorium, a nuclear material, into the SNS facility, numerous additional precautions were implemented to ensure safety and compliance with broader ORNL protocols for handling and storing nuclear materials. Even though thorium is not classified as a special nuclear material (SNM), it is categorized as a source material due to its modest proliferation risk when used to breed ^{233}U. While the thorium usage in NuThor does not pose a high-risk category, it does necessitate certain safety considerations. These precautions were promulgated from many rounds of discussion and negotiation between COHERENT representatives, SNS Radiation Control personnel and many other dedicated members of the ORNL safety apparatus. These risk mitigation strategies do not directly impact the physics processes being investigated in NuThor, but they do place a greater burden on the experimenter, requiring multiple layers of permissions and the presence of on-site personnel to facilitate any modifications to the NuThor instrument (Fig. 5.2).

Fig. 5.1 All 15 thorium metal plates stacked concentrically and tagged with individual sample numbers for bookkeeping. This is their orientation inside the aluminum canister prior to being sealed as a single encapsulated radioactive source

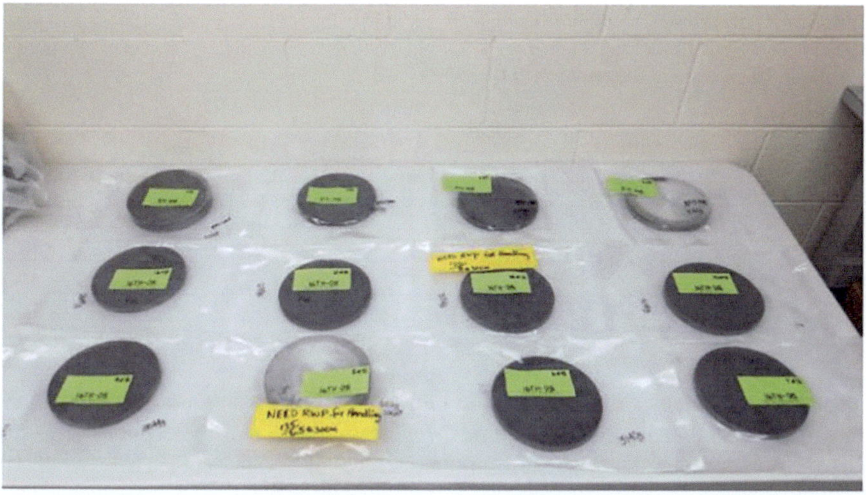

Fig. 5.2 12 of the 15 thorium metal plates sourced for usage in the NuThor detector as target mass. The thorium plates were coated with a $\sim \mu$m layer of aluminum to prevent corrosion. These plates were later installed into a custom aluminum canister and plastic lining

Some of these additional safety precautions include:

- The thorium was required to be turned into one large encapsulated radioactive source akin to a button source e.g. ^{137}Cs, ^{133}Ba. This was done with a 12 mil[1] custom-made plastic bag from the ORNL Bag Shop and a custom-made aluminum canister manufactured also at ORNL. The thorium plates were then stacked one on top of the other for installation into the bag and then the aluminum canister. Shown in Fig. 5.1.
- A tamper proof lock was required to be installed somewhere on the NuThor instrument on a hinged door. A radiation label was also added to the face of NuThor.
- The entirety of the apparatus needed to be covered in some form of cage to inhibit quick access to the contents of the instrument.
- The thorium stack had to be covered in a thick shield of lead then checked by radiation control personnel to verify the radioactive dosage immediately outside of the lead shielding. This determined whether Neutrino Alley needed to have a permanent radiation area–a step that was widely viewed as non-ideal.
- The thorium would be accounted for by the SNS Material Balance & Accountability (MBA) representative for regular inspections and verification of adherence to the above terms.

Because thorium is a material that must be tightly controlled and stewarded, a relatively precise measure of all the thorium plates' mass was recorded and utilized for event rate estimation. The masses of all 15 thorium plates are listed in Table 5.1.

[1] A mil is a unit referring to one thousandth of an inch.

Table 5.1 The masses of the 15 thorium metal plates used in the NuThor detector's target. The mass of the plates were accurately accounted for by ORNL NMC&A. The concentration of contaminants in the thorium plates was measured as 200 ppm

Piece ID	Mass [g]	Piece ID	Mass [g]	Piece ID	Mass [g]
1	2922	6	3140	11	3095
2	2930	7	2864	12	2849
3	3090	8	3148	13	2788
4	3084	9	3091	14	3152
5	6312	10	3083	15	3145

These thorium plates were used in previous measurements at Oak Ridge National Laboratory and the subsequent technical reports indicate that these plates are very pure thorium with contaminants present at the level of 200 ppm [1]. This was determined by chemical analysis of the scraps left over from the plates' fabrication.

5.1.1 *Radioactivity Survey*

Prior to using these thorium metal plates as a neutrino detector target mass, spectroscopic studies were performed to better quantify the radioactive backgrounds that the metal plates themselves would pose to the neutrino-induced nuclear fission experiment. Examination of the thorium plates was conducted using ORTEC portable, high-purity germanium (HPGe) radioisotope identification devices (RIIDs) at one of Oak Ridge National Laboratory's facilities [2]. One Micro-Detective-HX[2] and one Detective-200 Ruggedized, Ultra-High Sensitivity, Transportable HPGe Radionuclide Identification System.[3] Both were loaded with software for immediate detection of special nuclear materials, which help assuage any anxiety of possible ^{233}U contamination.

ORTEC's Micro-Detective is a 50 mm diameter x 33 mm P-type high purity germanium crystal held inside a cryostat and cooled down with a low-power Stirling cycle cryo-cooler. The Detective-200 is of similar design but boasts a larger HPGe crystal at 5675 sq mm. The two ORTEC spectroscopy instruments were placed far enough away from the bare thorium plate to not induce significant dead time for the measurement. This γ-ray spectroscopy is informative for what level of lead shielding around the thorium target mass is needed for (1) safety of experimenters in Neutrino Alley and (2) the successful operation of particle detectors instruments throughout the NuThor apparatus.

---

[2] https://analytical-online.com/uploads/3/4/8/8/34882511/micro-detective-hx.pdf.

[3] https://www.ortec-online.com/products/portable-radioisotope-identification-devices-riids/detective-200.

The energy spectra captured by the HPGe detectors were processed using Sandia's Interspec software [3] which performed the energy calibration via the observation of known radioactive peaks (e.g. ^{40}K and ^{208}Tl) then iteratively fit the observed photopeaks with gaussian models on either a linear or exponential background model. The fit peaks were compared with a catalog of measured γ-ray lines from the Thorium decay series and all fit peaks were matched with a known radioactive process in the decay series. The result of this analysis is shown in Table 5.2. The best fit position of the γ-ray line is given in the center column. The integrals of the fit model are given in the rightmost column (Fig. 5.4).

With a view toward future energy calibration using known photopeaks observable in the energy spectrum, it is important to note certain features. The ^{208}Tl photopeak is the most problematic background emitted from the thorium and occurs in the beta decay from ^{208}Tl to an exited state in ^{208}Pb then the subsequent 2.614 MeV γ-ray

Table 5.2 Table of radioisotopes, their energies, and counts per second (cps). These values were measured by the ORTEC HPGe detectors shown in Fig. 5.3. The individual decays were identified using the software, Interspec

Radioisotope	Energy (keV)	Cps
^{212}Pb → ^{212}Bi β decay	238.63 ± 1.886	1.668 ± 0.092
^{228}Ac → ^{228}Th β decay	338.32 ± 2.085	0.6463 ± 0.0612
^{228}Ac → ^{228}Th β decay	462.88 ± 2.085	0.4268 ± 0.0445
^{208}Tl → ^{208}Pb β decay	510.77 ± 2.417	2.141 ± 0.073
^{208}Tl → ^{208}Pb β decay	583.19 ± 1.895	2.707 ± 0.075
^{212}Bi → ^{212}Po β decay	727.33 ± 2.099	0.742 ± 0.0444
^{228}Ac → ^{228}Th β decay	772.29 ± 1.924	0.16 ± 0.0296
^{228}Ac → ^{228}Th β decay	794.95 ± 1.857	0.5573 ± 0.0375
^{228}Ac → ^{228}Th β decay	835.71 ± 1.230	0.129 ± 0.0229
^{208}Tl → ^{208}Pb β decay	860.56 ± 2.180	0.4372 ± 0.0339
^{228}Ac → ^{228}Th β decay	911.20 ± 1.987	2.742 ± 0.072
^{228}Ac → ^{228}Th β decay	964.77 ± 1.864	0.4843 ± 0.034
^{228}Ac → ^{228}Th β decay	968.97 ± 1.921	1.699 ± 0.057
^{228}Ac → ^{228}Th β decay	1459.14 ± 2.456	2.264 ± 0.064
^{228}Ac → ^{228}Th β decay	1495.91 ± 2.925	0.09982 ± 0.01688
^{228}Ac → ^{228}Th β decay	1501.57 ± 2.982	0.05386 ± 0.01526
^{228}Ac → ^{228}Th β decay	1580.53 ± 2.689	0.07394 ± 0.01601
^{228}Ac → ^{228}Th β decay	1588.20 ± 2.616	0.3182 ± 0.0267
Double escape peak	1592.24 ± 2.577	0.198 ± 0.0221
^{212}Bi → ^{212}Po β decay	1620.95 ± 2.724	0.1657 ± 0.0207
^{228}Ac → ^{228}Th β decay	1630.75 ± 1.669	0.13 ± 0.018
Single escape peak	2103.78 ± 3.192	0.37 ± 0.0286
^{208}Tl → ^{208}Pb β decay	2614.58 ± 2.987	2.847 ± 0.07

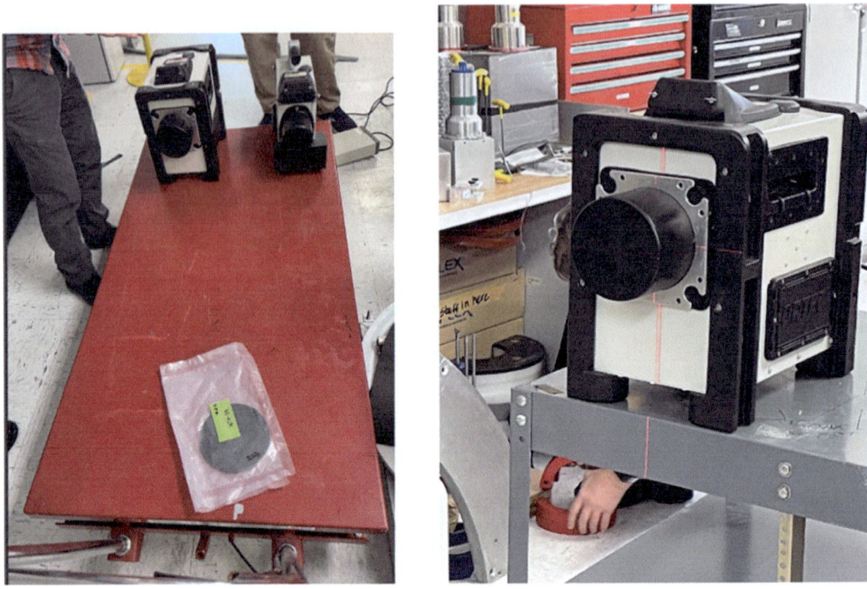

Fig. 5.3 (left) The survey setup at ORNL to assess the radioactive background posed by the thorium plates. The plates were studied in the position shown here as well as at a further distance to reduce the dead-time experienced by the HPGe detectors. (right) Laser alignment used to place the two HPGe detectors at equidistant positions from the target sample

to the ground state. However, a photopeak at 2.614 MeV is not the only vestige left by this beta decay. These higher energy γ-rays are able to both Compton scatter and produce escape peaks.

5.1.2 Double Escape Peak

When a photon propagates in a medium with nuclei present and possesses greater than 1.022 MeV, there is a chance of interaction via pair production [4]. There is also a chance that when this photon pair produces, both the electron and positron escape the detection apparatus without depositing their respective energies [5]. When this happens, the new photopeak can be observed at an energy equivalent to the difference between the originating photon energy and sum of the electron-positron rest masses–this is called a double escape peak. When only one of the pairs escape, that is a single escape peak. Both of these peaks can be used in an energy calibration, but it behooves the experimenter to know if there is additional activity in the immediate vicinity. In this case, immediately next to the double escape peak at 1.592 MeV is the γ-ray from the beta decay from ^{228}Ac $\rightarrow$ ^{228}Th at 1.588 MeV as shown in Fig. 5.5. An energy calibration using the Double escape peak would be receiving events from both the double escape peak and the ^{228}Ac β-decay.

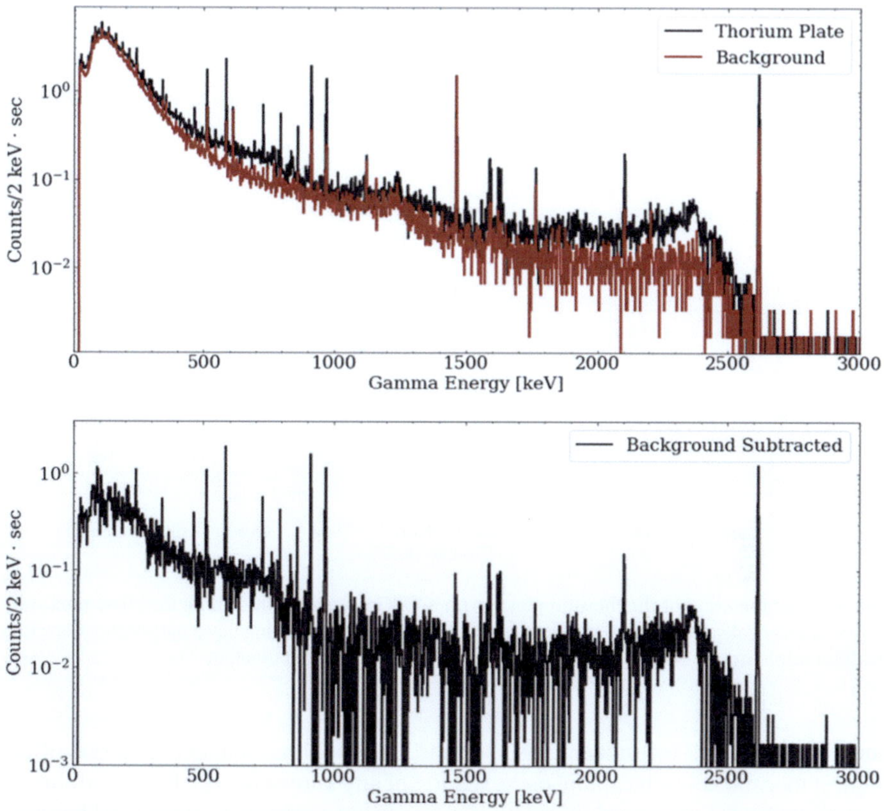

Fig. 5.4 Energy spectra collected using the ORTEC HPGe detectors in proximity to one of the ^{232}Th metal plates available at ORNL. Only one plate was surveyed in this manner because each of the 15 total plates have nearly the same history and masses. (Top) the thorium plate's radioactive energy spectra and the recorded background with the thorium plate removed. The two distributions are normalized by their respective live times for direct comparisons. (Bottom) the background subtracted energy spectrum of gamma rays observed in the HPGe detector from the thorium plate

5.2 Overview of the NuThor Design

The NuThor detector is an apparatus meant to detectorize the thorium plates introduced in the previous section. NuThor is a bespoke neutron counter. The facade of the detector is shown in Fig. 5.6. The NuThor detector design is comprised of a stack of ^{232}Th plates encapsulated by a neutron capture apparatus. Because the ideal signal for neutrino-induced nuclear fission is 3+ prompt neutrons, the detector design must be optimized to capture as many of these neutrons as possible. Our goal is therefore to construct what is known as a neutron multiplicity meter (NMM) [6, 7]. Of the cadre of efficient neutron capture materials (Li, B, Cd, Gd, etc.), gadolinium dissolved in water was chosen for the design of the NuThor detector. Not only is water an effective

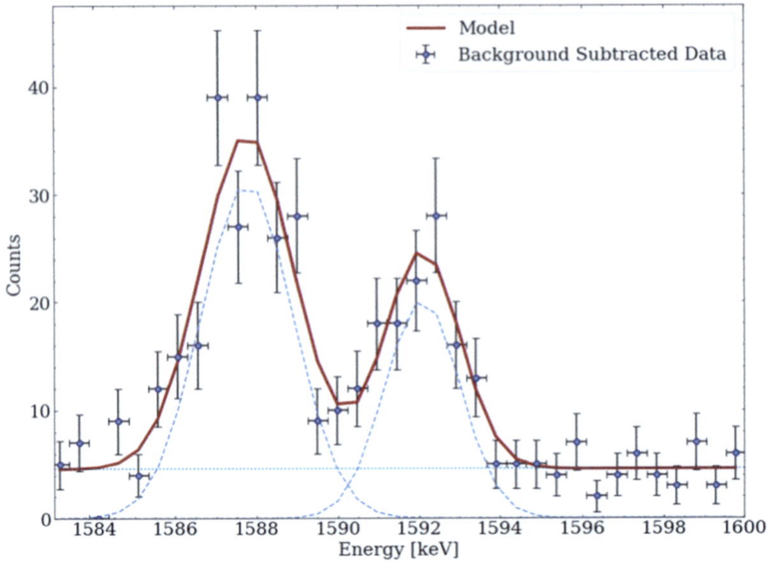

Fig. 5.5 Examination of the thorium plates with a HPGe detector illustrating the close proximity of two photo-peaks. The HPGe energy resolution is capable of resolving these but another detection medium might not be able to so it would appear as the two blurred together

and cheap material for moderating neutrons, but gadolinium also boasts one of the highest thermal neutron capture cross section of any known material.[4] Gadolinium is a popular tool for particle physics [8]; it is a possible contrast agent for MRI imaging [9]; it even maintains a shutdown state in CANDU reactors [9].

A relatively small amount of gadolinium dopant significantly increases the efficiency of neutron absorption, with neutron capture also resulting in the emission of approximately 8 MeV of γ-rays. Additionally, since NuThor is deployed in Neutrino Alley, it is subject to the constraint that no object can obstruct the service corridor, limiting its depth to 1 m. While this is unusual compared to the typical size of neutrino detectors worldwide, COHERENT has successfully constructed several neutrino detectors within this footprint. A compact neutrino-induced nuclear fission detector is made possible by the rapid thermalization and subsequently capture of neutrons in gadolinium-loaded water (Figs. 5.6 and 5.8).

Next, the design of the instrument will be reviewed in detail. Due to the nature of shielding and safety precautions, the outermost vantage of the instrument is not very illuminating. A sectional view of the detector is more instructive on how the detector works. Stepping through the detector's components moving from the center outwards:

---

[4] Gadolinium chief competitors for the title of best neutron capturing agent are neutron poisons created ephemerally in nuclear reactor cores.

Fig. 5.6 The façade of the NuThor detector in Neutrino Alley at the Spallation Neutron Source. The outermost portion of the detector is covered in a custom made cage along with a tamper proof lock at the topmost compartment of the instrument. These are precautions necessary to be in compliance with the SNS radiation control and material balance & accountability (MB&A) policies. An MB&A representative for this location in the SNS must routinely inspect the status of these precautions to stage in good standing. Additionally, because borated polyethylene falls in the list of combustible materials, a thin layer of aluminum sheet metal was installed on all outward pointing faces of the neutron shielding material in order to comply with directives from the Fire Marshall

- In dark grey is the 52 kgs of thorium metal inside the custom aluminum canister.
- The green immediately around the thorium is lead shielding to attenuate the radioactivity of the thorium target. The lead and thorium are their own seismically decoupled system called the "Inner core." Figure 5.7 depicts the lead and thorium inner core as well as the result of a seismic study examining the deflection under horizontal force.
- In cyan, are custom made polypropylene water bricks. These water bricks are doped with a solution of either 2.5% or 1% gadolinium nitrate hexahydrate by mass. Each brick is 3.75" × 2.25" × 10" and easily re-positioned within NuThor. The water is an effective neutron moderators and the gadolinium is a high efficiency neutron capture agent even at low concentrations.

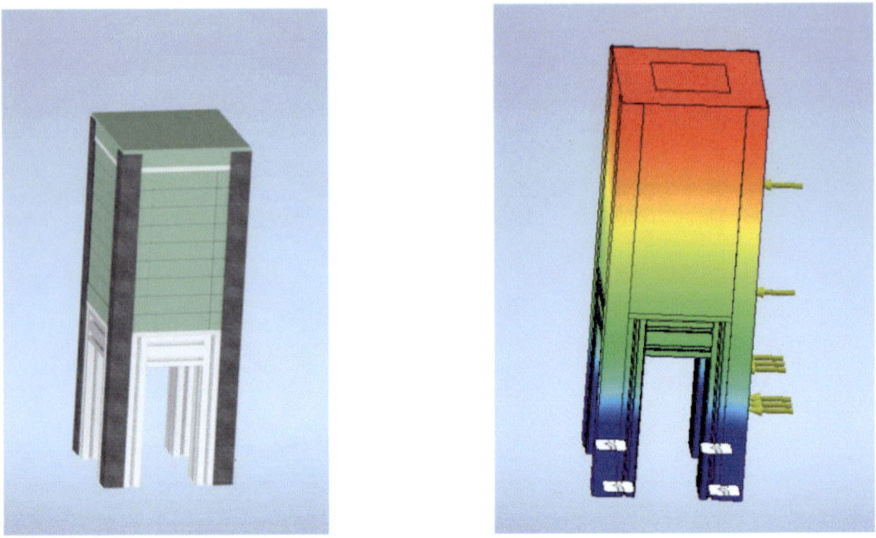

(a) CAD rendering (b) Seismic stability analysis

Fig. 5.7 The inner core of the NuThor detector. It is seismically isolated from the remainder of the NuThor detector apparatus

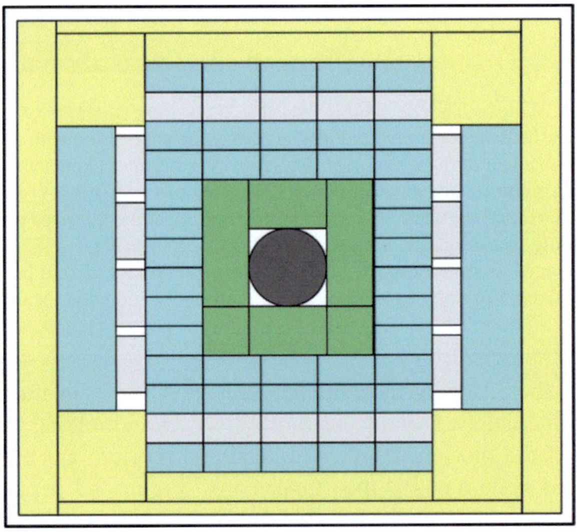

Fig. 5.8 Illustration of the cross sectional view of NuThor in it's production run #1 configuration. This is a bird eye's view looking downward from the top of the apparatus. Dark Grey = thorium target mass stack. Green = Lead shield. Cyan = Polypropylene water bricks doped with gadolinium nitrate hexahydrate. Light Grey = NaI[Tl] scintillator detectors. Yellow = Borated polyethylene

Fig. 5.9 Illustration of the cross sectional view of NuThor in it's production run #1 configuation. This is a point-of-view looking at the instrument from the vantage point of in Neutrino Alley with the SNS wall immediately behind it. Dark Grey = thorium target mass stack. Green = Lead shield. Cyan = Polypropylene water bricks doped with gadolinium nitrate hexahydrate. Light Grey = NaI[Tl] scintillator detectors. Yellow = Borated polyethylene

- In light grey are the 35 NaI[Tl] scintillating crystals. These are the only active components in NuThor. NaI is an effective γ-ray detection medium for its high stopping power.
- In yellow is the 30% borated high density polyethylene. Borated polyethylene is similar to the Gd-doped water in that the hydrogenous plastic is a good neutron moderator and the boron interspersed in its volume is a good neutron capture agent. The borated polyethylene is exclusively for shielding the detector from outside neutrons–though RCT surveys have not yielded measurable quantities of thermal neutrons in Neutrino Alley (Fig. 5.9).

5.2.1 *NaI Scintillators*

The active components of the NuThor detector are 36 NaI[Tl] scintillating crystals inherited from the Department of Homeland Security's Advanced Spectroscopic Portal Program. Roughly seven tons of these instruments were inherited by a combi-

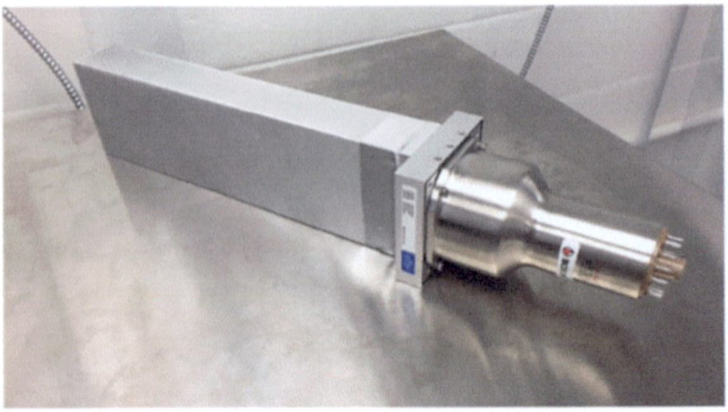

Fig. 5.10 Example of one of the NaI[Tl] scintillating crystals used for the NuThor experiment and several other neutrino detectors deployed to Neutrino Alley in previous campaigns

nation of universities and national laboratories. NaI is a common radiation detector material thanks to it's high photon stopping power and large light yield. It is used in numerous dark matter and neutrino experiments worldwide [10]. Further, a common practice is to dope the NaI crystals with Thallium to increase the light yield after energy depositions. This approach was used by PICO-LON [11], ANAIS [12], and DM-Ice [13] to name just a few.[5]

Each of the NaI[Tl] scintillators has 7.7 kgs of NaI with dimensions of 15.786" × 4.375" × 2.25". These NaI crystals are encapsulated in an aluminum housing and come instrumented with 3.5" diameter BURLE S832013F photomultiplier tubes with 10 stage dynode amplification (See Figs. 5.10 and 5.11).

5.2.2 *Neutron Multiplicity Meter*

There are several potential neutron capture agents that can be employed in the NuThor detector, each with varying degrees of effectiveness due to differences in their neutron capture cross sections. Figure 5.12 shows a comparative analysis of several commonly used neutron capture agents through their neutron absorption cross sections as a function of the neutron's kinetic energy. This information was collected via the JENDL 4.0 database [14]. As one can see from the head-to-head comparisons, gadolinium outperforms the other possible choices especially at thermal energies (Fig. 5.12).

Gadolinium's massive thermal neutron capture cross section allows experimenters to attain high neutron capture efficiency by doping a volume of water or liquid scintillator with a fairly small concentration of the chosen gadolinium compound. Several neutrino experiments like Super-Kamiokande [15] and Daya Bay [16] detect final

[5] Thallium dopants work to increase the observable light yield because after the energy depositions occur in the crystal medium, it creates the quasi-particle known as excitons. The excitons can encounter and energize a thallium ion that will the radiatively de-excite via the emission of a photon. Conversely, an exciton could instead scatter off a phonon propagating through the crystal's lattice. The now excited phonon would likely lose energy non-radiatively.

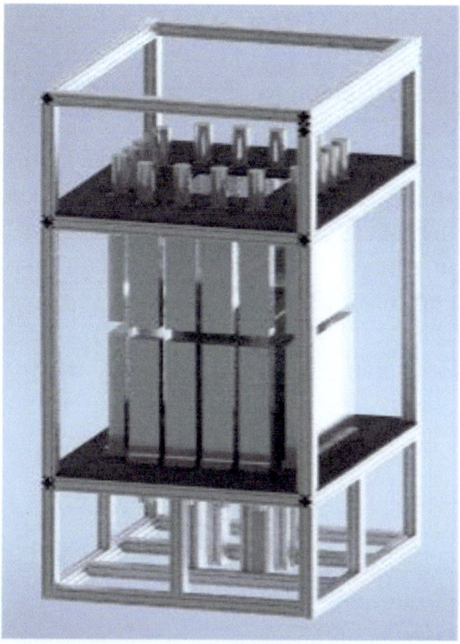

(a) CAD mock up of the Neutron multiplicity meter.

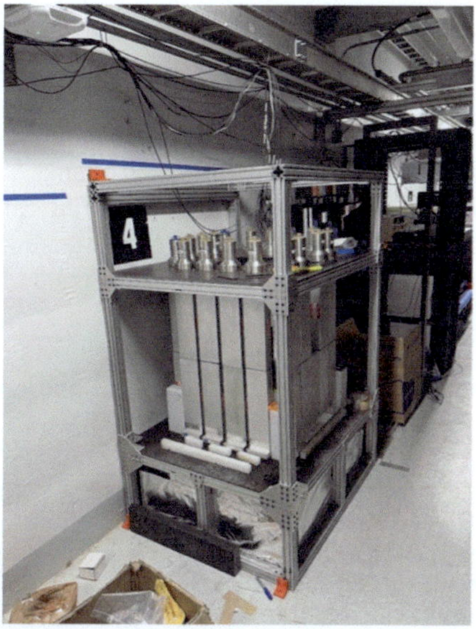

(b) The neutron multiplicity meter under construction in place in Neutrino Alley.

Fig. 5.11 The NaI[Tl] scintillator configuration within the Neutron Multiplicity Meter portion of NuThor

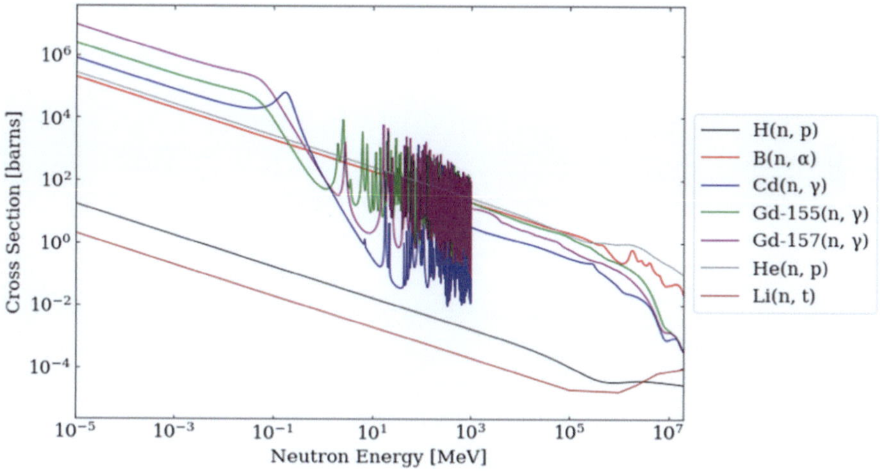

Fig. 5.12 The neutron absorption cross sections of common neutron capture agent isotopes as a function of the neutron's kinetic energy. This information was extracted from the JENDL 4.0 database

state neutrons by doping their detector volumes with 0.03% and 0.1% gadolinium by mass, respectively. The dark matter experiments LZ (0.1%) [17] and XENONnT (0.48%) [18] utilize gadolinium-doped outer detectors for tagging background neutrons.

The gadolinium was mixed with water inside small, modular water bricks that were custom made from Go-To-Tanks,[6] a commercial plastic tank vendor. This configuration was chosen for its flexibility, allowing for easy adjustments to the detector's inner components to accommodate new requirements. The chosen configuration of gadolinium-doped water bricks inside the neutron multiplicity meter is shown in Fig. 5.13.

Natural gadolinium has the following isotopic abundances and respective thermal neutron capture cross sections (Table 5.3).

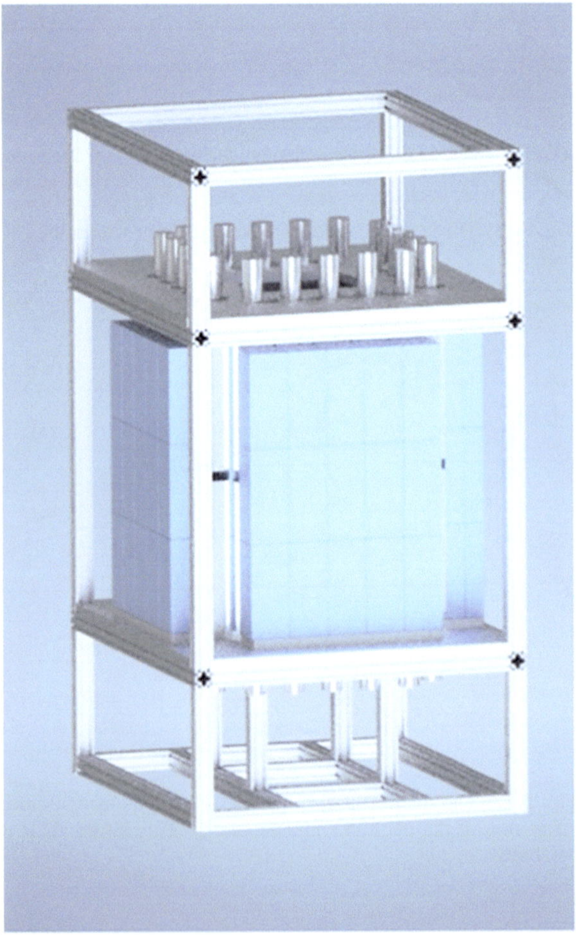

Fig. 5.13 Configuration of the custom-made polypropylene water bricks inside NuThor's neutron multiplicity meter. The water bricks are the light blue rectangles stacked around the vertically placed NaI detectors. There are additional water tanks on the inside

[6] https://gototanks.com/.

Table 5.3 The abundance of the naturally occurring gadolinium isotopes and their associated radiative thermal neutron capture cross sections [19]. The isotope with the greatest thermal neutron capture cross section is ^{157}Gd followed by ^{155}Gd

Isotope	Abundance [%]	Cross section [b]
^{152}Gd	0.200	735
^{154}Gd	2.18	85
^{155}Gd	14.80	60,900
^{156}Gd	20.47	1.8
^{157}Gd	15.65	254,000
^{158}Gd	24.84	2.2
^{160}Gd	21.86	1.4

The two largest contributors to gadolinium's effectiveness as a neutron capture agent are ^{157}Gd and ^{155}Gd which have the following reactions:

$$^{157}\text{Gd} + n \rightarrow ^{158}\text{Gd} + \gamma' s (Q = 7.937\text{MeV}) \tag{5.1}$$

$$^{155}\text{Gd} + n \rightarrow ^{156}\text{Gd} + \gamma' s (Q = 8.536\text{MeV}) \tag{5.2}$$

However, the Q-value of the neutron capture reaction is not uniformly emitted through a single γ-ray. Instead, the total Q-value[7] is distributed amongst several γ-rays. Figure 5.14 illustrates how these energies are typically distributed.

Gadolinium Compound

Two gadolinium compounds were evaluated: gadolinium nitrate ($Gd(NO_3)_3$) and gadolinium sulfate ($Gd_2(SO_4)_3$). Notably, both were assessed for usage in the campaign to add gadolinium to the Super-Kamiokande experiment (which also consid-

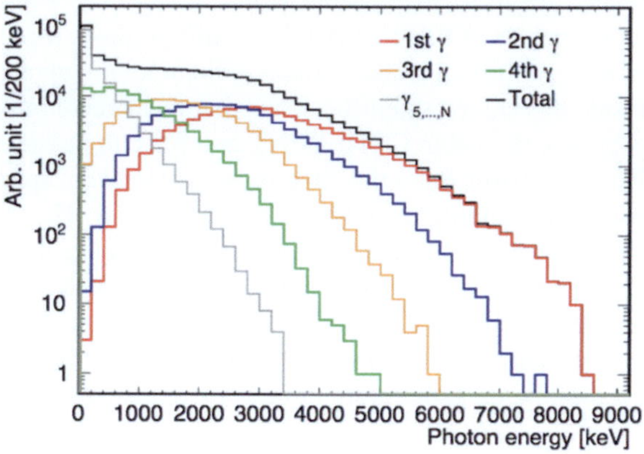

Fig. 5.14 Distribution of γ-ray energies broken into the components delineated by the gamma multiplicity [20]

[7] The Q-value is the total energy available in the nuclear reaction.

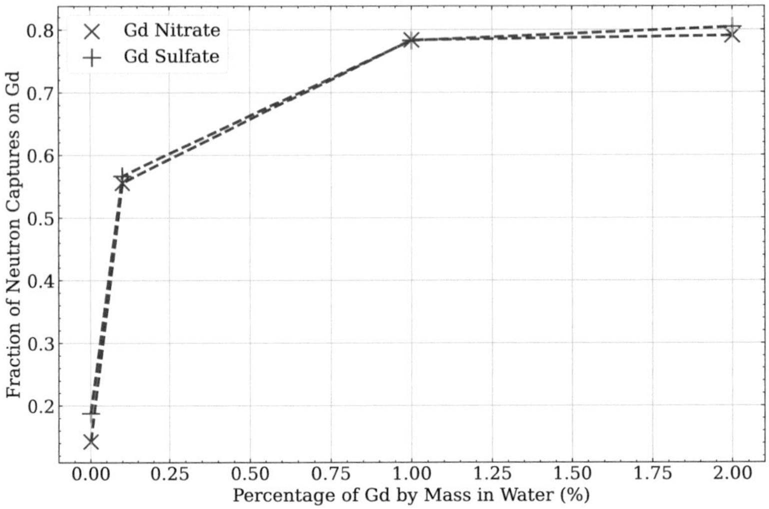

Fig. 5.15 Head-to-head comparison of gadolinium nitrate and gadolinium sulfate using MCNP-Polimi simulations at a range of gadolinium compound concentrations. With identical starting conditions except for the exact gadolinium compound, these two compounds were found to perform the same

ered Gadolinium Chloride[8]) [21]. Gadolinium nitrate is dissolved into water simply by mixing. Conversely, gadolinium sulfate requires both mixing and the addition of sulfuric acid to dissolve into the water. The two test gadolinium compounds were simulated in an early version of the NuThor detector design's geometry using MCNP-Polimi. The neutrons thrown in this simulation were initialized inside the thorium target mass and used MCNP's built-in thorium spontaneous fission source to generate a sampling of neutron energies following the Watt curve [22].

The goal of this detector is to capture as many neutrino-induced nuclear fission-related neutrons as possible while functioning within the fairly restrictive space requirements of Neutrino Alley. For this purpose, we sought to optimize the thickness of the gadolinium-water volume without taking up an unnecessary amount of space inside this shared scientific space that must accommodate numerous other neutrino experiments within COHERENT and for safe passage through the service corridor. Figure 5.16 illustrates the effectiveness plateaus at approximately 10 in. of Gd-loaded water surrounding the thorium target material. The study was performed with both gadolinium sulfate and nitrate to directly compare their effectiveness in each context.

Based on Figs. 5.15 and 5.16, the performance of the two gadolinium compounds are comparable in the window of concentrations tested. Because of this result and the fact that gadolinium sulfate requires the additional step of mixing in sulfuric acid for dissolving, gadolinium nitrate was selected as the ideal material.

---

[8] The Super-Kamiokande gadolinium demonstrator experiments evaluated gadolinium chloride, but was not examined to the same level of scrutiny because it was corrosive to the stainless steal tanks used in the experiment.

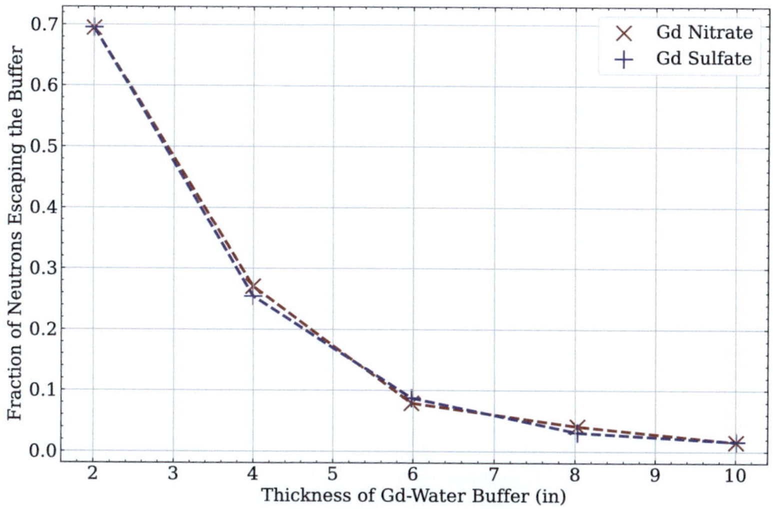

Fig. 5.16 Head-to-head comparison of the gadolinium nitrate and gadolinium sulfate using MCNP-Polimi simulations at a range of thicknesses of Gd-Water buffer

An additional optimization was performed to determine the ideal concentration of gadolinium nitrate to mix into the water in the water bricks. Although several other neutrino-related experiments used rather minuscule concentrations of gadolinium compounds (0.1–5%), those experiments were mostly multi-ton detectors where higher concentrations would be cost prohibitive. For NuThor, a small-scale detector, concentrations of up to 50% were simulated and the results are shown in Fig. 5.17.[9] The optimal concentration of gadolinium nitrate is 10% according to these studies, but with a rapid die-off in growth after approximately 2%.

The reduced efficiency of neutron capture on gadolinium can be better comprehended by examining neutron captures on other nuclei within the NuThor detector (See Fig. 5.18). The current understanding is that as the gadolinium concentration reaches extremes, there are fewer hydrogen nuclei for the neutrons to encounter thus reducing the efficiency of moderating the neutrons down to thermal energies where gadolinium's absorption cross section is far greater. Instead, the neutrons more readily escape from observation and either leave the detector all together or reach the borated polyethylene on the outer edge of the neutron multiplicity meter which has no mechanism for observing these neutrons.

Another consideration is that although the gadolinium compound may dissolve when initially mixed with water, it can still eventually precipitate out of the solution. To control for this hypothesis, several concentrations of gadolinium-water solutions were prepared and observed over a long period of time (several years). The vials with

[9] It's important to keep in mind that the chemistry of the solution is not considered this exercise. Although the 50% concentrated solution was simulated, it is highly unlikely that level of gadolinium nitrate would stay dissolved in the water without precipitating out.

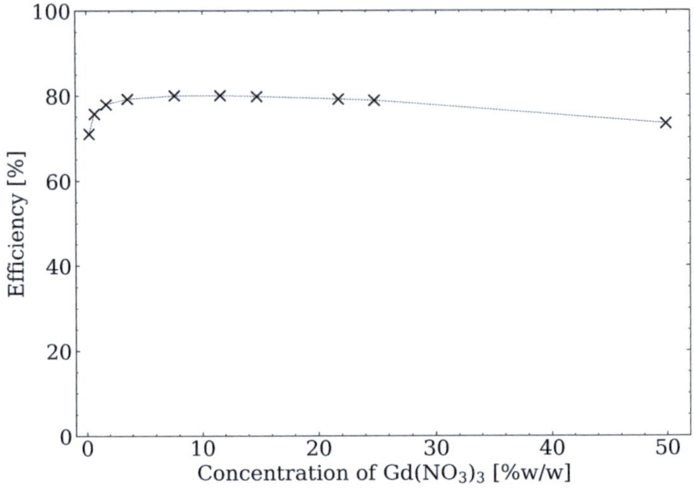

Fig. 5.17 The percentage of neutrons captured on one of the gadolinium isotopes in gadolinium nitrate with varying concentrations by mass. There is a diminished effectiveness after 10% concentration

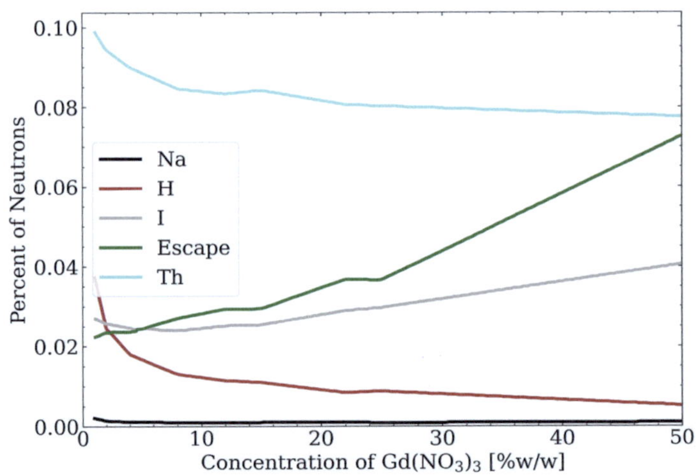

Fig. 5.18 Neutron capture percentages on non-gadolinium nuclei as the gadolinium concentration is increased

Table 5.4 Emission spectroscopy analysis results on the ProChem Inc. gadolinium nitrate hexahydrate. The results are reported in parts per million. ND indicates the element is not detectable

Ag	ND	Cd	0.19	Eu	<10
Al	0.89	Ce	<20	Nd	<20
As	<2	Co	ND	Ni	ND
Ba	<2	Cr	0.22	Pb	ND
Be	ND	Cu	ND	Pr	<20
Bi	ND	Dy	<10	Sb	ND
Ca	<1	Er	<10	Se	ND
Si	<2	Sm	<10	Sn	<0.5
Sr	ND	Tb	<25	Ti	ND
V	ND	Y	<10	Yb	<10
Zn	<0.2	Zr	ND		

10+% gadolinium nitrate had notable precipitant present at the bottom in the form of a yellow-ish residue. All the vials of solution below this concentration exhibited no precipitant after greater than two years of observation.

Because of these studies, a 2.5% concentration by mass of gadolinium nitrate hexahydrate is used for the innermost section of Gd-water bricks. Because simulations show that the majority of neutrons capture within the first few cm of water, the importance of the gadolinium concentration in water bricks further out is lessened. Thus those bricks receive a concentration of 1% by mass.

Five kilograms of gadolinium-nitrate hexahydrate was acquired from ProChem Inc.[10] where emission spectroscopy was performed on the purchased sample with the following result in Table 5.4.

The sample is 99.9% gadolinium nitrate with a measured Gd percentage of 34.85%. The spectroscopic study helps bring confidence that there is not another neutron capture agent in the sample that could divert neutrons from the detection channel we expect. The only contaminant found in the test sample sent from ProChem Inc. that is typically used as a neutron capture agent is cadmium and it is present at a mere 0.19 parts per million.

5.3 Deployment

Beginning in the Spring of 2022, the lengthy process of deploying the NuThor detector to the SNS began at Duke University. All individual components of the detector were manufactured in-house at Duke/TUNL or commissioned with local

[10] https://prochemonline.com/.

(a) The neutron multiplicity meter 80-20 frame and Inner core being test built at Duke University.

(b) The neutron mulitplicity meter 80-20 frame and Inner core being installed into Neutrino Alley in the Summer of 2022.

Fig. 5.19 The early stages of the NuThor deployment in 2022

manufacturers such as Benco Steel Inc. who fabricated and plasma cut the steel plates used to affix the NaI scintillators in place (Fig. 5.19).

The deployment of NuThor happened in stages starting with phase 0 and finally reaching phase 2 which constitutes NuThor reaching maturity and taking science data suitable for incorporation into the analysis.

5.3.1 *NuThor Phase 0*

The detector was commissioned in Phase 0. This included installing all necessary components in Neutrino Alley, completing seismic safety studies, ensuring fire safety compliance, and verifying the functionality of all electronics and hardware required for the successful operation of the experiment (Fig. 5.20).

Another key aspect of Phase 0 was the absence of thorium, as the necessary permissions from SNS and ORNL had not yet been obtained. During this period without the thorium, the data acquisition system was set up and background data were taken. This data is limited in usefulness to the final analysis because only 26 of the 36 NaI crystals were installed, and the front shielding was incomplete. Both were intentionally left unfinished to facilitate the future installation of thorium through the detector's front face (See Fig. 5.21).

Additionally, the data collected during this phase used an internal trigger with a level threshold, followed by online energy integration using the digitizer's onboard accumulators—referred to as 'List Mode.'

(a) NuThor reaching full construction awaiting the installation of the thorium target in the center. The front instrumentation and shielding had yet to be included to allow for an easy inclusion of the thorium.

(b) Close up of the NuThor inner cavity showing where the thorium would be installed once the necessary SNS & ORNL permissions were achieved.

Fig. 5.20 NuThor Phase 0 status. At this stage, the thorium was no yet procured, but the majority of the hardware was deployed to Neutrino Alley

(a) The thorium target arriving at the SNS and being unpacked by the SNS research mechanic team led by David Proveaux.

(b) The thorium target encapsulated in its custom canister and packaged in a DOT standard radioactive material drum and custom cribbing constructed by ORNL carpenters.

Fig. 5.21 Thorium target mass being delivered to the SNS for NuThor Phase 1. This operation was conducted with a number of ORNL research technicians, radiation control technicians, and COHERENT personnel

5.3.2 NuThor Phase 1

NuThor Phase 1 was the first official data-taking campaign and took place between December 2022 and March 2023. The data collected from this phase is also called Science Run 1. This phase also marked the first data taken with the thorium target mass installed inside the detector. Within the first week, the data acquisition method was switched from List Mode (internal trigger with level threshold and accumulator integration) to an external trigger with recorded waveforms, driven by the desire for a 'zero-threshold' analysis.

5.3.3 NuThor Phase 2 and On

Science Run 1 illuminated several key areas to upgrade for future campaigns starting in Science Run 2. During Science Run 1, it was apparent that the dynamic range of the digitizer was not being sufficiently exploited. So the first alteration made was to increase the gain on all detector channels from their initial 900 V value. Several voltages were tested in "List Mode" in CoMPASS The results are depicted in Fig. 5.22.

Another upgrade performed between Science Run 1 and 2, was the addition of extra lead shielding around the inner core of NuThor. The Science Run 1 dataset confirmed that the intrinsic radioactivity of the thorium target mass was significant and needed to be mitigated with additional high-Z, dense passive shielding material. Consequently, two inches of lead were added to the existing lead stack around the inner core in the configuration shown in Fig. 5.23.

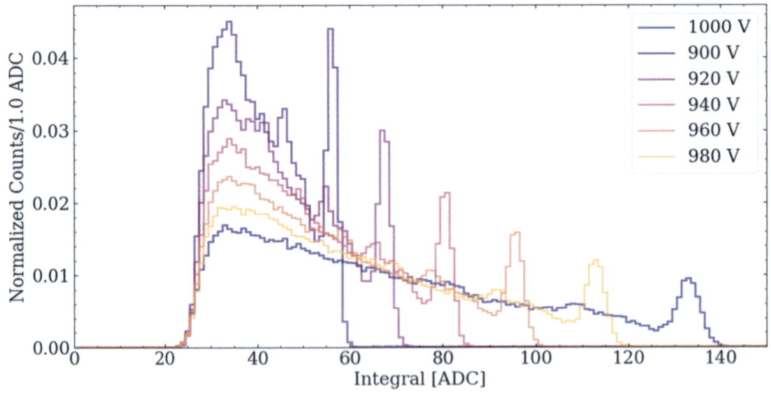

Fig. 5.22 Test of multiple voltages applied to the PMT bases in NuThor to illustrate the effect of varying voltages on the background energy spectrum (in ADC) recorded in CoMPASS's List Mode. Increasing the applied voltage proportionally increases the integral of the notable features in the energy spectrum as one can see the migration of the Tl-208 gamma ray peak at 2614 keV

Fig. 5.23 A schematic illustration of the newly installed lead shielding places around the existing inner core lead shielding in NuThor

This added 341 kg to the existent payload of lead of 536 kg for a total of 877 kg. This is nearly equivalent to the lead target mass deployed in the Lead Neutrino Cube referred to in Chap. 3. This, in turn, increases NuThor's potency as a lead NIN measuring instrument, but also increases the NIN background for a dedicated neutrino-induced nuclear fission measurement by populating the end product with beam-correlated neutron events (Fig. 5.23).

The additional lead proved highly effective in reducing the background rate from the thorium target's radioactivity on the NaI scintillators. The effects of the lead upgrade on the NaI trigger rates is illustrated in Figs. 5.24 and 5.25. Note that the observed attenuation of the thorium-related radioactive background is degenerate with the increase in PMT gain from 900 V to 1000 V. Had the voltage remained constant between production runs 1 and 2, the level of attenuation would have been greater than what is demonstrated here.

5.4 Electronics and Data Acquisition System

NuThor's 36 NaI scintillating crystals are equipped with PMTs connected to custom dual-gain PMT bases, originally designed by Dr. Sam Hedges and Lorenzo Fabris for NaIvETe [23] and constructed by a dedicated team led by Dr. Adryanna Major, with members from various COHERENT institutions. The original intent of these bases were primarily for the NaIvETe subsystem but they were gracious enough to supply NuThor with this equipment which is greatly appreciated (Fig. 5.26).

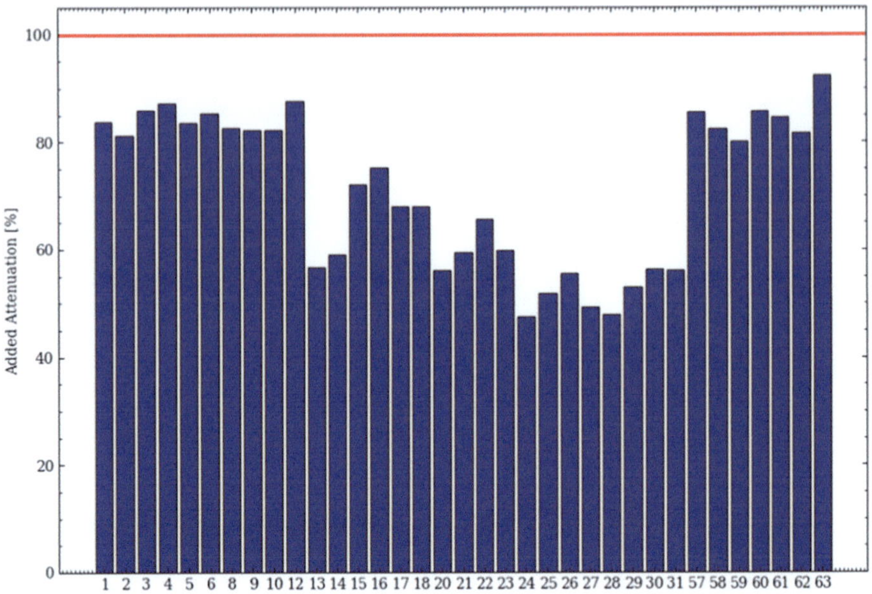

Fig. 5.24 The effect of installing the additional 341 kgs of lead shielding the NuThor's inner core around the existing stack of lead shielding by channel. The channels closest to the thorium mass on the bottom section of the neutron multiplicity meter received the greatest attenuation (approximately 80%) while those furthest away on the top enjoyed an attenuation around 50–70%. It is important to note that the attenuation of the gamma backgrounds on the NaI detector shown here are degenerate with the increases in gain shown in the previous subsection meaning that these attenuation of background rate are achieved despite increases in background associated with increased gain

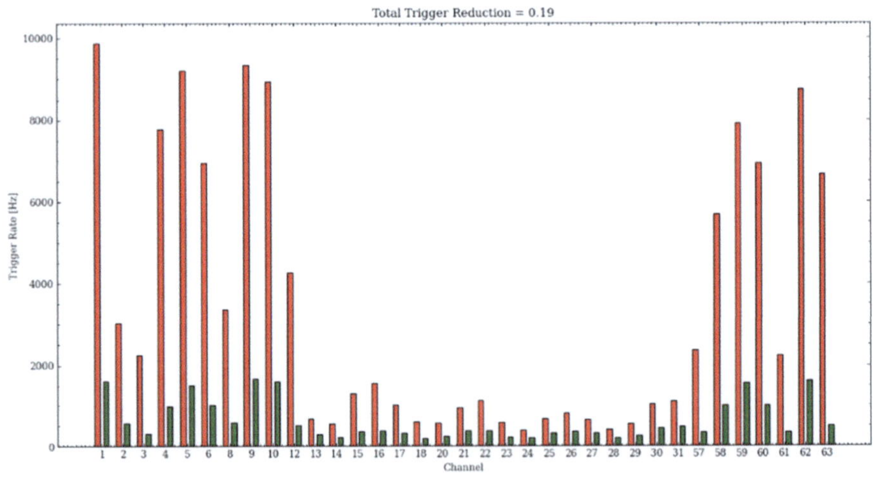

Fig. 5.25 The trigger rates of each NaI detector channel before (red) and after (green) installing the 341 kgs of additional lead shielding around NuThor's inner core. This attenuation of trigger rates is degenerate with increases to the detector PMTs' gains which would partially negate the effectiveness of the shielding to trigger rates

Fig. 5.26 PMT bases installed on the top section of NaI detectors in NuThor. A complementary set of bases is installed in the lower compartment of the NuThor detector

The bases are powered by a combination of high and low voltage. Two low voltage supplies are used to produce +15 and −15 V to power the amplifiers onboard the base's printed circuit board (PCB). The low voltage lines are distributed amongst the 36 PMT bases with a distribution board designed by ORNL and North Carolina Central University (Fig. 5.27).

The high voltage powering the PMT bases is from a CAEN SY5527LC high voltage supply with a 12-channel CAEN AG538D card. This high voltage is distributed through a high voltage distribution board constructed by the University of Washington.

Each PMT base is connected with a BNC cable to a transition card (shown in Fig. 5.28) outfitted with two high-density ERNI connectors. The transition card is connected to a CAEN 1740D 64-channel, 62.5 MSample/sec digitizer.[11] The 1740D digitizer is loaded with the CAEN DPP-QDC firmware, connected to a DAQ computer with an optical cable plugged into a PCIe controller with the A3818 optical-to-USB adapter. The acquisition software utilized here is CAEN's CoMPASS (Fig. 5.29).

[11] The 1740D digitizer is prone to cross talk between input channels within the same grouping of 8. For future reference, large amplitude signals like the SNS timing signals will induce substantial artifacts across the channel group(s) they are a part of. If possible, large amplitude signals should be assigned to their own groups physically separated from the other detector channels.

Fig. 5.27 Low voltage distribution board fabricated by NC Central University which distributes the $\pm$ 15 V to each of the 36 NaI detector PMTs

Fig. 5.28 Transition card designed by ORNL which allows individual detector channels to be read out into the CAEN 1740D digitizer with the goal of reducing cross talk typically observed with off-the-shelf CAEN A746B patch panels connected with two ribbon cables

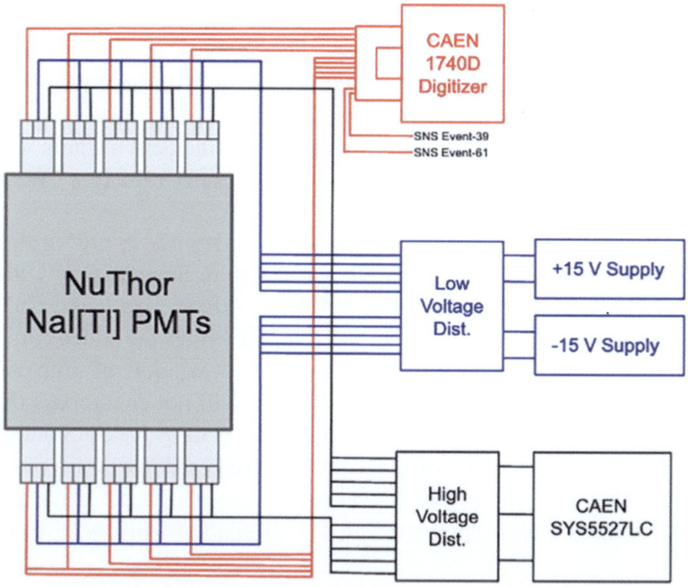

Fig. 5.29 Hardware diagram for NuThor showing the simplified NuThor detector with PMT bases with signal readout connected to a CAEN 1740 digitizer, low voltage lines supplied through a custom distribution board and two voltage supplies, and high voltage lines supplied through a custom distribution board and a CAEN SYS5527LC

5.4.1 Trigger Scheme

The aforementioned phases of the NuThor data taking campaigns have different triggering schemes as the detector reached maturity and as priorities crystallized with initial data taking, health checks, and analysis. The proceeding section lays out the variations in each of the phases.

Phase 0 Trigger Scheme

Each NaI detector channel and SNS timing channel is individually internally triggered with a hardware level threshold set in the CoMPASS software. Data taking used "List Only" mode, which records only the values of the accumulators for signal events that exceed the threshold. An accumulator is an electronic component that temporarily stores charge and converts the input analog signal into a digital value. This allows for online integration of the waveform signals.

Phases 1 and 2—Current Configuration

The CAEN 1740D is externally triggered by the SNS timing signal (Event 39) to allow for a zero-threshold analysis. This means that there is no minimum signal peak-height or size required for the digitizer to record data–thus, signals that would otherwise not exceed these thresholds are recoverable later during analysis. However, CoMPASS has limitations in its data throughput. The original Event 39 signal is passed through a function generator to produce several copies staggered in time to trigger the digitizer multiple times between beam spills. During Science Run 1, there were 1 coincident trigger and 9 anti-coincident triggers for a total of 10 triggers per beam spill–a 600 Hz trigger rate. For Science Run 2, the number of anti-coincident triggers was reduced from 9 to 4–a 300 Hz trigger rate. Data collection was performed in "Waves" mode to record waveforms.

With 36 detector channels (one non-functional) and two SNS timing signals combined with the somewhat high trigger rates (600 Hz in Science Run 1 and 300 Hz in Science Runs 2, 3), the maximum, possible waveform size using CoMPASS is 15,360 ns (960 samples). This maximum wavelength brings the added complications that not all sought-after signals will occur within that window of time. As one can see in Fig. 5.30, the 15,360 nanosecond waveform will not encompass the entirety of the expected neutron capture times relative to the start of the SNS neutrino beam pulse. This is further exasperated by the externally triggered waveforms having a pre-trigger region where one should not expect neutrinos yet. The pre-trigger effectively reduces the fraction of neutron captures in the acquisition window (Fig. 5.30).

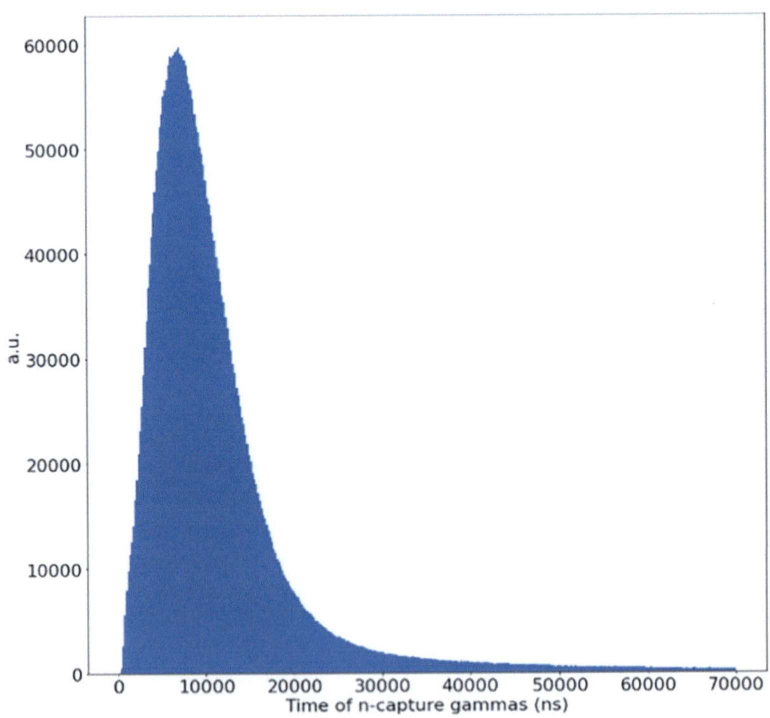

Fig. 5.30 The timing structure of neutron-capture γ-rays from the gadolinium nuclei inside nuThor. This is generated from a MCNP–Polimi simulation using the Phase 1 detector geometry

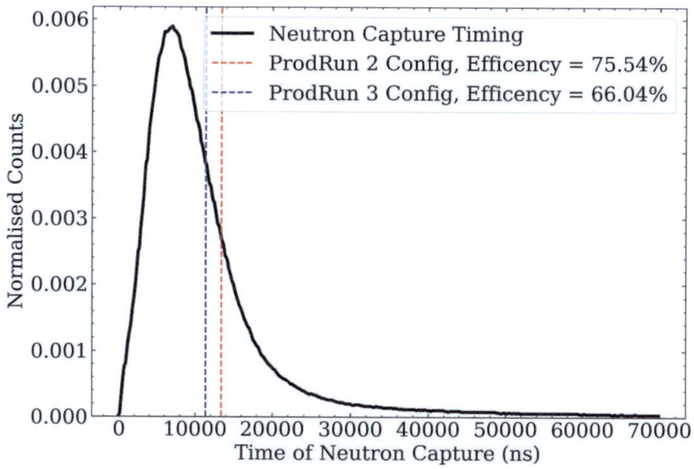

Fig. 5.31 Effect of 2 microsecond delayed SNS timing signal relative to the arrival of neutrinos

Phase 3 Trigger Scheme

Future Science Runs beyond what is reported in this work will need to account for an additional 2 microsecond delay introduced between Science Run 2 and 3 (See Fig. 5.31).

5.5 Signal and Backgrounds

NuThor is set up to pursue neutrons as its primary signal and leverage the high neutron multiplicity to distinguish neutrino-induced nuclear fission from other sources of backgrounds. However, consider the many possible signal candidates in the context of a fully formed detector design. In the same vein, the most pertinent background sources are discussed.

5.5.1 Signal

There are several possible candidates for the neutrino-induced nuclear fission signal in NuThor. A non-exhaustive list and brief discussion is summarized in Table 5.5. Some of these signals could be more advantageous with another apparatus and that will be discussed in a later section of this work.

Table 5.5 Summary of neutrino-induced nuclear fission signal candidates and a brief discussion of each one's eligibility as a strong signal observable specifically in NuThor at the SNS

Signal candidate	Efficacy as NuFission confirmation
Prompt neutrons	The primary signal. Neutrino-induced nuclear fission yields a high multiplicity of neutrons (See Fig. 4.19), thus the neutrino-induced nuclear fission observation becomes a neutron counting exercise in which one must measure an excess of high multiplicity neutron detections in coincidence with the SNS beam pulse
Delayed neutrons	Only approximately 5% of nuclear fissions exhibit delayed neutrons thus is an unreliable signal given the rarity of the neutrino-induced nuclear fission event. The timing structure of the delayed neutrons vary widely and can be broken into six categories–some of which extend out to as large as 200 s. For a beam spill-based experiment, this timing could prove disqualifying
Fission fragments	The mean free path of fission fragments is only a few microns so is unlikely to escape the thorium target mass. A target with the thickness of a foil would likely be necessary to successfully capture these fragments with good efficiency
Charged-current electron	Although the high-energy electron from the CC-interaction is approximately 20 MeV, MCNP simulations suggest it is unlikely to escape the target mass. This possibility is driven to near zero would accounting for the inner core's lead shielding
Fission γ-rays	The lead used to block the intrinsic radioactivity massively attenuates photons attempting to escape the inner core. Even without the lead shielding, the radiogenic γ-rays from the thorium would swamp any signal γ-rays

Prompt Neutrons

As computed in Fig. 4.21, there are, on average, a large number of neutrons emitted per neutrino-induced nuclear fission event. Additionally, neutrons lose a small fraction of their kinetic energy per elastic scatter with both thorium and lead due to the large disparity in the projectile's mass versus the target nucleus. This means that neutrons are readily able to scatter out of the inner core and into NuThor's neutron multiplicity meter for observation. The collection of prompt neutrons is the strongest candidate for the neutrino-induced nuclear fission signal observable.

Another possible technique for distinguishing prompt fission neutrons is through reconstructing the angles between the emitted neutrons. Because fission fragments are launched in the opposite directions via their electrostatic repulsion, the neutrons emitted by these fragments are also boosted in that original direction. This anisotropy could serve as an indicator of fission [24]. For this to be a viable signal approach, the neutrons must scatter very few or no times before reaching the detector; otherwise, the directional information will be lost or too blurred to be useful. The current geometry of nuThor is not ideal for this method of event discrimination because neutrons are likely to scatter several times before leaving the thorium and lead inner core (Fig. 5.32).

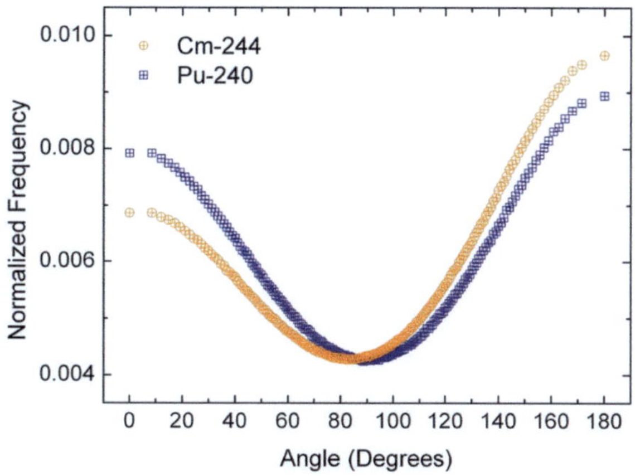

Fig. 5.32 The angle between prompt fission neutrons from ^{244}Cm and ^{240}Pu as simulated from FREYA2.0.5 without any energy threshold or detector effects applied, from Ref. [24]

Charged-Current Electron

The charged-current electron in the final state is a possible signal to demonstrate neutrino-induced nuclear fission. However, there is a low probability that the electron successfully escapes the thorium and an even more diminished probability it can propagate through the lead shielding of the inner core. For added context, the stopping power of thorium metal is plotted in Fig. 5.33.

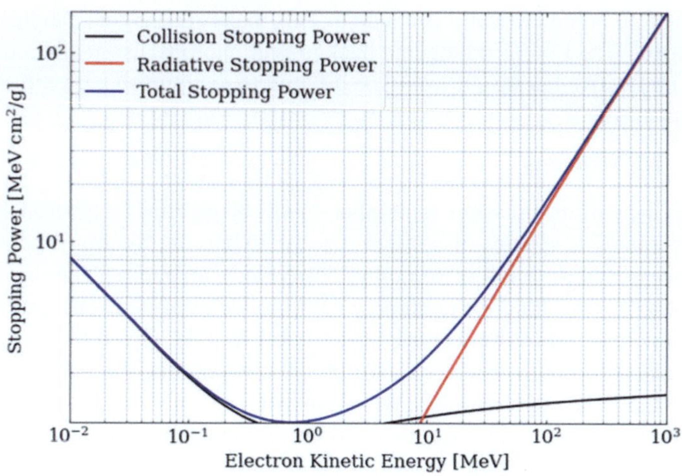

Fig. 5.33 Stopping power of electrons in Thorium. Data from NIST XCOM Database [25]. In the tens of MeV range, where the charged-current electron are expected, the cross section from radiative stopped power is extreme

Table 5.6 Mean range of fission fragments from ^{235}U, tabulated by Ref. [26] which extracted the values for the references [27, 28]

	Air [mm]	Hydrogen [mm]	Helium [mm]	Argon [mm]	Xenon [mm]
Light fragment	25.4	21.1	28	23.9	23
Heavy fragment	19.5	17.7	23	19.4	18
(Total range)	44.9	38.8	51	43.3	41

Fission Fragments

The critical issue with using the fission fragments is the high stopping power in dense media (and even air) of these highly charged, energetic particles. This behaviour is well summarized by the information presented in Table 5.6. One can easily see that if the mean range of fission fragments is of the order of tens of millimeters for even the gaseous materials in the table, there is an exceedingly minuscule probability of a fission fragment successfully escaping the thick thorium metal plate which has a density of 11.72 g/cm^3.

Prompt Gamma Rays

The angular momentum of the fission fragments is carried away with the emission of prompt fission γ-rays. Because the design of NuThor involves a thick layer of lead inner shielding between the thorium target mass and the active components of the detector (the NaI detectors), it is highly unlikely that these γ-rays escape through the shielding before being critically attenuated. This makes this signal candidate a very weak prospect. The rate of radiogenic background photons also renders this signal candidate ineffective because it would be difficult to distinguish between signal and background events.

Beta-Delayed Neutrons

Beta-delayed neutron emission is another mode of ejecting a neutron following nuclear fission. This species of neutron emerges when the fission fragments beta-decay to get to stability and in so doing, decays into an excited state with energy above the energy threshold for emitting a neutron. Delayed-neutrons' potential as a signal candidate suffers from its rarity as seen in Fig. 5.34. Given that the neutrino-induced nuclear fission is already a rare process, depending on $\sim$5% of events to generate a beta-delayed neutron is a weak prospect.

Additionally, the timing of beta-delayed neutrons is determined by the half-lives of the fission fragments which are largely on the scale of several seconds [29]. This is not conducive with the rapid 60 Hz frequency of the SNS neutrino pulses.

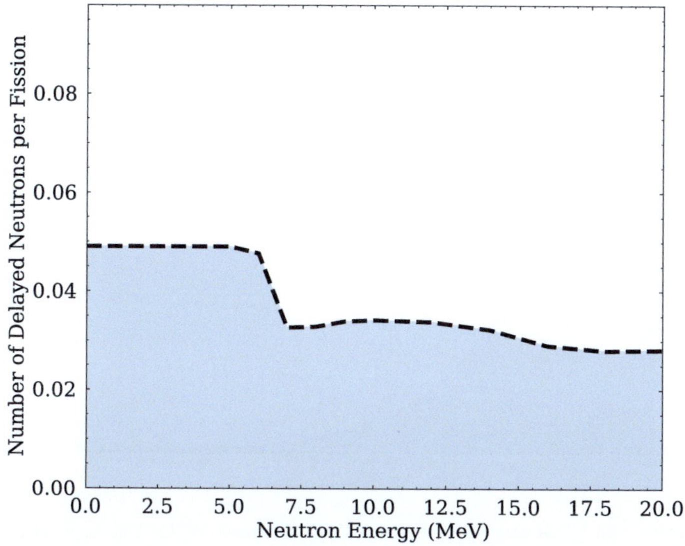

Fig. 5.34 The delayed-neutron probability from neutron-induced nuclear fission on ^{232}Th data [14]

5.5.2 *Backgrounds*

In rare-event searches, a thorough understanding of experimental backgrounds is just as important as knowledge of the signal itself. NuThor faces backgrounds that are both completely random and correlated with the beam producing the neutrino-induced nuclear fission events. The characterization of each of these backgrounds is of paramount importance. The mitigation strategy of each of these sources of backgrounds are discussed in detail throughout this work.

Neutrino-Induced Neutrons

This category of neutron backgrounds is particularly difficult to approach because they are created in coincidence with the SNS beam spill. Any material inside the NuThor apparatus is a potential target to create a NIN background event. The potency as a NIN target scales with the specific nucleus's neutron excess $(N - Z)$ and its neutron separation energy. As such, the target nuclei identified as key, potential culprits are lead, thorium, iron, and iodine.

Several neutrino experiments view NINs as their primary signal rather than a background to reduce. With that in mind, one could alternatively view the detection of these neutrino events as a fruitful result as well.[12]

[12] To put it colloquially, "One man's trash is another man's treasure."

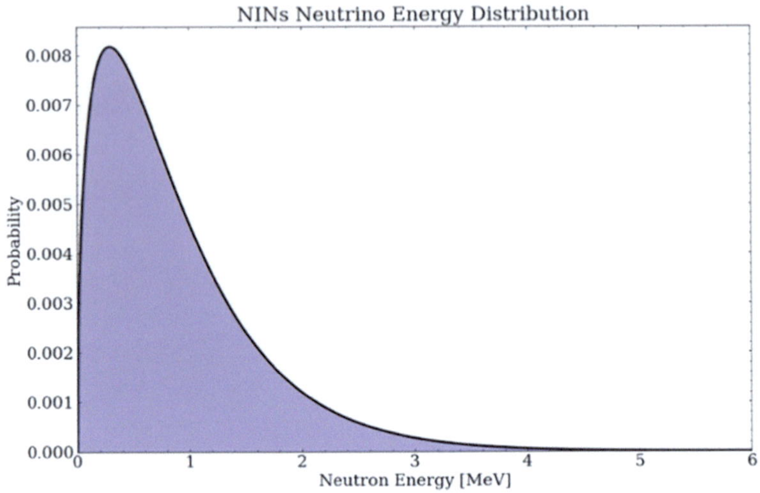

Fig. 5.35 The ^{232}Th NIN energy distribution calculated from INCL+ABLA. A smooth curve is generated by fitting a Watt curve to the simulation output

Thorium NINs

The 52 kg of thorium is a neutrino-induced nuclear fission target, but also is readily able to emit NINs following neutrino interactions. The NIN energies are defined as a Watt Spectrum with parameters: a = 0.59255 MeV and b = 0.01501 MeV^{-1}. These were extracted from a fit to simulated data yielded by INCL/ABLA–see Fig. 5.35. One can also model the thorium NIN energy spectrum using another evaporative model, such as a Maxwellian distribution, with similar results.

Appropriate care must be taken to understand the Th NIN behaviour because amongst all the potential NIN target nuclei, its daughter nucleus has the lowest neutron separation energy at 5.5 MeV. This brings with it, an enhancement of the higher multiplicity NIN events. Despite this, the average neutron multiplicity is still approximately 2–below the average for neutrino-induced nuclear fission.

The branching ratios expected from $^{232}\text{Th}(\nu_e, e^-)^{232}\text{Pa}$ de-excitations are tabulated using INCL/ABLA. A transfer matrix with bases in excitation energy and fission probability is formed and is then used to establish the branching of fission versus neutron emission.

Lead NINs

The lead NINs are generated uniformly within the lead inner shielding inside the neutron multiplicity meter. NuThor's Science Run 1 had 536 kg of lead. Upon observing high background rates in the NaI detectors, additional lead shielding was added to the inner shielding stack for a total of 881 kg of lead. The neutron energies are computed through MARLEY's output shown in Fig. 5.36a. MARLEY is used for this nucleus because there is no need to compute the competition between de-excitation trough

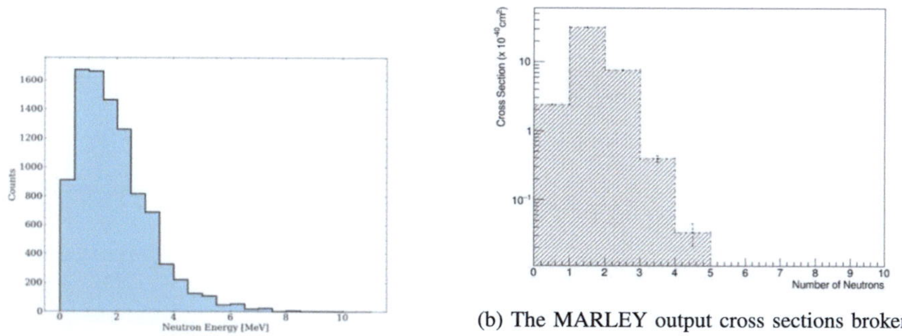

(a) Neutron energies from electron neutrino capture on ^{208}Pb from MARLEY.

(b) The MARLEY output cross sections broken into neutron multiplicities. No correction is yet added to consider the Neutrino Cube constraints in the rates.

Fig. 5.36 Lead NIN predictions made with the MARLEY software. The inputs for this assume $g_A = 1.27$ and takes in charge-exchange reaction data from experiment

nuclear fission and neutron emission (the fission barrier for lead is greater than the expected excitation energies at the SNS).

Figure 5.36 depicts the flux-averaged cross sections per neutron multiplicity for neutrino-induced neutron emission on ^{208}Pb. The cross section can be roughly scaled by the N–Z [30] of each of the individual naturally occurring isotopes of lead (weighted by the natural abundances) to make an estimate of the number of events expected within NuThor. Those rates are corrected according to the combined fit between the Eljen Cell detector and the Lead neutrino cube detector (See Fig. 3.8) which reported a constraint on the Pb NINs rate relative to the MARLEY prediction as $0.29^{+0.17}_{-0.16}$.

Iodine NINs

The iodine NINs are generated uniformly in the 36 NaIs. There is approximately 235 kg of ^{127}I in the detectors. The normalization of the simulated Iodine NINs backgrounds is based on the combination of the MARLEY computed cross sections (shown in Fig. 5.37) and then corrected by the measured constraints evidenced by the NaIvE-185 publication also performed by the COHERENT collaboration.

However, the iodine NINs are a special case, because they are accompanied by the electromagnetic component of the charged-current interaction inside the NaI detectors. For this reason, there is a reasonable probability of veto-ing or at least tagging ^{127}I NINs using their proximity in time to a high-energy deposition in the NaI crystals within the beam window.

A proposed mitigation strategy for this species of NINs ranges to exploit the fact that the region-of-interest in energy for this search is from 50 keV to 10 MeV per neutron event (more discussion of this in the subsequent chapters). 98.95% of ^{127}I charged-current events are above this region, thus would not be incorporated into the analysis pipeline if this criteria was applied (Fig. 5.38).

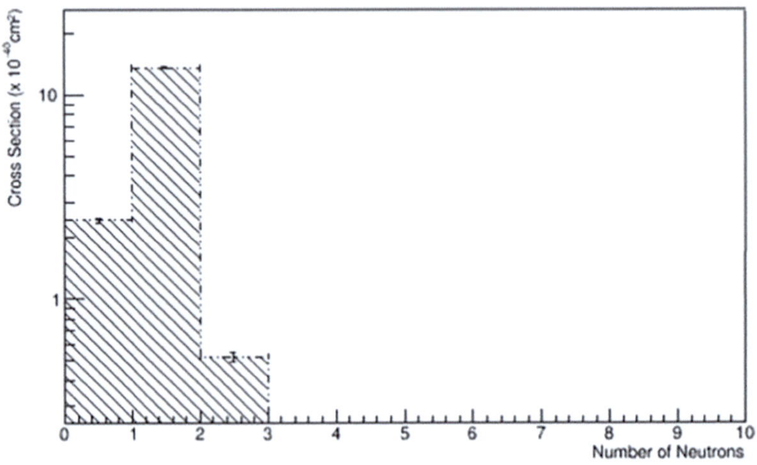

Fig. 5.37 ^{127}I NINs exclusive cross sections per neutron multiplicity. This computation was completed in MARLEY

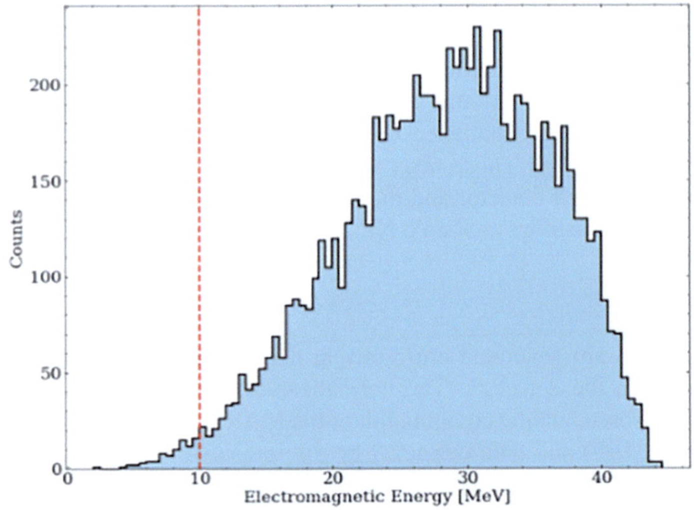

Fig. 5.38 The electromagnetic energy emitted from CC interactions on ^{127}I. The majority (98.95%) of events occur above the NuThor energy ROI

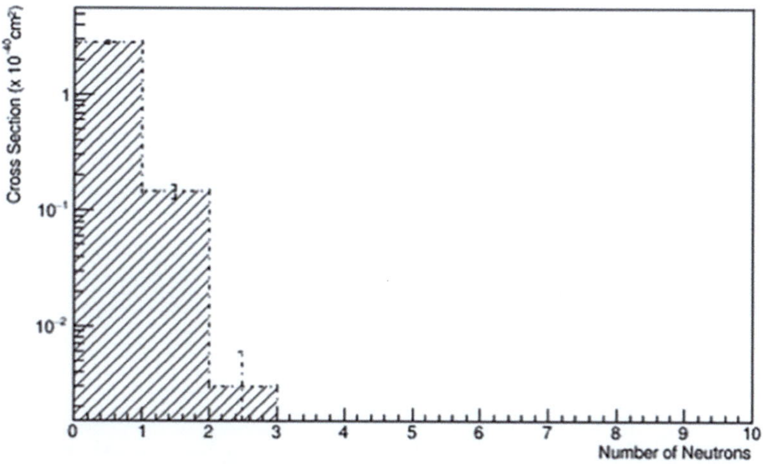

Fig. 5.39 The NINs cross sections broken into true neutron multiplicity for CC interactions of ^{56}Fe. The cross sections are notably lower than the other isotopes tested in this note

Iron NINs

The iron NINs are generated uniformly in the two steel sheets on the top and bottom of the neutron multiplicity meters. There are approximately 172 kg of steel in the design.

However, the steel plates in the NuThor geometry are on the periphery of the neutron capture materials so there is an inherent reduction in capture efficiency—roughly half of the already sparse Fe NINs will scatter in the direction away from the neutron multiplicity meter's capture region. Further, because the charged-current neutrino interaction cross section scales with N–Z, iron has a low cross section compared to the other NIN target materials. Figure 5.39 illustrates the smaller exclusive cross sections. This is further demonstrated by the severely statistically limited neutron energy spectrum in Fig. 5.40 where there are too few simulated neutrino events which actually result in the emission of a neutron.[13] One could continue to run the event generator until the statistical fluctuations subside but because this background is negligibly small, that was not done. Studies meant to elucidate the nature of Fe NINs likely will not be possible in this incarnation of the detector.

Beam-Related Neutrons

The SNS is primarily a facility for producing neutrons. The laboratory space known as "Neutrino Alley" is the product of both considerable forethought and serendipity in finding a location with a relatively low neutron fluence. However, neutrons

[13] Further increasing the number of simulated events would eventually populated the statistically jittery histogram, but given the small importance of the background, it is neglected in this work.

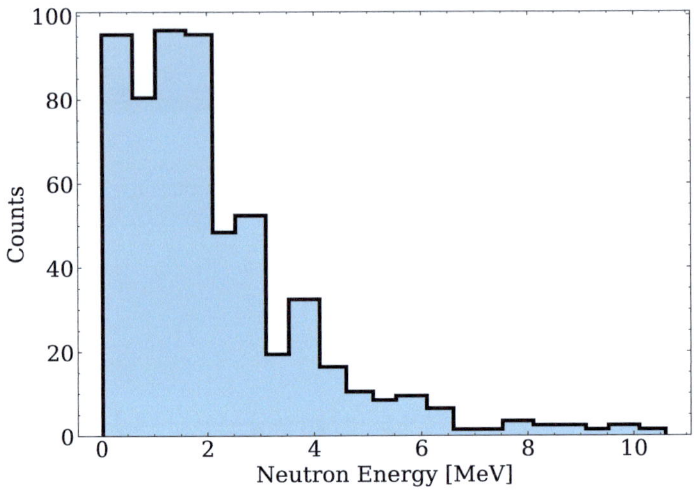

Fig. 5.40 Neutron Energies from Fe NINs in MARLEY

originating from the SNS beam still occasionally manage to find their way to Neutrino Alley. These neutrons are aptly dubbed "Beam-related Neutrons"–hereafter referred to as BRNs (Fig. 5.41). Figure 5.42 illustrates the flow of beam-related neutrons from the SNS liquid mercury target using a neutron camera.

Given the nature of the experiments dwelling in Neutrino Alley (which heavily rely upon nuclear recoils), it is of paramount importance for the COHERENT collaboration to understand this BRN background. Concomitant to COHERENT's aims, BRNs are of great interest to the nuThor detector because (1) neutrons are the primary signal for the detector and (2) there is a possibility that some of these high-energy BRNs could burrow into the core of the detector and either induce a fission on thorium or spall on high-A materials. To assess the propensity for this to occur, existing initiatives for characterizing the BRN background are examined. Namely, the MARS and SciBath detectors. In addition to these detectors, there is a dedicated effort within the COHERENT collaboration to simulate the energy and multiplicity of BRNs as they enter the experimental space. These simulations have been performed in both MCNP and GEANT4 (Fig. 5.43).

The flux of BRNs was measured by both multiple COHERNET background monitoring subsystems. The MARS detector [31] and the SciBath detector reported fluxes of $2.51 \pm 1.18 \times 10^{-6}$ n m^{-2} spill^{-1} and $2.73 \pm 0.37 \times 10^{-5}$ n m^{-2} spill^{-1}, respectively. We attain an estimated number of BRNs per year by combining the aforementioned BRN flux with the size of the detector and the total number of beam spills per year. This results in approximately 3,780 BRNs per year incident on the NuThor detector. This calculated normalization is not what is used for the eventual event rate estimates. The above rate is the an estimate of the number of BRNs which enter the surface around nuThor but many of these will not result in a detectable signal.

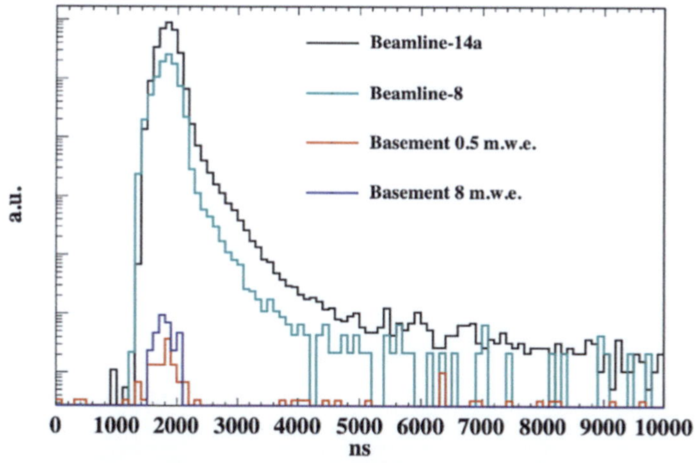

Fig. 5.41 Beam-related neutron populations observable at several locations in the SNS facility. This study was performed using the Scatter Camera detector and illustrates the importance of location to the success of COHERENT measurements given that the Beamline-14a and -8 locations are plagued by not just a vastly larger normalization to the BRN rate, but also a significant presence of neutrons in the delayed region of the beam pulse where previous neutrino measurements conducted by coherent have needed a tightly constrained BRN component in time to fit out the electron neutrino or muon anti-neutrinos in the delayed region of the beam pulse. Figure courtesy of the COHERENT collaboration

Fig. 5.42 Neutron camera depicting the propagation of beam-related neutrons emanating from SNS target (the red hot stop). Image courtesy of Dr. Jason Newby of ORNL and Professor Yuri Efrimenko at the University of Tennessee–Knoxville

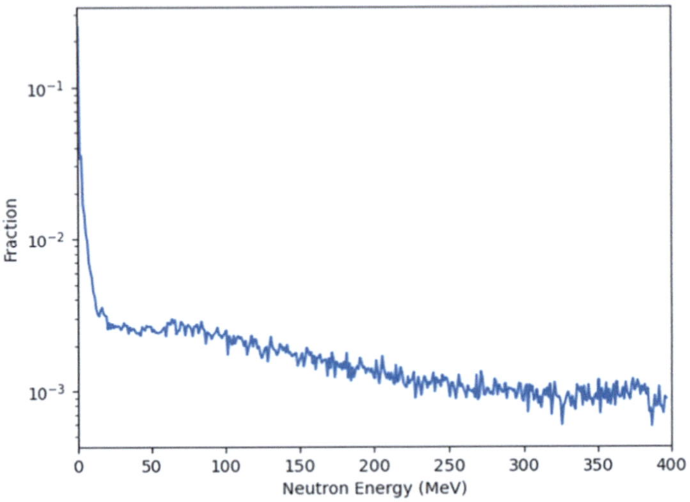

Fig. 5.43 Beam-Related Neutrons from the SNS expected in Neutrino Alley as simulated by Brandon Becker

Based on the number of expected BRNs and neutrino-induced nuclear fissions per year, there is an exceedingly small probability of coincidences between the two. The accidental coincidence rate given by Eq. 5.3

$$R_{\text{acc}} = R_{\nu F} R_{\text{BRN}} w = 1.4 \times 10^{-7} \text{year} \tag{5.3}$$

where w is the average neutron capture time inside the NuThor detector, $R_{\nu F}$ is the neutrino-induced nuclear fission rate, and R_{BRN} is the BRN rate.

GEANT4 simulations performed by Brandon Becker yielded a BRN energy spectrum that we then normalize and scale using the value calculated in the previous section [32].

Because high-energy (> 20 MeV) neutrons often lack cross section measurements in MCNP's libraries, it is important to benchmark simulations with real world data. Simulations with the NuThor detector geometry and energy spectrum shown in Fig. 5.43 suggested that BRNs incident on the detector geometry would produce violent and highly observable collisions with many of the detector channels in a very narrow time window. Notably, because BRNs are created in tandem with the protons-of-target (and does not correspond to the radioactive decay of the muon like the electron neutrinos), there would be an easily identifiable set of criteria to demarcate a BRN event from a bona fide signal event. The BRN would engage a large number of detector channels and be tightly constrained to the prompt region of the neutrino beam to coincide with the protons-on-target. These shower-like events from BRNs were independently identified to occur in the NaIvE-185 detector as depicted in Fig. 5.44. This timing will come into play in the final analysis (Fig. 5.45).

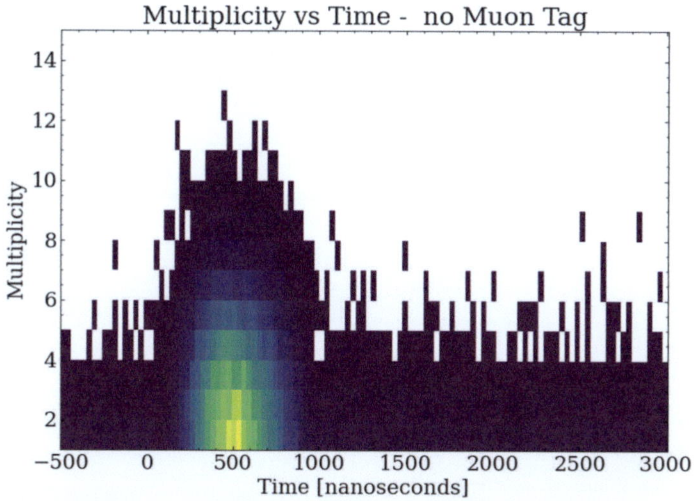

Fig. 5.44 The multiplicity of NaI detectors struck versus time of events relative to the beam pulse extracted from NaIvE-185 unblinded data. The energies of these BRN events are from 10–55 MeV in the NaIvE-185 region of interest. These events are those that do not have a corresponding muon tag in the muon veto system around the detector

Fig. 5.45 Distribution of thorium spontaneous fission neutron multiplicity derived from the Eq. 5.4

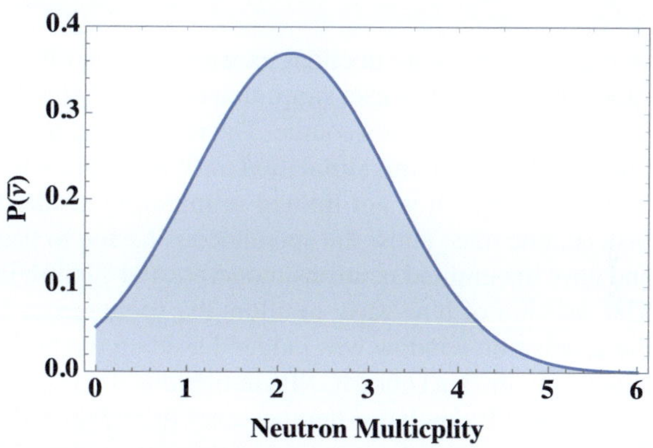

Thorium Spontaneous Fission

Despite the fact that thorium was selected as the target material for its low spontaneous fission (SF) rate, the rate is not zero. The spontaneous fission rate coupled with a relatively large quantity of thorium means there is potentially a measurable signature of spontaneous fission even in the neutrino beam coincident window. Thankfully, the spontaneous fission rate of ^{232}Th is well explored. The neutron multiplicity can be expressed with Eq. 5.4.

$$P(\bar{\nu}) = \frac{1}{\sqrt{2\pi\sigma^2}} e^{(\nu-\bar{\nu})^2} / 2\sigma^2 \tag{5.4}$$

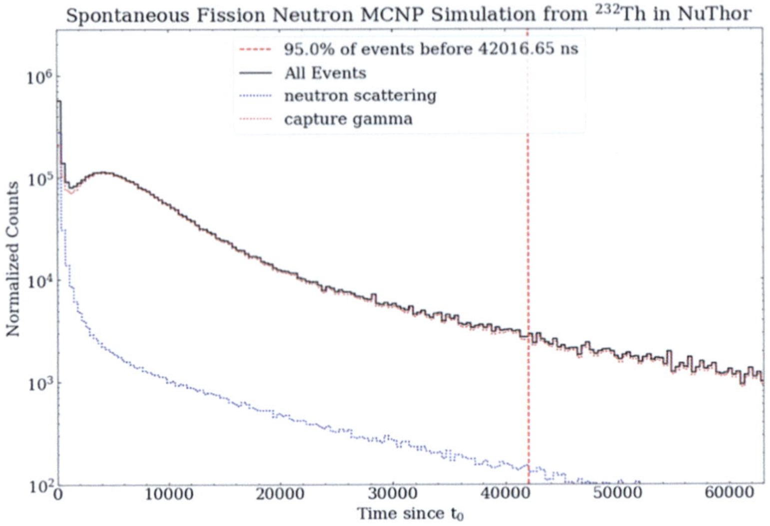

Fig. 5.46 The timing structure of ^{232}thorium spontaneous fission neutrons simulated in the NuThor geometry originating in the thorium target mass. Notably, 95% of the neutron-related detections in the NaI detectors occur within 42 microseconds of the start of the simulation event

where $\sigma = 1.08$ for all nuclides except ^{252}Cf, and $\bar{\nu} = 2.14$ [26]. Simulating thorium spontaneous fission is also simplified by the fact that MCNP already possesses a thorium spontaneous fission source. The result of such an MCNP simulation is shown in Fig. 5.46. However, this simulation must account for the fact that this is a steady-state background which is not limited to initiating inside the DAQ acquisition window. Instead, one must allow the spontaneous fission to begin before the external trigger and have the emitted neutrons moderate and capture inside the acquisition window. The decision of how early to allow the spontaneous fission events to begin before the acquisition window was judged based on the timing structure of the SF neutron captures in the NuThor MCNP simulation.

The MCNP simulation demonstrates that 95% of SF neutron captures occur within 42 microseconds of the start time of the thrown history. Thus, the accidentally coincidence rate between spontaneous fission (and the scaling factor of the SF background PDF) is computed using 42 microseconds, a beam pulse frequency of 60 Hz and the literature value of the spontaneous fission half-life for ^{232}Th of 1.2×10^{21} years [33].

References

1. Olsen DK, Ingle RW, Portney JL (1980) Precise measurement and analysis of neutron transmission through 232Th. Tech. rep. CONF-800942–8. United States
2. Klann RT, Shergur J, Mattesich G (2009) Current state of commercial radiation detection equipment for homeland security applications. Nucl Technol 168(1):79–88. https://doi.org/10.13182/NT09-A9104
3. Johnson W et al (2021) InterSpec v. 1.0.9. [computer software]. https://doi.org/10.11578/dc.20221016.3
4. Motz JW, Olsen HA, Koch HW (1969) Pair production by photons. Rev Mod Phys **41**(4), 581–639. https://doi.org/10.1103/RevModPhys.41.581. https://link.aps.org/doi/10.1103/RevModPhys.41.581
5. Johnson JR, Mann KC (1973) Single and double escape peaks in Ge(Li) gamma-ray spectra. Nucl Instrum Methods 112(3):601–602. ISSN: 0029-554X. https://doi.org/10.1016/0029-554X(73)90186-9. https://www.sciencedirect.com/science/article/pii/0029554X73901869
6. Hennings-Yeomans R, Akerib DS (2007) A neutron multiplicity meter for deep underground muon-induced high-energy neutron measurements. Nucl Instrum Methods Phys Res Sect A Accel Spectrometers Detect Assoc Equip 574(1):89–97. ISSN: 0168-9002. https://doi.org/10.1016/j.nima.2007.01.137. https://www.sciencedirect.com/science/article/pii/S0168900207001738
7. Wilhelmy JB et al (1972) Angular momentum of primary products formed in the spontaneous fission of ^{252}Cf. Phys Rev C **5**(6), 2041–2060. https://doi.org/10.1103/PhysRevC.5.2041. https://link.aps.org/doi/10.1103/PhysRevC.5.2041
8. Akimov DY et al (2010) The ZEPLIN-III anti-coincidence veto detector. Astropart Phys 34(3):151–163. ISSN: 0927-6505. https://doi.org/10.1016/j.astropartphys.2010.06.010. https://www.sciencedirect.com/science/article/pii/S0927650510001222
9. Fraum TJ et al (2017) Gadolinium-based contrast agents: a comprehensive risk assessment. J Magn Reson Imaging 46(2):338–353. https://doi.org/10.1002/jmri.25625. eprint: https://onlinelibrary.wiley.com/doi/pdf/10.1002/jmri.25625
10. Adhikari G et al (2019) Search for a dark matter-induced annual modulation signal in NaI(Tl) with the COSINE-100 experiment. Phys Rev Lett 123(3):031302. https://doi.org/10.1103/PhysRevLett.123.031302. https://link.aps.org/doi/10.1103/PhysRevLett.123.031302
11. Fushimi K et al (2021) PICOLON dark matter search project. J Phys Conf Ser 2156(1):012045. https://doi.org/10.1088/1742-6596/2156/1/012045
12. Amaré J et al (2014) Preliminary results of ANAIS-25. Nucl Instrum Methods Phys Res Sect A Accel Spectrometers Detect Assoc Equip 742:187–190. In: 4th Roma international conference on astroparticle physics. ISSN: 0168-9002. https://doi.org/10.1016/j.nima.2013.09.019. https://www.sciencedirect.com/science/article/pii/S0168900213012539
13. Pettus WC (2015) DM-Ice: current status and future prospects. arXiv:1510.00378 [physics.ins-det]
14. Shibata K et al (2011) JENDL-4.0: a new library for nuclear science and engineering. J Nucl Sci Technol 48(1):1–30. https://doi.org/10.1080/18811248.2011.9711675
15. Fukuda Y et al (1998) Evidence for oscillation of atmospheric neutrinos. Phys Rev Lett 81(8):1562–1567. https://doi.org/10.1103/PhysRevLett.81.1562. https://link.aps.org/doi/10.1103/PhysRevLett.81.1562
16. Cao J, Luk K-B (2016) An overview of the Daya Bay reactor neutrino experiment. Nucl Phys B 908:62–73. Neutrino oscillations: celebrating the nobel prize in physics 2015. ISSN: 0550-3213. https://doi.org/10.1016/j.nuclphysb.2016.04.034. https://www.sciencedirect.com/science/article/pii/S0550321316300724
17. Mount BJ et al (2017) LUX-ZEPLIN (LZ) technical design report. arXiv: 1703.09144 [physics.ins-det]
18. Collaboration X et al (2024) The XENONnT dark matter experiment. arXiv: 2402.10446 [physics.ins-det]

19. Rosman KJR, Taylor PDP (1998) Pure Appl Chem 70(1):217–235. https://doi.org/10.1351/pac199870010217
20. Hagiwara K et al (2019) Gamma-ray spectrum from thermal neutron capture on gadolinium-157. Prog Theor Exp Phys 2019(2):023D01
21. Marti L et al (2020) Evaluation of gadolinium's action on water Cherenkov detector systems with EGADS. Nucl Instrum Methods Phys Res Sect A Accel Spectrometers Detect Assoc Equip 959:163549. ISSN: 0168-9002. https://doi.org/10.1016/j.nima.2020.163549. https://www.sciencedirect.com/science/article/pii/S0168900220301315
22. Watt BE (1952) Energy spectrum of neutrons from thermal fission of U^{235}. Phys Rev 87(6):1037–1041. https://doi.org/10.1103/PhysRev.87.1037. https://link.aps.org/doi/10.1103/PhysRev.87.1037
23. Hedges S (2024) Measurement of quenching factors in NaI[Tl]. In: Low energy neutrino-nucleus interactions at the spallation neutron source. Springer Nature Switzerland, Cham, pp 45–65. ISBN: 978-3-031-68110-3. https://doi.org/10.1007/978-3-031-68110-3_4
24. Yoon S et al (2021) Energy-dependent characteristics of prompt neutron anisotropic emission by ^{244}Cm and ^{240}Pu according to Monte Carlo simulations. J Radioanal Nucl Chem 330(2):481–491. https://doi.org/10.1007/s10967-021-07893-8
25. Berger M et al (1999) XCOM: photon cross section database (version 1.2)
26. Hyde E (1964) The nuclear properties of the heavy elements. Prentice-Hall, Inc
27. Bøggild JK, Minnhagen L, Nielsen OB (1949) A cloud-chamber study of fission fragment ranges in Air. Phys Rev 76(7):988–989. https://doi.org/10.1103/PhysRev.76.988. https://link.aps.org/doi/10.1103/PhysRev.76.988
28. Bøggild JK, Arrøe OH, Sigurgeirsson T (1947) Cloud-chamber studies of electronic and nuclear stopping of fission fragments in different gases. Phys Rev 71(5):281–287. https://doi.org/10.1103/PhysRev.71.281. https://link.aps.org/doi/10.1103/PhysRev.71.281
29. Keepin GR, Wimett TF, Zeigler RK (1957) Delayed neutrons from fissionable isotopes of Uranium, Plutonium, and Thorium. Phys Rev 107(4):1044–1049. https://doi.org/10.1103/PhysRev.107.1044. https://link.aps.org/doi/10.1103/PhysRev.107.1044
30. Osterfeld F (1992) Nuclear spin and isospin excitations. Rev Mod Phys 64(2):491–557. https://doi.org/10.1103/RevModPhys.64.491. https://link.aps.org/doi/10.1103/RevModPhys.64.491
31. The COHERENT Collaboration (2022) Monitoring the SNS basement neutron background with the MARS detector. J Instrum 17(03):P03021. https://doi.org/10.1088/1748-0221/17/03/P03021
32. Becker BJ (2022) Measurement of neutrino-induced neutron production in lead. PhD Dissertation, University of Tennessee. https://trace.tennessee.edu/utk_graddiss/7105
33. Holden NE, Hoffman DC (2000) Pure Appl Chem 72(8):1525–1562. https://doi.org/10.1351/pac200072081525

Chapter 6
Signal Processing

Abstract The NuThor detector contains 36 NaI scintillating crystals which are externally triggered by the SNS beam pulse as well as a few times between beam spills to collect information about the beam anti-coincident backgrounds. The data acquisition system records 15 μs long waveforms. A signal processing pipeline is developed to pass all recorded waveforms through in order to lower energy thresholds, optimize energy resolution, and maintain sensitivity to pile-up events. After processing the signals for their respective energy depositions in the NaI crystals, each channel is calibrated every two weeks and has the gain tracked over the span of the months-long data taking campaigns to monitor for abnormalities. Events are then clustered to create N_{100} event pairs as well as super-clusters which correlate every detector event to its respective beam spill from the SNS.

With the NuThor detector constructed and deployed (Sect. 5.3) at the Spallation Neutron Source (Sect. 3.1), it is necessary to interpret and analyze the signals recorded by the data acquisition system (Sect. 5.4), which reads out from the 36 NaI detectors in the neutron multiplicity meter (Sect. 5.2.2). This chapter outlines the nature of the detector signals, the procedure for extracting information from those signals, the optimization of that extraction, and the handling of the data ultimately used in the search for neutrino-induced nuclear fission.

The Waveforms

The waveforms read out from the digitizer must be analyzed offline to extract key values using a custom signal processing scheme tailored to NaI scintillation pulses specific to these crystals and voltage settings. Due to the fact that the digitizer is externally triggered, a large portion of the recorded data will not contain physics information, with most of it consisting of noise characteristic of the detector channel. Both the signal-containing waveforms and the noise will be utilized. A small subset of the noise-only waveform traces is saved for further study, while the rest are excluded from later analysis steps–though they are not deleted. Waveform traces identified as signal have key values calculated.

T. Johnson, *The First Indication of Neutrino-Induced Nuclear Fission*,
https://doi.org/10.1007/978-3-032-01936-3_6

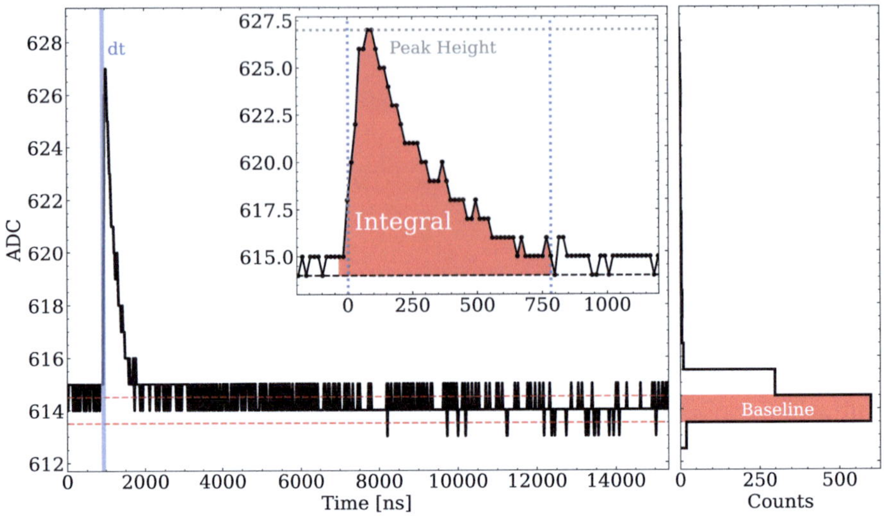

Fig. 6.1 Example of a NuThor waveform with the key values that are computed and stored. The inset is the waveform zoomed-in

The list of extracted values from the waveforms that contain signals are as follows (Fig. 6.1):

- **Timestamp of the onset**: The estimate of the time when the signal exceeds a certain threshold relative to the digitizer's clock, which begins at the beginning of each CoMPASS run. This is expressed in nanoseconds.
- **Integral**: The sum of the samples before and after the onset of the signal pulse minus the baseline estimate at every sample bin.
- **Peak High Value**: The value in ADC of the highest point in the waveform
- **Baseline**: A simple estimate of the baseline of the waveform calculated as the mean. This is a poor baseline estimate but is only used for monitoring detector health and an initial guess in the CMA filtering algorithm explored in a later section.
- **dt**: The number of samples before or after the external trigger. Given the waveform acquisition length is set to 960 samples and the pre-trigger length is set to 62 samples, the values of dt span from -62 to 898.
- **Strobe**: A number associated with the nature of the recorded event.
  0 = A detector channel,
  1 = An SNS timing signal coincident with the beam pulse. These occur at 60 Hz on only on channels 54 and 55,
  2 = An SNS timing signal anti-coincident with the beam pulse. These occur at 240 (540) Hz for production run 2 (1).

- **Entry Number**: This is the number in the ROOT TTree associated with this signal event. Because waveforms are so memory intensive, the output files after the signal processing do not contain the waveforms themselves. Rather, this number refers back to the waveform, such that the original waveforms can be queried without duplication of terabytes of data.

6.1 Timing Signals

In addition to the detector signals, there are timing signals provided to the COHERENT collaboration by the SNS: Event 39 and 61. Physics events in the COHERENT detectors can be associated with the arrival of the neutrino pulses thanks to these timing signals.[1]

Event 39 is a signal recorded on NuThor's channel 54 and transmitted to all SNS operations via the Experimental Physics and Industrial Control System (EPICS) [1]. This signal, operating at a 60 Hz frequency, is synced to an atomic clock, providing precise timing. It serves as a reference for predicting the arrival time of neutrinos to the instruments in Neutrino Alley. The physical source of Event 39 originates from the kicker magnet and neutron choppers, and it is closely tied to the timing of the proton packets [2] (Fig. 6.2).

Event 61 is a signal recorded on NuThor's channel 55 also sent to all SNS operations through the EPICs archiving system. This signal is tied to the delivery of protons to the SNS target via the ring-to-target beam transport (RTBT). As such, at nominal beam operations, Event 61 has a frequency of 60 Hz—albeit with a "skip" every 10 seconds. However, when the beam is off or is being tuned, the frequency of event 61 can be zero or simply not consistently 60 Hz. These moments are not incorporated in the Golden run list for the final analysis.

The NuThor data acquisition system is externally triggered by the event 39 timing signal and the duplicate triggers sent to the digitizers between the true beam pulses, thus anti-coincident and valuable for evaluating the steady-state backgrounds specific to beam operations. The number of anti-coincident triggers is set by the pulser. In Science Run 1 (winter 2022), there were nine anti-coincident triggers. In Science Run 2 onwards, there were four anti-coincident triggers. It is noteworthy that the event 39 TTL signal sent through the front face of the digitizer in the TRIG IN is delayed compared to the original Event 39 it is copied from. This results in the digitized Event 39 (on Channel 54) arriving at the digitizer 392 nanoseconds before the trigger.

Each SNS timing signal is evaluated event-by-event as either a true, coincident trigger or a duplicated, anti-coincident trigger. The evaluation is a simple level threshold from the baseline. The baseline estimate is simpler for these channels because

[1] There is no significance to the number of each SNS timing signal type. It is simply the bit address in the binary word sent to the neutrino alley via FPGA.

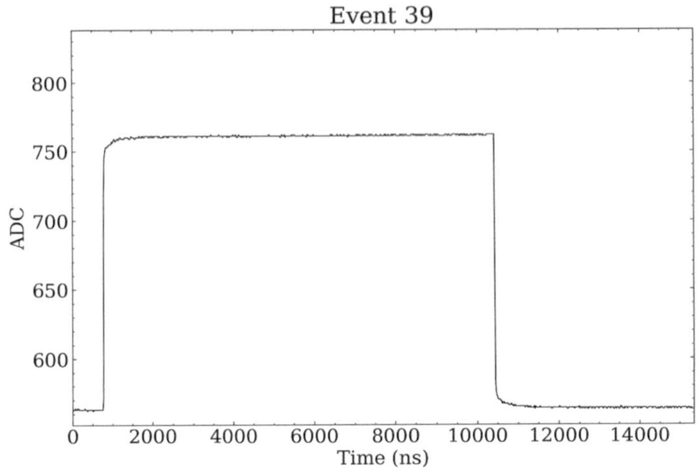

Fig. 6.2 Example of event 39 SNS timing signal digitized by NuThor. The signal is a TTL logic pulse

the onset of the signal always lies in the same sample in the waveform, so simply taking the average of the pre-trigger window is a sufficient approximation. A signal is present if, at any point, the waveform exceeds 5 sigma from the baseline with sigma is calculated as the mean average deviation of the pre-trigger window. Timing signals judged to be beam coincident are assigned a strobe value of 1. Conversely, beam anti-coincident timing signals are assigned a strobe value of 2.

6.2 Detector Signals

The detector channels must be treated specially to meaningfully extract the information listed above. There were two iterations of signal processing schemes. The first version is called "DSPv1" and was used primarily to judge the health of the NuThor DAQ and identify shortcomings of a more simplistic approach to pulse identification that used a level threshold to detect signals. Using DSPv1, several goals for the follow-up signal processing scheme crystallized:

1. **Low Energy Threshold**: Because the signal readout from the PMT bases is on the low gain output, a level threshold-based pulse ID approach can only achieve an energy threshold of approximately 100 keV. The design of the chosen signal processing scheme should prioritize lowering the energy threshold to combat the minimal amplification of the base.
2. **The ability to handle pile-up of signals in the same waveform trace** (Fig. 6.3) The radioactivity of the thorium target makes the NaI detectors highly susceptible

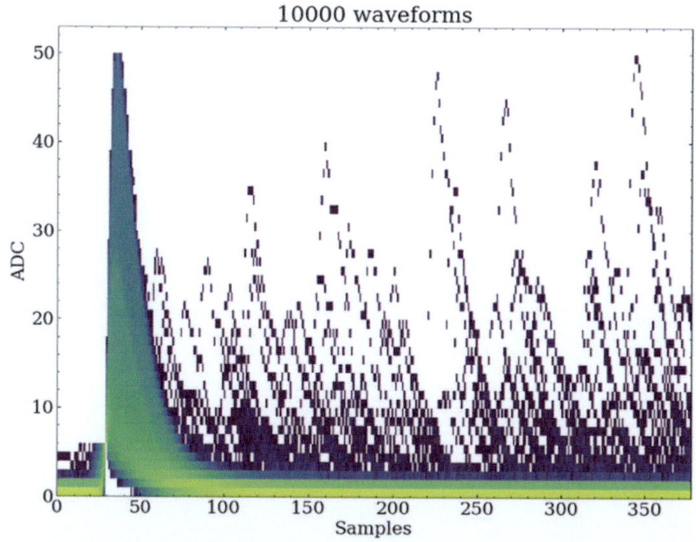

Fig. 6.3 Demonstration of NuThor pile-up. Pulses triggered and beginning at a fixed bin show many examples of an additional scintillation pulse also inside this same acquisition window on the same detector channel

to pile-up. Often, signal processing schemes discard events that contain multiple signals. This analysis seeks to include these pile-up waveforms.

An overview of the signal processing is shown in Fig. 6.4. Each component of the processing pipeline will be discussed in greater detail in the subsequent sections.

There are a few prerequisites to the DSPv2 analysis pipeline. First, a large sum of noise pulses were harvested and stored in a repository. Waveforms with maxima less than a level threshold set to 2σ from the baseline are judged to only contain noise (Fig. 6.5). A set of a few thousand of these signal-less waveform traces are saved into a database for later usage.

The frequencies present in the noise waveforms harvested for the noise database are examined in FFT-space as a power spectrum density and shown in Fig. 6.6 [3]. This information is fertile ground to explore possible signal processing approaches if any of the features present in the power spectrum density are viewed as undesirable. At lower energies, signal amplitudes can be comparable to the high-frequency noise present in the waveform trace. Employing a band-stop to remove a specific frequency band or low-pass filter to remove frequencies greater than a threshold may be advantageous for reducing high-frequency noise and lowering the energy threshold. However, studies conducted on NuThor NaI signals revealed significant overlap between the signal and noise across frequency bands, resulting in a reduction of both signal and noise that did not aid the goal of lowering the threshold.

Alternatively, the collected noise serves as the basis for generating a repository of toy waveforms. In this context, toy waveforms refer to artificially generated wave-

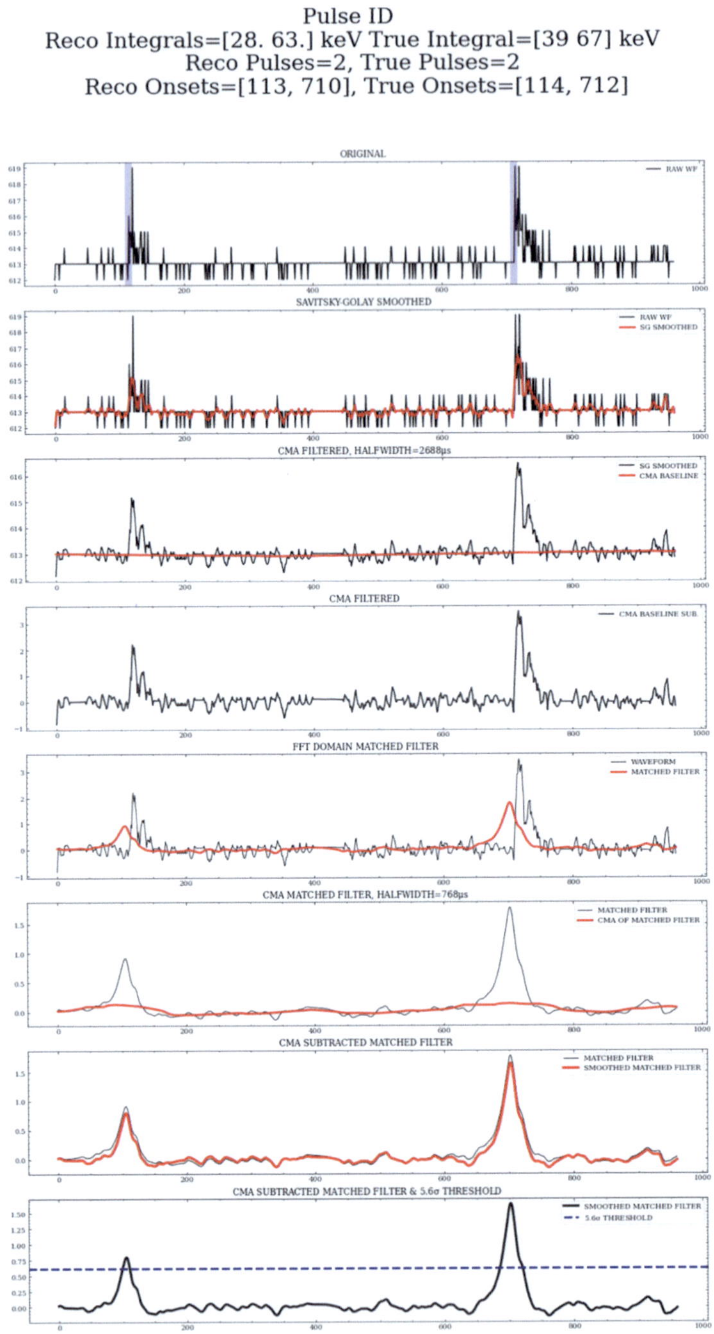

Fig. 6.4 Overview of the NuThor DSPv2 analysis pipeline applied to a toy waveform with two pulses with integrals of 28 and 63 keV

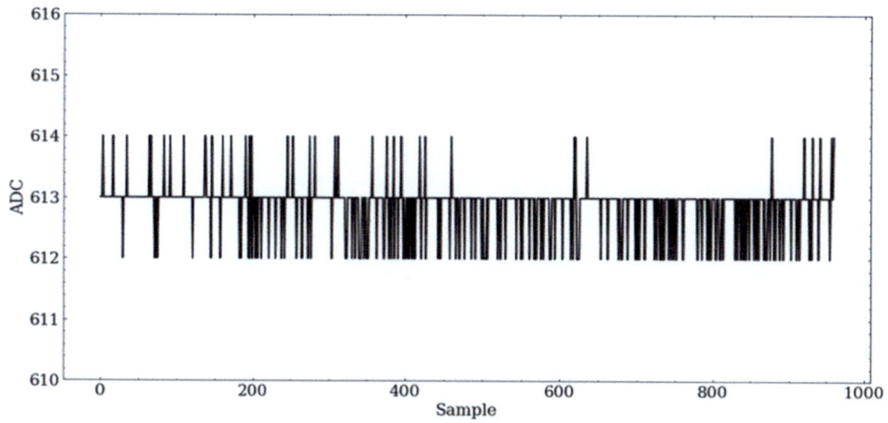

Fig. 6.5 Example noise waveform with no signal present from the NuThor Science Run 2 dataset

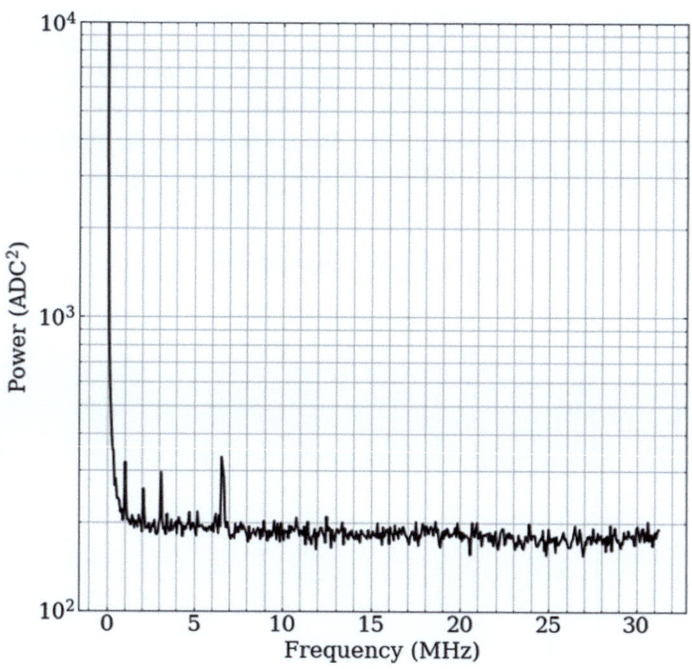

Fig. 6.6 The average power spectrum density (PSD) of 1500 noise signals from recorded NuThor data. The random fluctuations on a waveform-by-waveform basis averages out leaving behind the persistent noise frequency bands

forms, created in software, that closely mimic the observed signal pulses in the NaI crystals. A collection of data-driven toy waveforms enables the design and optimization of signal processing to be tailored to the specific characteristics of the NuThor hardware. The more accurately the toy waveforms replicate the actual waveform signals, the greater the understanding of the signal processing effects. Rather than attempt to fabricate the noise in the waveforms, the collected, empty waveform traces already exemplify the desired behaviour.

Also instrumental in the creation of toy waveforms is an understanding of the behavior of the signal pulses that will be seeded on top of the harvested noise waveforms. NaI scintillation has a very well understood timing structure [4]. One can use the waveforms displayed in Fig. 6.3 and take the average of each individual sample to get an average waveform that is useful as a waveform template.

The average shape of the waveform shown in Fig. 6.7 is area-normalized to one in order to construct a probability density function (PDF). This PDF is the basis of the signal component in the toy waveforms. Then to produce a toy waveform, one randomly selects a noise waveform from the database, and randomly samples the waveform PDF N-times, where N is the desired, true integral of the signal pulse. This iterative, random-sampling of the NaI scintillation PDF assures an accurate reflection of the discrete nature of the digitized waveforms recorded in hardware. Any number of pulses can be seeded onto the toy waveform to study how different signal processing approaches are affected by the presence of pile-up. An example of

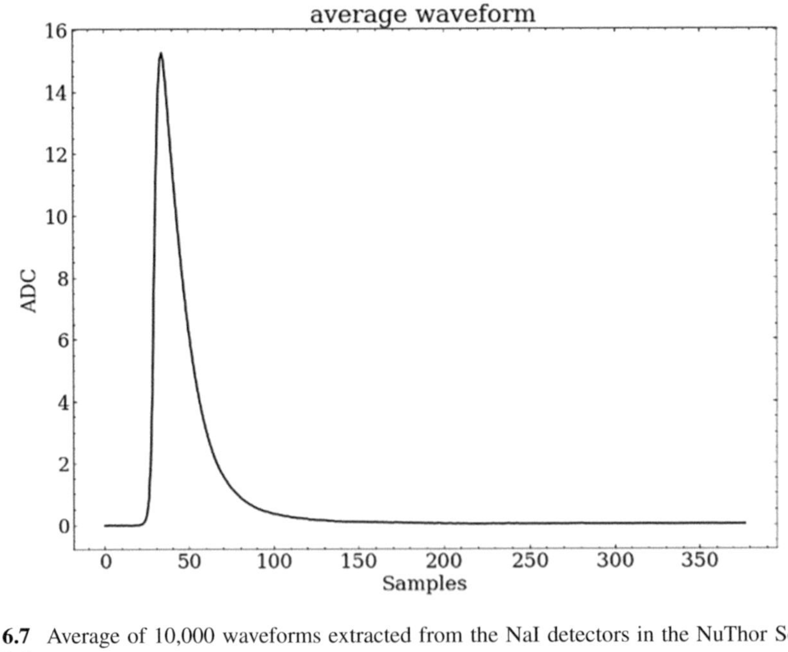

Fig. 6.7 Average of 10,000 waveforms extracted from the NaI detectors in the NuThor Science Run 2 dataset

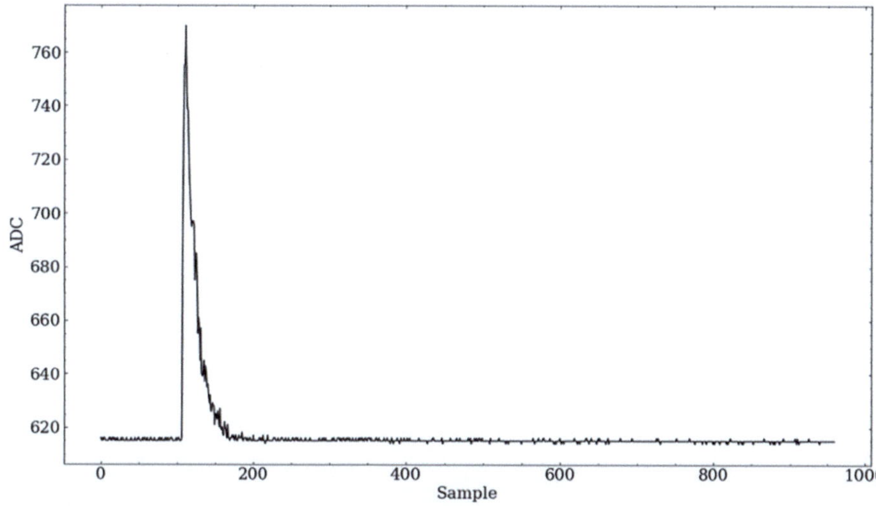

Fig. 6.8 Example of a toy waveform with integral of 2.614 MeV (the energy of the Tl-208 gamma ray)

a toy waveform through the above procedure is shown in Fig. 6.8 and shows good agreement with the waveform from data in Fig. 6.1.

After generating a large assortment of toy waveforms of a user-defined integral and onset time, a signal processing scheme can be created and then iteratively tested to control for unexpected quirks and inefficiencies.

6.3 Savitzky-Golay Smoothing

The high-frequency noise exhibited in the signals (and depicted in Fig. 6.6) complicates the baseline estimation needed for a later step in the signal processing pipeline and can confuse the peak finding routine when looking at low-integral signals that have peak high values close to the baseline. To combat this, a smoothing algorithm, that is called a Savitzky-Golay (SG) Filter, is employed [5].

The Savitsky-Golay algorithm scrubs a moving window of a specified width along the waveform's length. At each incremental step across the waveform, the content of the window is fit with a polynomial of some degree using least-squares minimization. In the case here, a 3rd order polynomial is used. After the polynomial is fit, the value at the center of the window is replaced with the best fit polynomial at that location. This process is repeated until the entire waveform has been smoothed.

The Savitsky-Golay filter employed for this work is a 3rd order polynomial (k = 3) with a moving average window parameter (m) of 11 which equates to a window length of $W = 2 \times m + 1$. A set of coefficients is computed at the outset

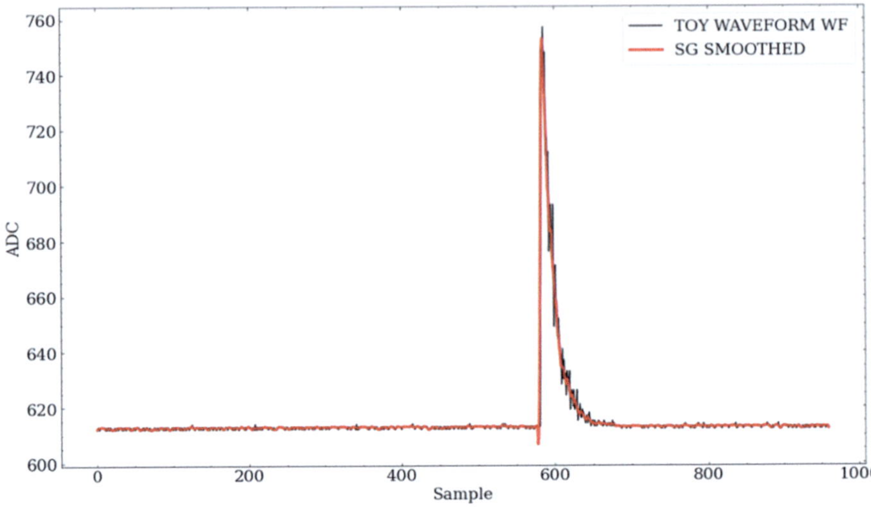

Fig. 6.9 Example toy waveform (black) with Savitzky-Golay Smoothing Filter (red) overlaid on top

and is applied to every iteration of the filtering. The coefficients for this context can be generated by computing the pseudo-inverse of a $W \times k$ matrix called the Design matrix. The steps are not shown in this work, but the result is the following vector of 11 coefficients (Fig. 6.9):

$$[-0.08391608, 0.02097902, 0.1025641, 0.16083916, 0.1958042, 0.20745921, 0.1958042, 0.16083916, 0.1025641, 0.02097902, -0.08391608]$$

As seen in Fig. 6.10, the SG Filter predominantly affects the higher frequency portion of the power spectrum density. Further, the histogram of each waveform sample's content in Fig. 6.11 indicates that the baseline is more tightly centered around a central value. Note that the binning is not uniform is this comparison of the pre- and post-filtered baselines; however, it is clear that the variation around the central value is reduced with the Savitsky-Golay Filter. Allowing the waveform samples to have non-integer values following the smoothing, yields a more accurate baseline estimate. Under many contexts, a integer versus non-integer waveform sample is a negligibly small difference. However, in the case of the NuThor waveforms of this PMT gain, the signal pulses tend to have low amplitudes, and both discrepancies and artifacts emerge if constrained to integer values. Particularly, the subsequent step in the signal processing chain, the conditional moving average filter creates errors from insufficient estimates of the waveform's baseline.

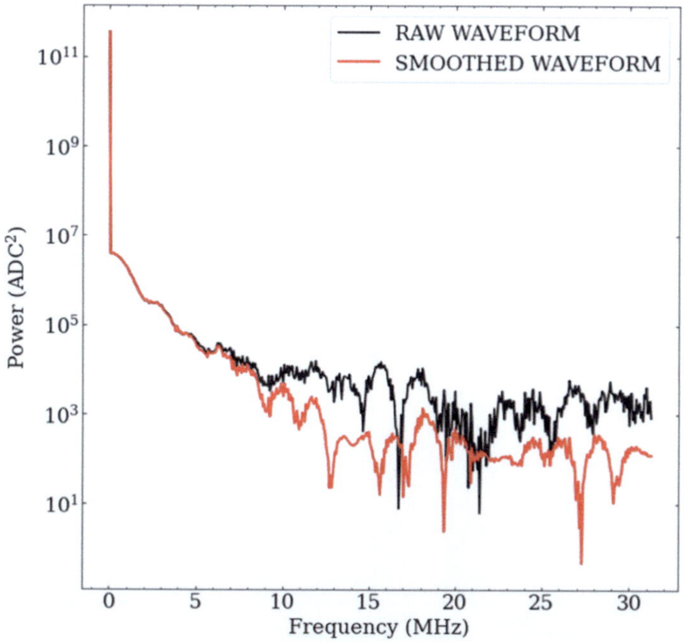

Fig. 6.10 Example toy waveform power spectrum density raw (black) and smoothed (red). The higher frequencies are affected while the lower frequencies are kept largely intact

6.4 Conditional Moving Average Filter

The conditional moving average (CMA) filter is employed here to remove the low frequency oscillations from the waveform trace while minimally impacting the signal itself [6, 7]. The CMA filter computes a baseline that is then subtracted from the original waveform. Rather than a single, flat baseline, the CMA-computed baseline is a vector with a length matching the waveform's number of samples. Two parameters called the reject threshold and the half width allow the algorithm to trace out the undulations without disturbing the signal. The filter works by instantiating a moving average window of length = half width × 2. The moving-window is iterated across the span of the waveform trace. Each new sample in the waveform trace will either be incorporated into the moving average or rejected/ignored if its value lies above the reject threshold.

This algorithm has three inputs: the reject threshold, the half width and the preload value.

- **Half Width**: The half-width is half the length of the moving average window that is stepped through the length of the waveform. This value must be less than the length of the waveform.

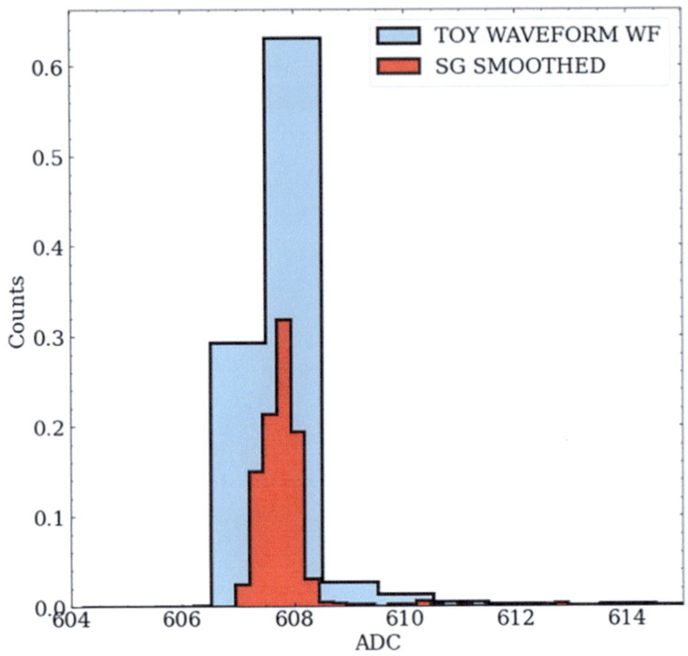

Fig. 6.11 The effect of the Savitzky-Golay smoothing filter on the baseline of the toy waveform. The two histograms depict the value of all the samples in a given waveform trace with (red) and without (light blue) the Savitsky-Golay smoothing. The binning between the two histograms displayed is not consistent, but that discrepancy is used to illustrate the more tightly centered bin content after the smoothing. Part of the utility of this smoothing is that it yields non-integer values

- **Reject Threshold**: As the moving window scrubs across the length of the waveform, the reject threshold determines if the new sample is incorporated into the window or rejected. A each step, the difference between the sample value and the current moving average is calculated. If that difference is greater than the reject threshold, that bin is not assimilated into the moving average. If that difference is lower than the reject threshold, the value is added to the moving average window. If adding that value increases the length of the moving average window to greater than 2×half width, the oldest sample is removed to conserve the length.
- **Preload Value**: The preload value is an initial guess of the baseline value for this waveform trace. An array, with a length matching the half-width, is then created with all elements set to the preload value. This value is especially important for low peak height pulses because slight deviations from the correct value will require the filter to migrate to the correct value. This migration is not instantaneous, thus an artifact will emerge in the output waveform in which an incline or decline will appear. Specifically, the first half width worth of samples will appear to have an

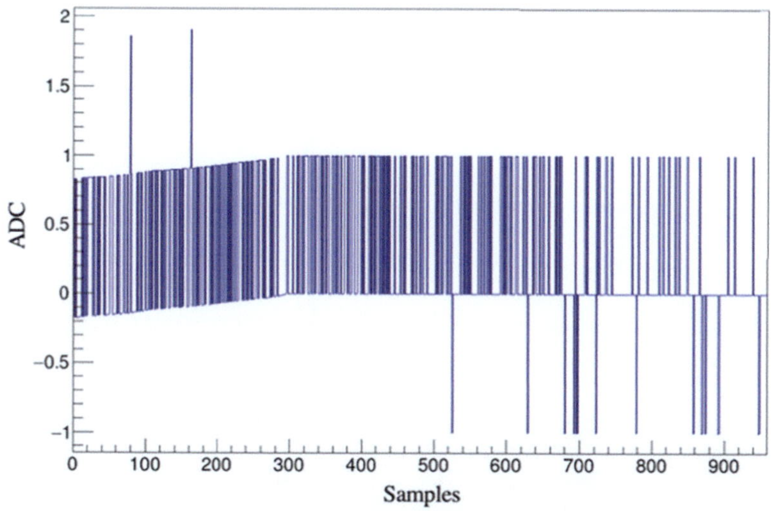

Fig. 6.12 Artifact introduced by an incorrect preload baseline estimate in the CMA algorithm. Although the artifact is visually slight, it can affect the trigger efficiency in that portion of the waveform trace when a pulse ID scheme is looking for signals of similar height to the noise

unphysical incline/decline (See Fig. 6.12). This defect is slight but can have an impact on the pulse ID efficiency in this span of samples.[2]

An example of the CMA filter computing a baseline to subtract from the waveform trace is shown in Fig. 6.13.

The effect of the CMA filter is evident by viewing the power spectrum density distribution before and after applying the CMA filter. Note that the contents of Fig. 6.14 is after the Savitsky-Golay filter has already been applied to the toy waveform.

Impact on Energy Reconstruction Efficiency

The CMA filter improved the overall energy reconstruction efficiency of the signal processing. This study speaks to the power of having a reliable repository of toy waveforms. The ability to test the an algorithm, the CMA filter, on a toy waveform with known, user-defined integral enables the iterative adjustment of the CMA parameters to most reliably return the "true" values. It also illuminates biases in the output values. In Fig. 6.15, several approaches are shown. Of the three tested approaches, the CMA produces the tightest population around one, where one represents a perfect reconstruction.

---

[2] A pulse ID algorithm that uses a level threshold will be directly inhibited from lowering its threshold if the CMA artificially inflates the number of triggers in the beginning region of samples.

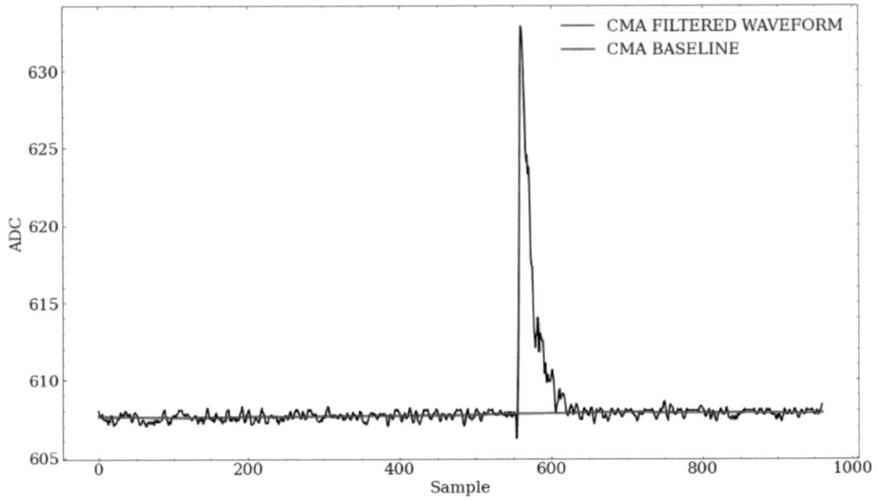

(a) the CMA baseline estimate

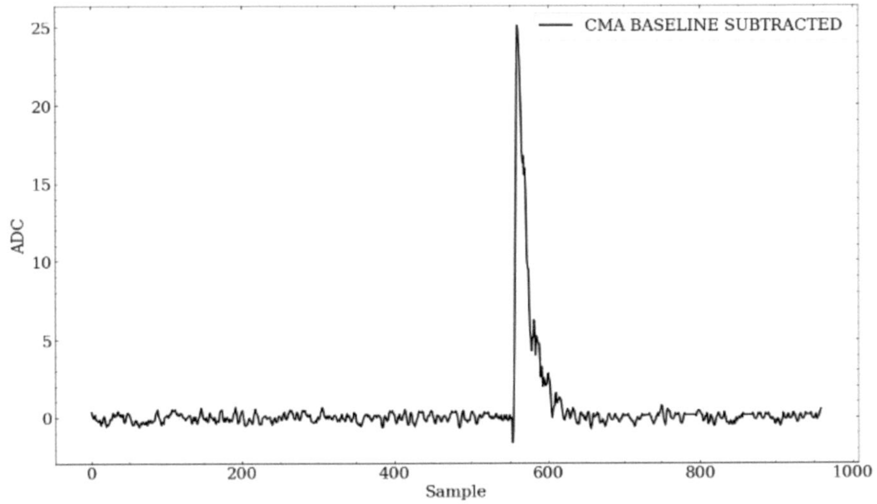

(b) The original waveform after subtracting off the CMA baseline

Fig. 6.13 The conditional moving average (CMA) filter applied to the toy waveform

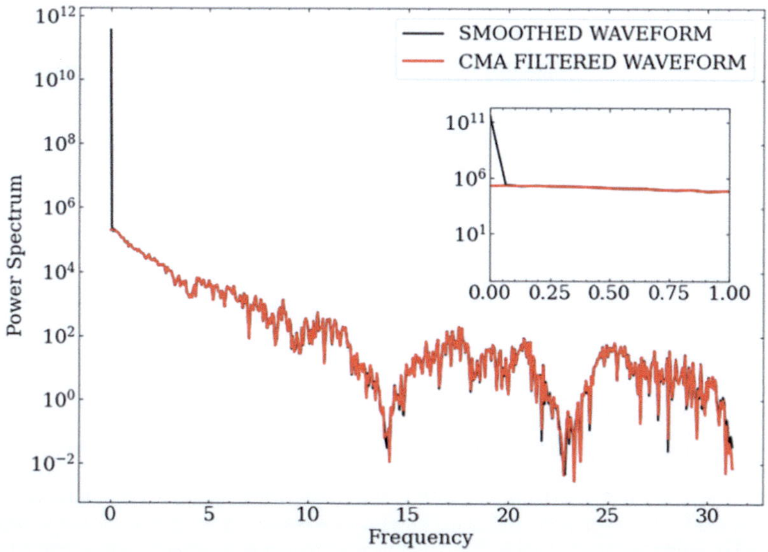

Fig. 6.14 The Power Spectrum Density before and after the CMA Filter is applied. The Savitzky-Golay Filter had already been applied. The Low Frequency noise is removed as illustrated in the inset plot

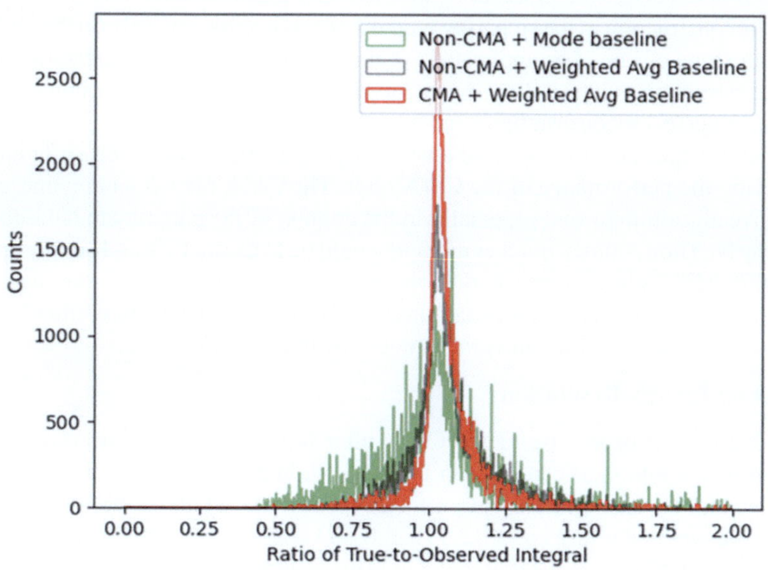

Fig. 6.15 The effect of various treatments of the waveform trace before integration. The performance is judged based on how tightly the various histograms fall near a ratio of true-to-observed integral of 1.00. The CMA + Weighted Average baseline computation exhibit the best performance by this metric

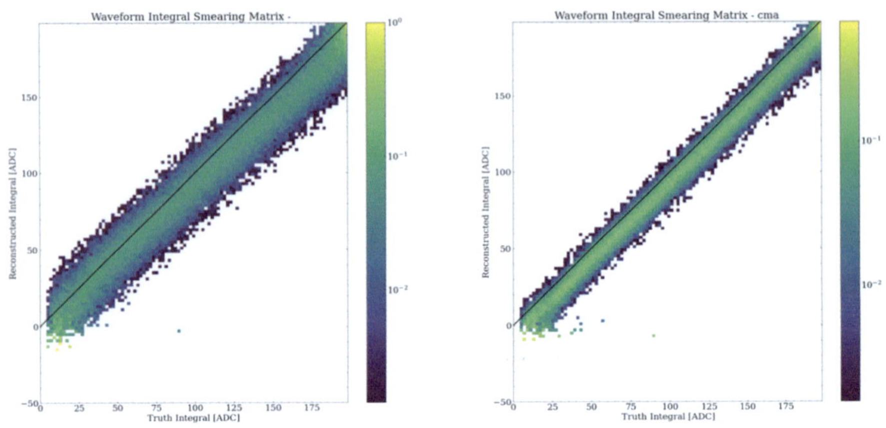

(a) No CMA applied to the toy waveform dataset.

(b) CMA applied to the toy waveform dataset.

Fig. 6.16 Smearing matrices with and without the CMA filter applied. The application of the CMA filter leads to a superior performance for energy reconstruction, as evidenced by the tighter population of points around the black diagonal line used to denote a perfect reconstruction for each true energy value

This is better displayed in smearing matrices created in the same way (Fig. 6.16). A perfect reconstruction in the smearing matrix context occurs precisely on the diagonal line with slope of one. The application of the CMA filter improves performance of the reconstruction evidenced by the reduced spread of data points around the line showing a perfect reconstruction.

A natural question to ask is whether a high-pass filter would be sufficient for mimicking the performance of the CMA filter. The CMA filter is a computationally expensive algorithm to run, especially in the context of the gargantuan datasets produced by NuThor. A much quicker method would be to use an FFT and simply apply a threshold in FFT space, where all the low-frequency contributions are removed from the waveform trace. Perhaps unsatisfyingly, this FFT-based high-pass filter leads to a poor reconstruction of the integral as depicted in Fig. 6.17.

Impact on Energy Resolution

An added benefit of the use of the CMA filter is that there is an improvement to the energy resolution, as can be seen in Fig. 6.18. When the photopeak is fit with a Gaussian and the background with a linear model, the width of the peak with the CMA Filter applied returns $\sigma = 18.113 \pm 0.533$ whereas the fit to the peak without the CMA filter gives $\sigma = 19.909 \pm 0.621$. This is approximately a 9% improvement in the energy resolution at 2.614 MeV.

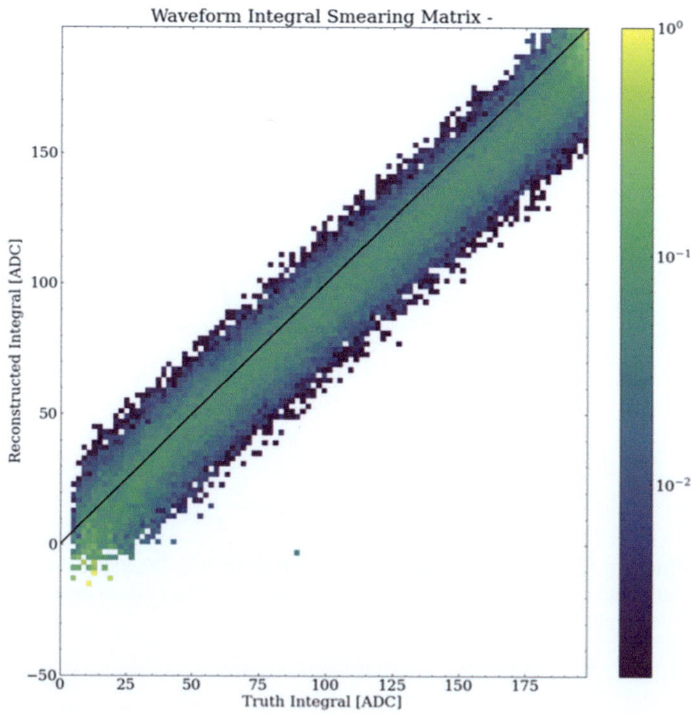

Fig. 6.17 Smearing matrix after a high-pass filter in FFT-space is applied to the toy waveforms. The high-pass filter upper threshold was set to 5 MHz. The performance is not

6.5 Frequency-Domain Matched Filter

A matched filter is now employed to detect signals inside the waveform trace [8]. A template is provided to aid this effort. The template in this case is the average waveform shape used to produce the toy waveforms. The template is convolved with the inspected waveform trace in the frequency domain using the Fast Fourier Transform (FFT) for its speed [9]. The same filter can be accomplished in the time domain but I had less success with this approach. Specifically, I used the template as a window function similar to a triangle or trapezoidal filter, but the length of the window function itself limited the ability to find pulses in the last portion of the waveform trace–the length of the template (Fig. 6.19).

As can be seen in the product of the matched filter, there is a characteristic offset between the peak of the matched filter product and the onset of the waveform. This offset is always the same and emerges from the presence of a pre-trigger in the waveform template (Fig. 6.7). The onset of the waveform is taken to be the position of this peak corrected for this global offset.

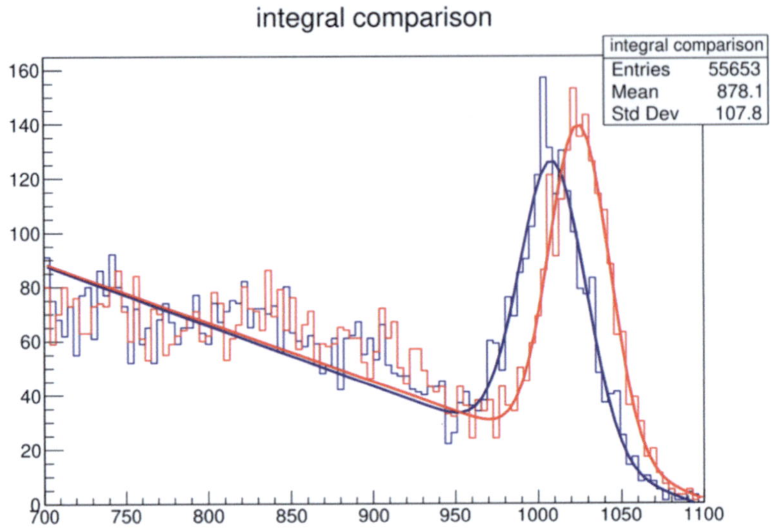

Fig. 6.18 (Blue) the integral spectrum with no CMA Filter applied. When a Gaussian is fit to the photopeak with a linear background, the best fit is $\sigma = 19.909 \pm 0.621$. (Red) the integral spectrum with CMA Filter applied. The best fit is $\sigma = 18.113 \pm 0.533$

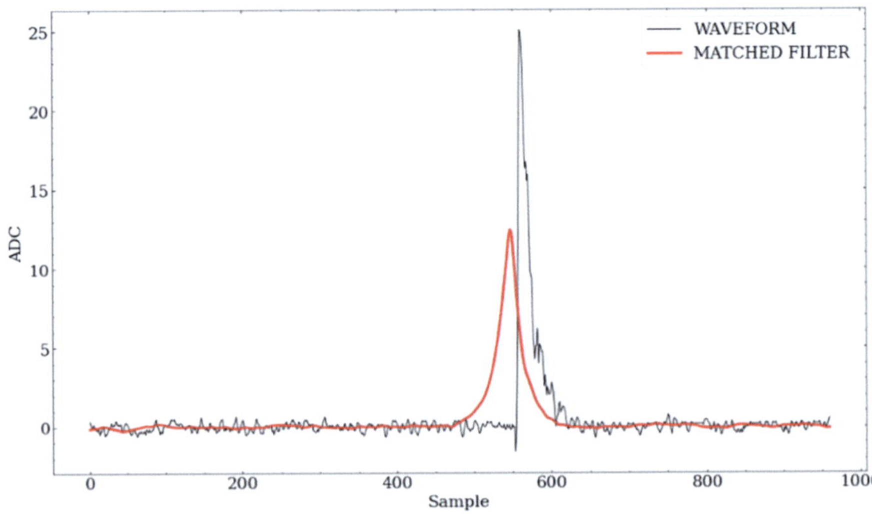

Fig. 6.19 The matched filter applied to the toy waveform (at this point, both the Savitzky-Golay and Conditional moving average filters have been applied). Note that the matched filter is askew compared to the toy waveform. The shift is proportional to the pre-trigger samples in the waveform template

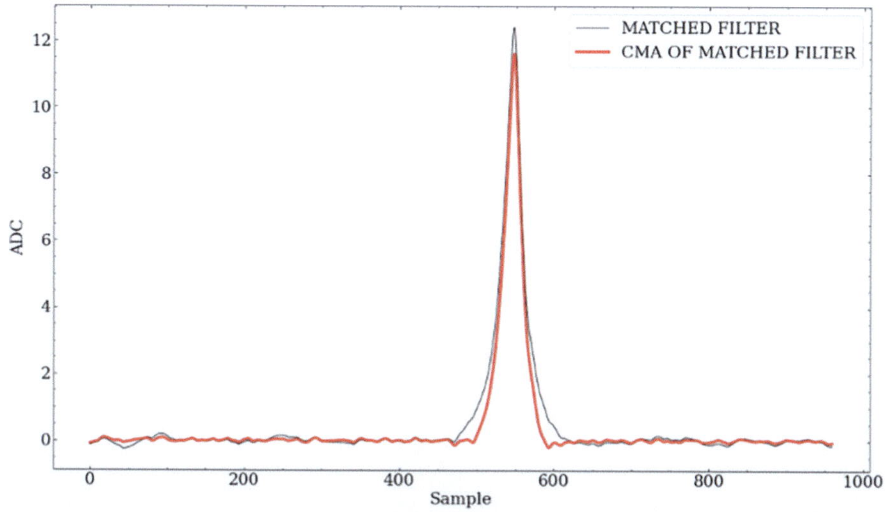

Fig. 6.20 The second CMA Filter applied to the output of the Matched Filter. A smaller half width is used for this second CMA filtering to better track the matched filter output, which has higher frequency baseline oscillations compared to the raw waveforms

Second Conditional Moving Average Filter

The peak finding algorithm benefits from applying a second CMA filter to the output of the matched filter (See Fig. 6.20). Because a simple level threshold is taken to identify the presence of signals, further removing random undulations of the baseline allows the signal processing scheme to find peaks closer to the baseline (lower in integral) thereby pushed down the achievable energy threshold. Figure 6.21.

Level Threshold On Match Filter Output

Finally, pulses are identified if the output of the matched filter exceeds a level threshold. The value of said level threshold will be discussed in subsequent sections. Features within the waveform trace that resemble a NaI pulse in shape result in a matched filter output to resembles a peak. The height of these peaks corresponds to how much the item in question "matches" the template it's being compared to (Fig. 6.22).

6.6 Integration Optimization

After signals are identified by the peak finding algorithm discussed in previous sections, one must then determine what the integral of the pulse is. This requires a decision of how far before and after the onset of the pulse one should integrate. The decision in this context was determined by running a 2D grid search allowing the samples before and the sample after the onset to vary. The samples before the onset

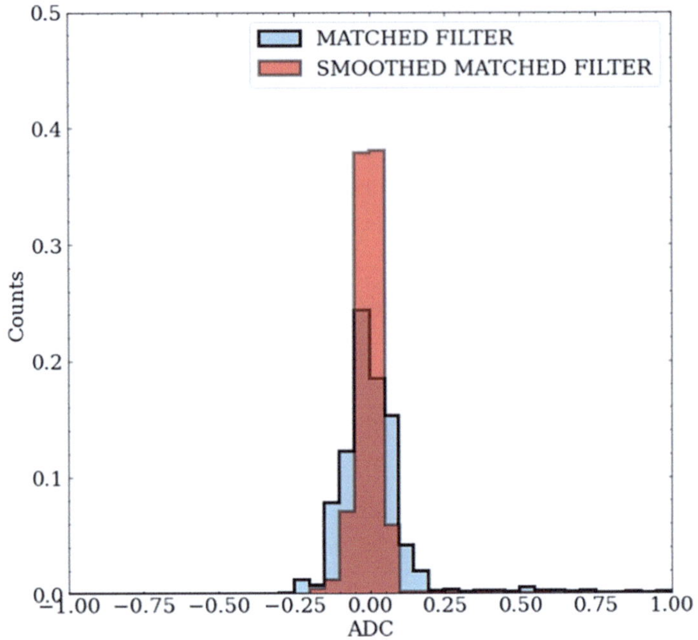

Fig. 6.21 Impact of the CMA filter of the Matched Filter output on the baseline. Plotted here is all the sample bin content in the waveform trace. The CMA applied to the matched filter output decreases the spread of the distribution

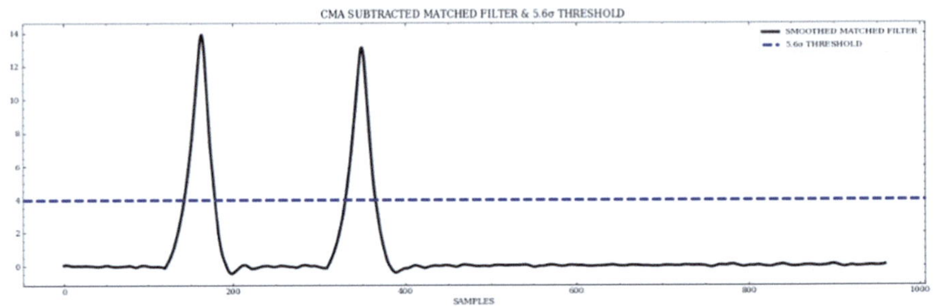

Fig. 6.22 Example of level threshold applied to the output of a matched filtered toy waveform. Pulses are identified only if they exceed the threshold in this step

varied between 6 and 18 samples (96–228 ns) and the samples after the onset varied between 40 and 200 samples (4,016–3,200 ns). Integrating too long leads to the an excessive addition of noise–thereby blurring out the structure. Failing to integrate long enough leads to the loss of potentially helpful information.

The optimal combination of integration parameters was found to be: Samples Before Onset = 6, Samples After Onset = 60 (Fig. 6.23).

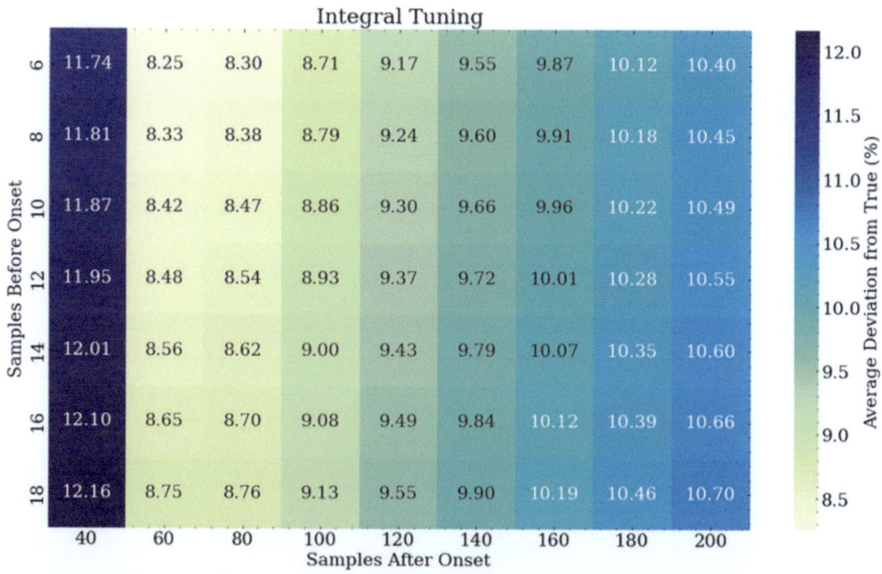

Fig. 6.23 2D grid search to determine the optimal integration parameters. The quality of each parameter set is judged by the average deviation from the True value

6.7 Differential Evolution Optimization

The hyperparameters for the signal processing shown above were educated guesses. They can likely be adjusted to optimize their efficiency at correctly reconstructing the toy waveforms' true values. To execute on this hyperparameter tuning, 10,000 toy waveforms were generated with integrals between 25 and 200 keV. The limited range of true integrals is because reconstruction above 100 keV has near 100% efficiency so about half of this sample contains waveforms in this regime. The aforementioned toy waveforms were generated with no pileup. Because pileup of signals in a single waveform trace is a sought after feature, that is optimized separately after the primary optimization.

The objective function being minimized here is shown below in Eq. 6.1. A total of sixth parameters are adjusted to achieve this minimization. The parameters are as follows:

1. The Savitsky-Galoy Smoothing Width
2. The waveform CMA Filter Reject Threshold
3. The waveform CMA Filter Half Width
4. The matched filter output CMA filter Reject Threshold
5. The matched filter output CMA filter Half Width
6. The matched filter output level threshold.

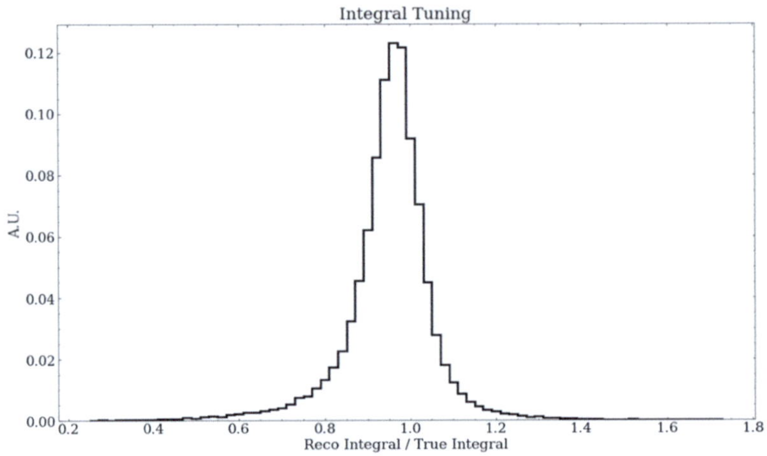

(a)

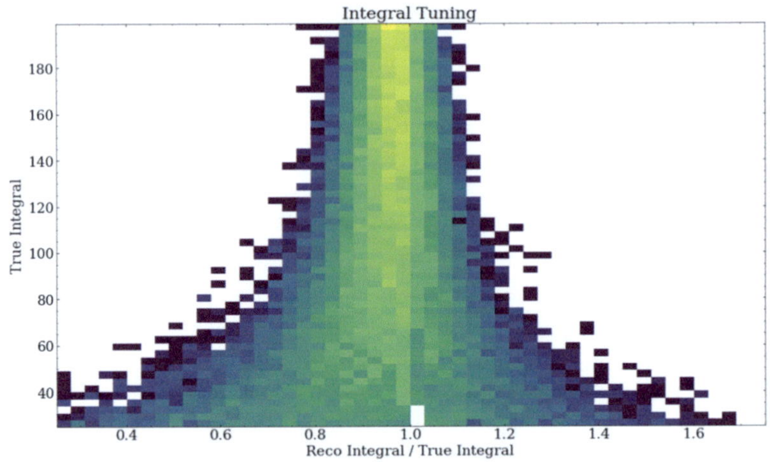

(b)

Fig. 6.24 Performance of the selected optimal set of integration parameters. A perfect integral reconstruction would be equal to 1

Note: Parameters 2, 4, and 6 are multiplied by the mean absolute deviation (MAD) on a waveform-by-waveform basis rather than a fixed level threshold for maximum accuracy. MAD is used rather than standard deviation because its numerator is linear rather than quadratic thus outlier samples (signal) won't skew the computation as severely.

Because the parameter space being optimized is N = 6, an ordinary grid search would be too computationally expensive to complete in a reasonable time frame. Instead, I turned to the stochastic minimization tool provided by Scipy called Dif-

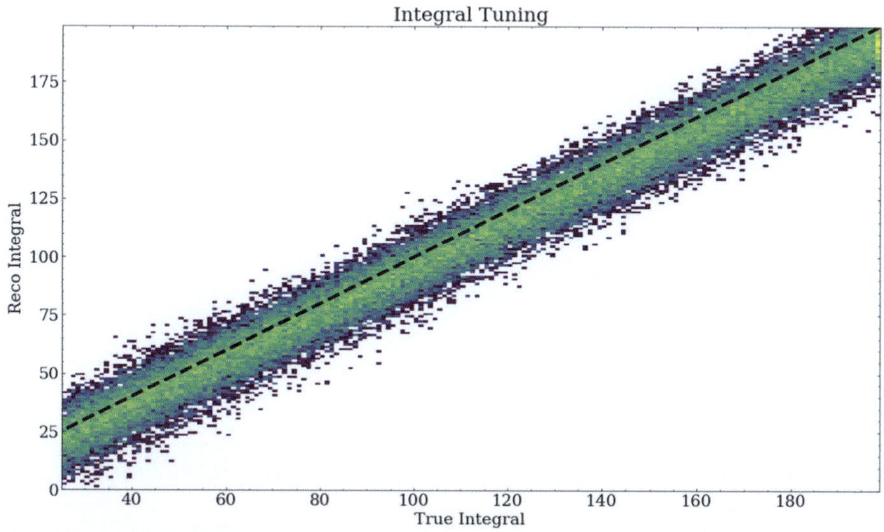

Fig. 6.25 Smearing matrix for the integral reconstruction after integration parameter optimization

ferential Evolution [10]. This tool works well to locate the global minima of a multi-variable space. The performance of the selected parameters is illustrated in Figs. 6.24 and 6.25. Part of the appeal is its usefulness for noisy and multi-modal objective functions. The objective function used in this work equally weighted the reconstruction efficiencies for energy, onset time, and pile-up (Fig. 6.26 and Table 6.1).

$$\text{Score} = \frac{\sum_i^N \left| 1 - \frac{E_{\text{reco}}}{E_{\text{true}}} \times \frac{T_{\text{reco}}}{T_{\text{true}}} \times \frac{N_{\text{reco}}}{N_{\text{true}}} \right|}{N} \tag{6.1}$$

Threshold Optimization for Pile-Up

To manage pileup of multiple signals occupying the same waveform trace, an additional dataset was generated where between 0 and 3 signals were placed inside a trace. This second round of optimization only varies the Matched Filter Threshold.

As demonstrated in Fig. 6.27, as the level threshold is decreased the peak finding algorithm incorrectly believes there is a signal when the true number of signals in the waveform trace is zero. This keeps the threshold from going too low in the optimization. The balance point between correctly identifying zero pulses and when there are pulses in the waveform is found to be at 5.62σ from baseline.

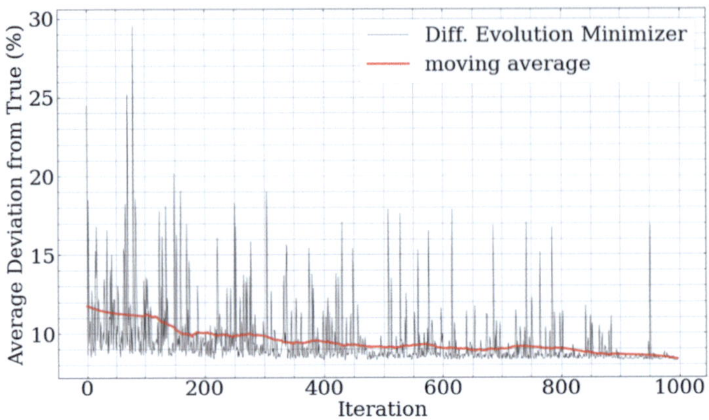

Fig. 6.26 The progression of the differential evolution minimizer over the course of 1000 iterations. An initial objective function score of 24.48% difference is diminished to 8.25% difference

Table 6.1 The adjustment between the initial parameter guesses and the final selections optimized through the differential evolution minimizer on a repository of toy MC waveforms. The resulting optimized parameters were used for the analysis of all datasets

Parameter	First guess	Final optimization
SG smoothing window	4	11
Waveform CMA reject threshold	2	2.43
Waveform CMA half width	195	168
Matched CMA reject threshold	2	2.62
Matched CMA half width	40	48
Matched filter level threshold	5	6.70

6.8 Data Cleaning

Signal processing often comes with unwanted artifacts that must be sorted out or removed from the analysis stream. Filters that move along the length of a waveform often experience edge effects where aberrant behaviour observed. The following two cuts are made on the processed data to address both the noise pulses allowed into the data stream and the edge effects observed at the tail end of the waveform traces.

Noise Removal

Despite applying an optimized level threshold on the output of the matched filter, there is still a certain amount of noise pulses that leak in. Noise can be identified with key features like (1) a high amplitude and (2) a low total integral. This is well encapsulated by the quantity A/E which is the peak height divided by the integral of the pulse. Figure 6.28 illustrates the A/E plotted at a range of energies. A clear population of noise is identified below approximately 50 keV.

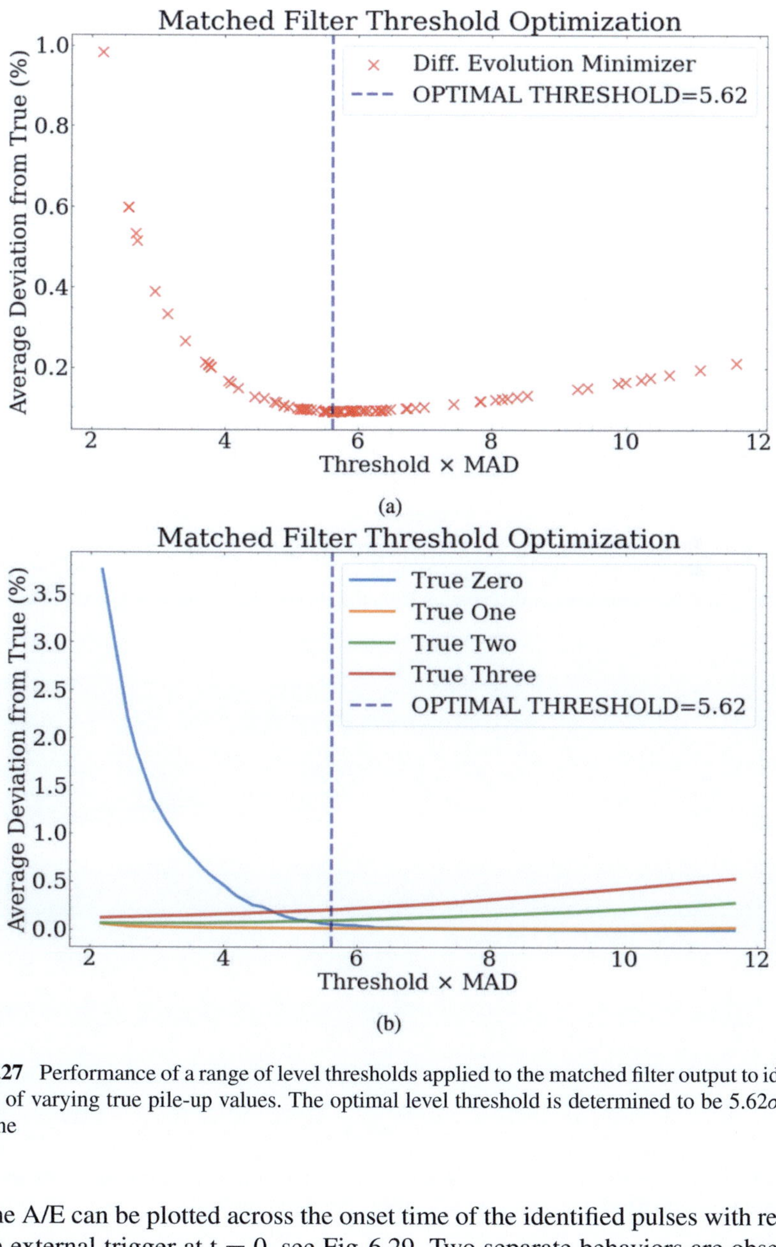

Fig. 6.27 Performance of a range of level thresholds applied to the matched filter output to identify pulses of varying true pile-up values. The optimal level threshold is determined to be 5.62σ from baseline

The A/E can be plotted across the onset time of the identified pulses with respect to the external trigger at $t = 0$, see Fig. 6.29. Two separate behaviors are observed. On the early side of the waveforms, a population of noise pulses is observed and mimics the A/E distribution seen in Fig. 6.28. One can address the noise by applying a software threshold of 50 ADC (100 ADC) in Science Run 2 (Run 1) to clean the trace of these events. Since an integral threshold was already planned, this comes at no additional cost.

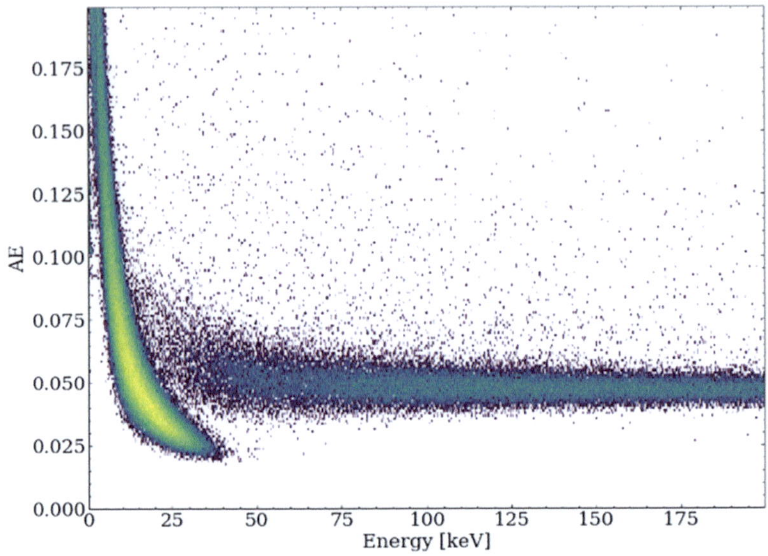

Fig. 6.28 The A/E versus Energy distribution from the NuThor data after energy calibration in this case

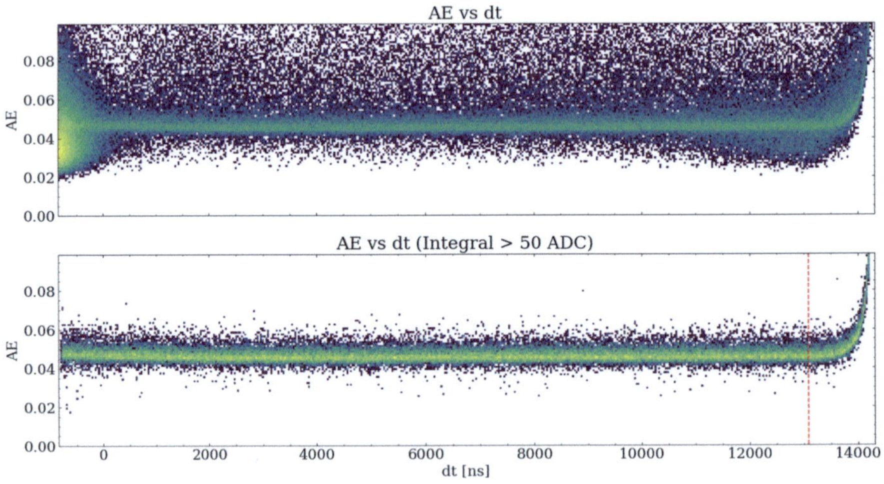

Fig. 6.29 (top) The A/E plotted across the onset time of the identified signal pulses, dt. (bottom) The same as the top panel, but with a 50 keV energy threshold applied

Second, there is an upward trend in A/E at the tail end of the waveform. The behavior at the end of the waveform trace is caused by events that are not fully contained within the trace leading to incomplete integration and skewing the A/E ratio into the noise-like region. This second issue is mitigated by cutting events that occur less than one integration length from the end of the waveform.

The effectiveness of these data cleaning cuts is displayed in Fig. 6.30. The two aforementioned interventions rectify the troublesome behavior in the onset sample distribution as evidences by the now uniform shape (Fig. 6.31).

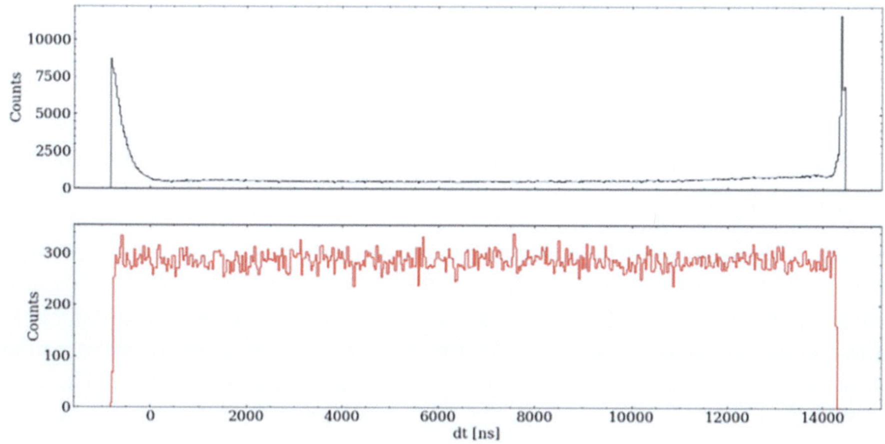

Fig. 6.30 (top) The onset sample for events examined in the beam OFF dataset. There are clear indicators of artifacts induced by the signal processing routine given the non-uniform shape to this steady-state background. (bottom) the result of applying a software threshold of 50 ADC and excluding events at the tail end of the waveform trace

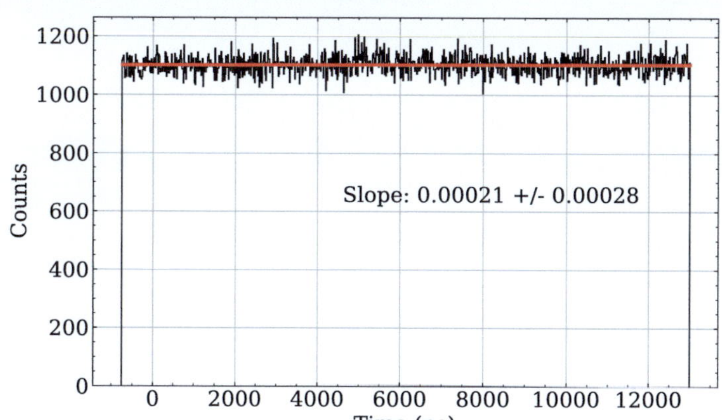

Fig. 6.31 The uniformity of the tagged pulses in time is checked by fitting the onset times with a linear function. The best fit value for the slope is consistent with zero

6.9 Energy Calibration

The NaI[Tl] contain intrinsic background-producing contaminants like ^{208}Tl and ^{40}K which previous experiments using these NaI[Tl] crystals used to calibrate their energies. However, the thorium target inside NuThor prevents using this approach exactly. Because thorium still emits copious amounts of the 2.614 MeV ^{208}Tl γ-ray line, that remains in the calibration method. However, the 1.460 MeV ^{40}K γ-ray line is masked by the lower energy radiation from the thorium. In its place, the 511 keV γ-ray emitted from the combination of the hot-off-gas pipe and the thorium's pair production serves as the second calibration peak for a two-point calibration to be performed–See Fig. 6.32. There are other possible calibration peaks such as the single- and double-escape peaks at 1.592 MeV and 2.103 MeV, respectively. However, for the purposes of this analysis, the aforementioned ^{208}Tl and 511 keV peaks are used (Fig. 6.32 and Table 6.2). Figure 6.33 shows the result of the energy calibration on all functioning detector channels individually and summed together.

Gain Tracking

Energy calibrations must be performed fairly regularly to account for incremental changes in the gain which can originate from a plethora of sources. For this reason, energy calibrations are performed weekly to account for any changes to the detectors. However, as can be seen from Fig. 6.35, the detector gains are stable over the course of the entire 12-week long Science Run 1 and 9-week long Science Run 2. The gain for each NaI detector is tracked via the mean and sigma of the Gaussian model fit

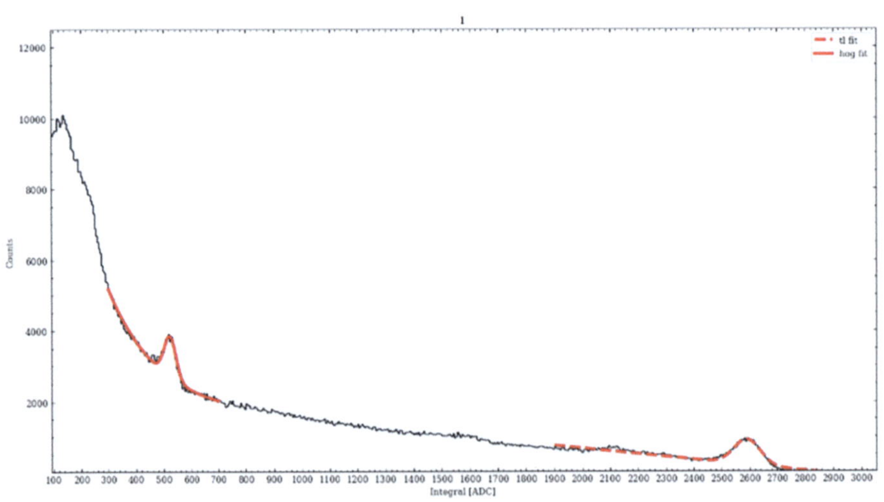

Fig. 6.32 Example of a two-point calibration on channel 1 in NuThor fitting peaks at 511 keV and 2.614 MeV. Both peaks are fit with Gaussians though with different background assumptions. For the 2.614 MeV peak, the background is fit with a linear model. For the 511 keV peak, the background is fit with an exponential model

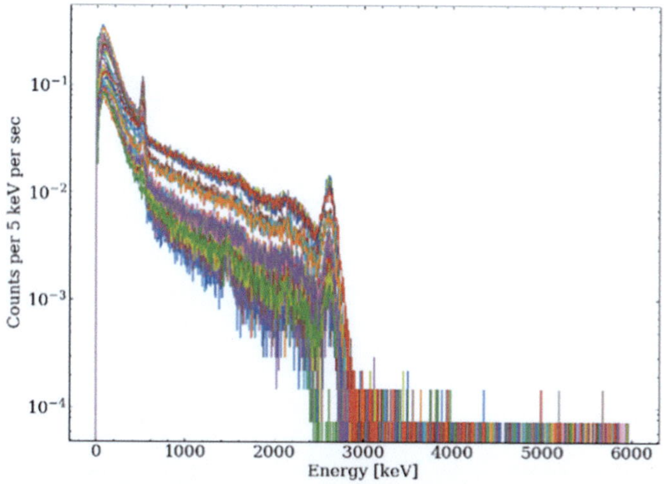

(a) Calibrated energy spectrum of all 35 NaI[Tl] channels plotted individually in counts per 5 keV per second.

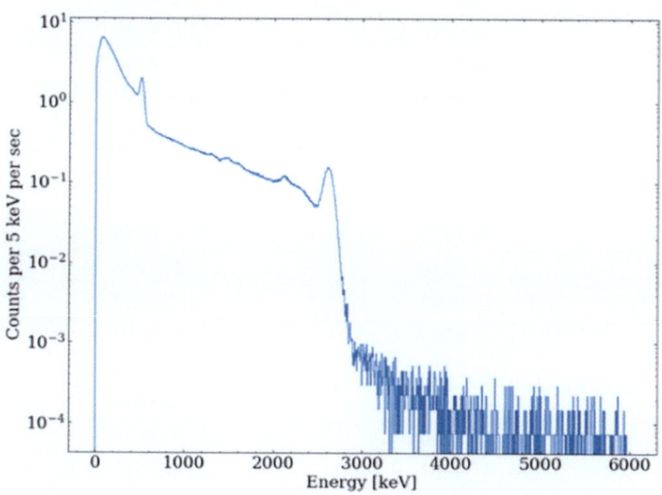

(b) Calibrated energy spectrum of all NaI[Tl] channels summed together in counts per 5 keV per second bins.

Fig. 6.33 The calibrated energy spectra from the NaI detectors plotted both separately (above) and summed (bottom)

Table 6.2 Overview of the energy resolution FWHM in keV across the 35 functioning NaI detector channels. The results were calculated using the photo-peaks at 511 and 2614 keV

Channel	511 keV	2614 keV
1	41.3545	94.3167
2	44.6184	90.6207
3	48.2117	103.3352
4	43.6943	104.3250
5	42.3527	101.9714
6	46.8465	112.1647
8	44.5926	87.6212
9	45.3011	99.2543
10	45.1893	99.1694
12	42.6346	95.4440
13	44.0632	82.3535
14	46.5462	88.7685
15	44.2167	90.8658
16	42.5862	88.6622
17	48.1021	123.7367
18	48.6319	136.6736
20	44.6485	101.1636
21	44.9127	91.0000
22	49.4925	120.4189
23	46.2566	88.2385
24	44.3910	98.5936
25	38.6062	93.4247
26	43.5939	103.6915
27	42.0850	99.1930
28	46.8066	107.0352
29	44.8209	91.6025
30	45.7516	88.8786
31	45.2860	88.7890
57	40.8340	90.8140
58	47.5167	104.3583
59	45.6565	93.5956
60	49.7817	102.9372
61	63.6438	108.9750
62	52.9325	128.4777
63	47.0517	106.5833

to the ^{208}Tl peaks. Despite the consistency observed in the detector channels, each week of the campaign receives an individual set of calibration parameters for greater accuracy (Fig. 6.34).

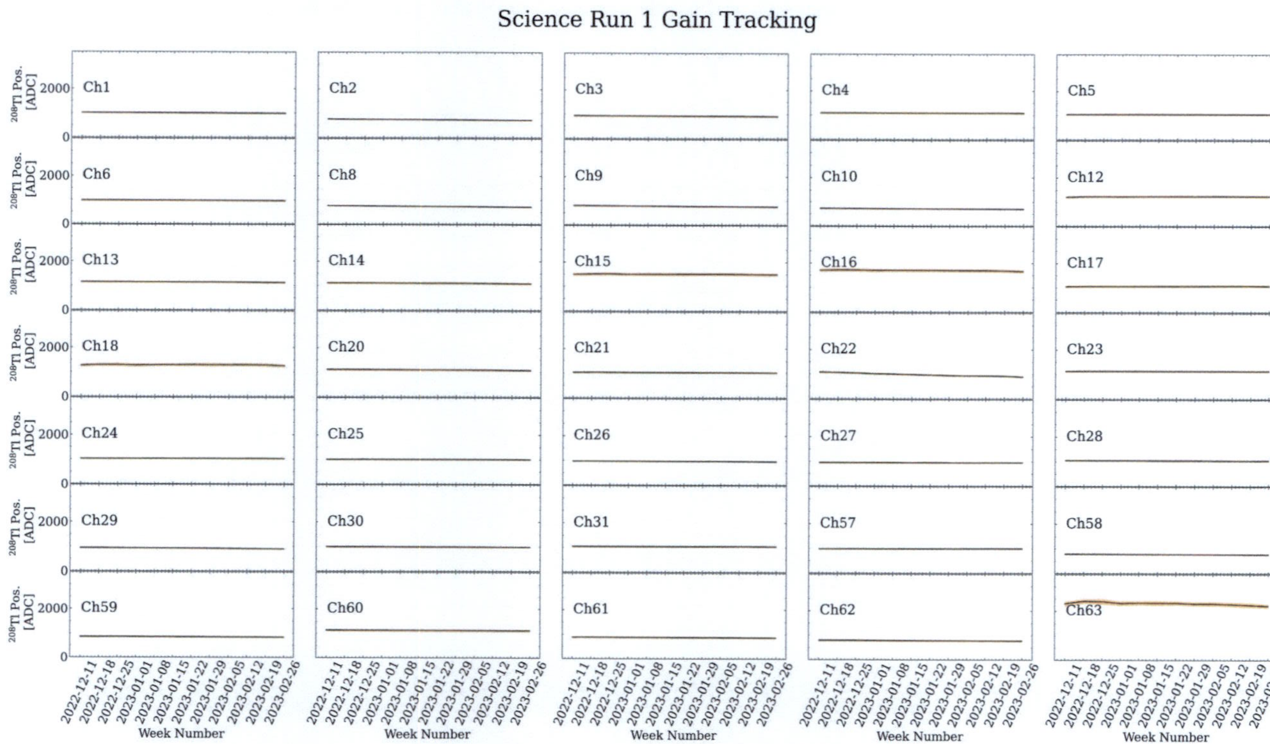

Fig. 6.34 Gain tracking performed during the 12-week long Science Run 1 campaign. The mean position of the Gaussian, 2.615 MeV ^{208}Tl γ-ray peak (and its sigma) are plotted over time

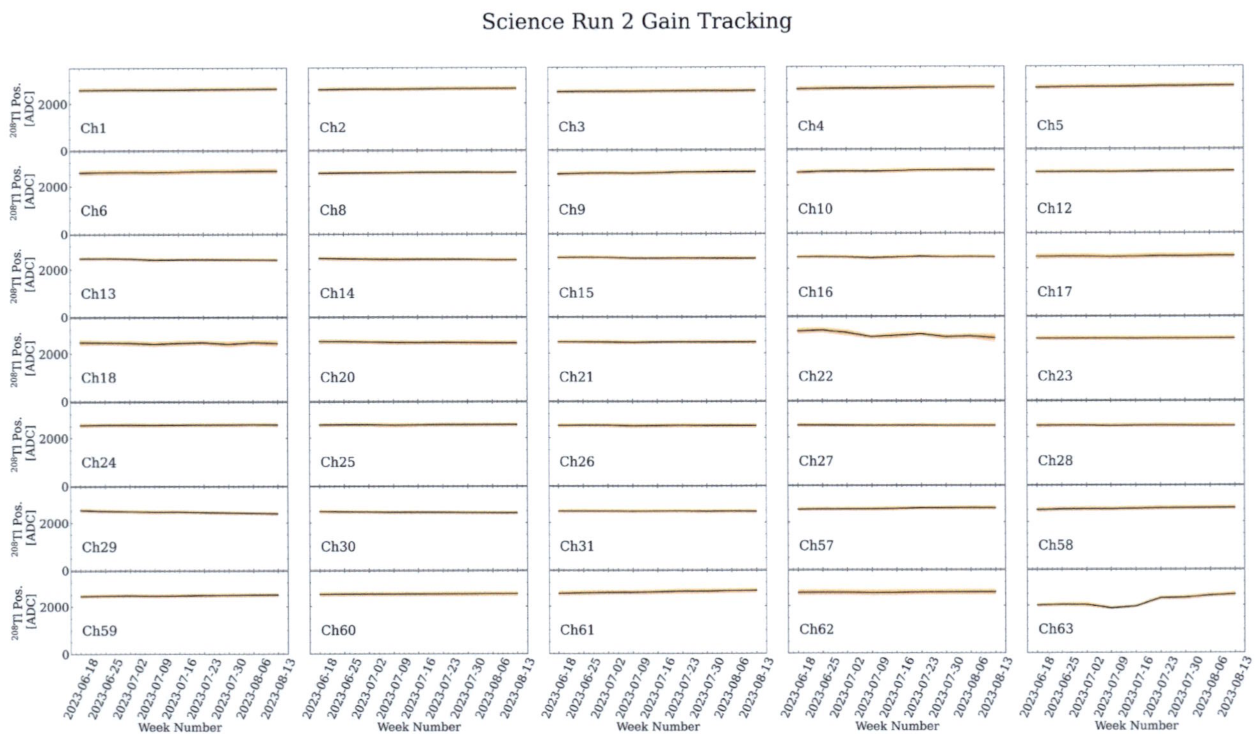

Fig. 6.35 Gain tracking performed during the 9-week long Science Run 2 campaign. The mean position of the Gaussian 2.614 MeV ^{208}Tl γ-ray peak (and its sigma) are plotted over time. There is very little variation across all of the NaI channels with the exception of slight deviations for channels 22 and 63

6.10 Health Checks

After data collection, the combination of the SNS beam's meta-information and detector quality assessments are used to select which data runs should be utilized for the final analysis. The SNS beam related health check conditions used in this analysis pertain to the SNS beam power and beam energy. Unlike, previous measurements made by COHERENT in Neutrino Alley, the radiation level emitted from the hot-off-gas pipe monitored by the environmental radiation monitor called "Radstack" is not used as a health check. Because the radioactive background from thorium is already high, the contribution from the hot-off-gas pipe relatively minor. The health check criteria is listed in Table 6.3.

A database is maintained to store all the meta-information for the approximately 60,000 data runs between Science Run 1 and 2. In this database, the UTC time that each data run occurs is corrected for the 4 (or 5) hour shift associated with time zones and time changes.

Beam Power

Although NuThor primarily collects data during period of sustained beam operations, there are times before, after, and even during these beam periods where the beam is in a condition that is not optimal for inclusion in the NuThor analysis. There are days during the beam periods where the operators are tuning the beam or performing accelerator studies. These are cut from the analysis by requiring the beam power to be greater than 800 kW. The distribution of average beam power for NuThor data runs is shown in Figs. 6.36 and 6.37.

Beam Energy

Similarly to the beam power, the beam energy is also subject to a quality cut for periods where the beam is being tuned or is simply not operational. The minimum beam energy utilized in this analysis is 950 MeV. Relative to the beam power, the beam energy is quite stably maintained for longer periods of time so is the less stringent cut condition. The distribution of average beam energy for NuThor data runs is shown in Figs. 6.38 and 6.39.

Table 6.3 The NuThor Golden Run criteria utilized to adjudicate the data run files that are used in the final analysis. These are in line with conditions set for previous COHERENT experiments

Quantity	Requirement
Average beam power	$\geq$800 kW
Average beam energy	$\geq$950 MeV
NaI detectors	All functioning

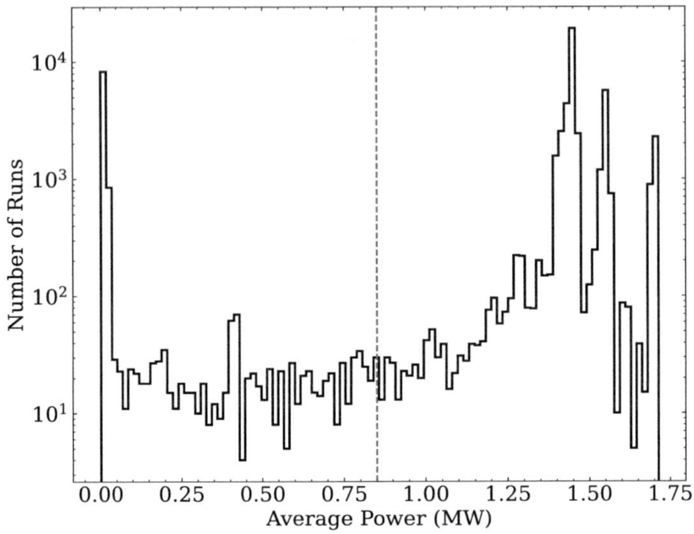

Fig. 6.36 Distribution of average beam powers for NuThor runs. The minimum average beam power of 800 kW is depicted with the vertical line

Detector Checks

All 35 NaI detectors must be functioning during data runs in order to be included in the final analysis. The 36th detector channel is already excluding due to a faulty BNC cable connecting the PMT base to the digitizer.

Beam Exposure

After applying the health check conditions, the remaining data runs compose the "Golden Run" list. These are all suitable for inclusion into the final analysis. A database of each Golden run was created and the combination of the run's duration, average beam power, and average beam energy was used to estimate the number of expected neutrinos per cm^2 at NuThor's distance from the SNS target. The beam exposure for Science Run 1 and 2 is 2.25 and 1.57 GWHrs, respectively (See Fig. 6.40 and Table 6.4).

6.11 Event Building

This section outlines how the signals across the 35 NaI detector channels are grouped together with each other, correlated to the SNS timing signals, and the region-of-interest is blinded. The neutrino-induced nuclear fission signal in NuThor is expected to be visible on several detector channels with an event topology that must be recorded in vivid detail in order to discriminate against the backgrounds posed by the thorium target mass (Fig. 6.41).

(a) Average beam power for Science Run 1 data runs.

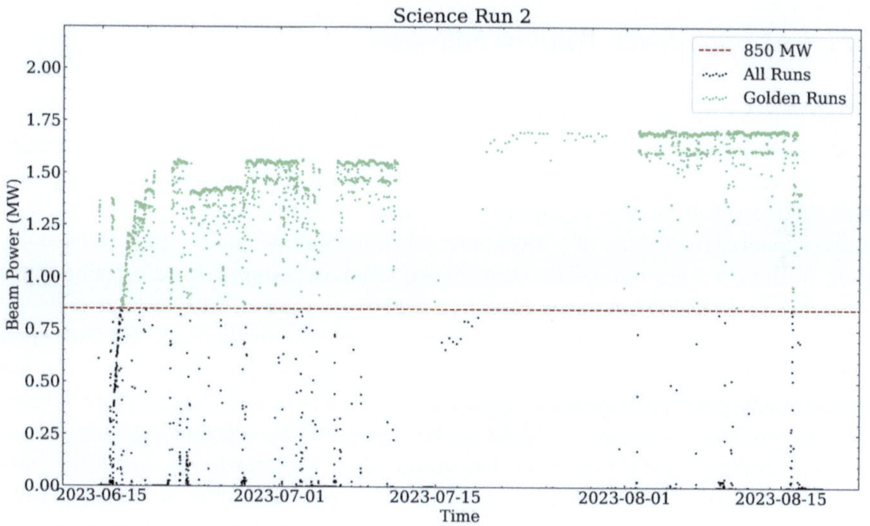

(b) Average beam power for Science Run 2 data runs.

Fig. 6.37 The beam power monitoring during both NuThor's Science Run 1 and 2. Green scatter points indicating which data runs passed the conditions for inclusion in the Golden run list

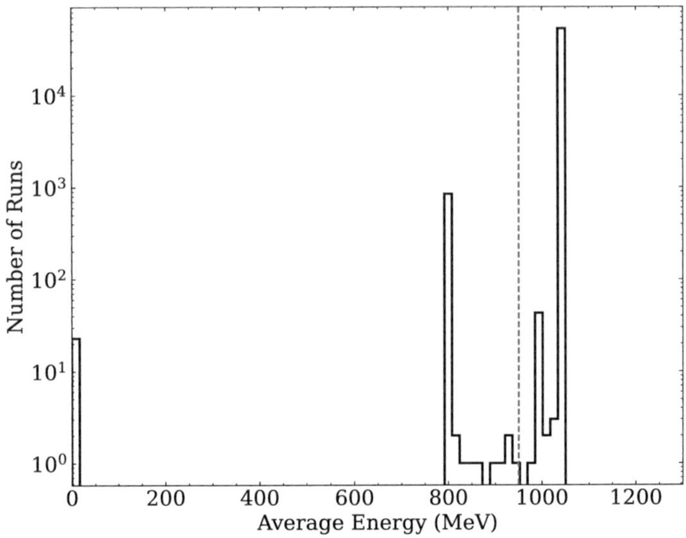

Fig. 6.38 Distribution of average beam energies for NuThor runs. The minimum average beam energy of 900 MeV is depicted with the vertical line

6.11.1 Coincidence Window Selection

Recall that the crux of the NuThor detection approach is neutron counting. Neutrons produced simultaneously within the thorium volume will moderate at different rates then subsequently capture on gadolinium nuclei at staggered intervals, creating time gaps between each successive capture. Each neutron capture on a gadolinium nucleus is accompanied by a spray of γ-rays outwards hitting some number of NaI detectors close in time to each other. The coincidence window assigned for clustering these γ-ray hits from a single neutron capture is 100 ns.

This coincidence window was determined quantitatively with NuThor data (Fig. 6.42) where the difference in time between detector hits was used to elicit the typical spacing between correlated and uncorrelated events in the detectors. Examples of correlated events are multiple scatters between two adjacent detectors where the same underlying physics event. Uncorrelated events refer to two random events like a muon and a ^{208}Tl γ-ray hitting two crystals near each other in time. The time scales of these two event populations can be represented by exponential models with different decay constants. The data points are fit to reveal where the correlated event population diminishes along the time axis. One can then extract an appropriate coincidence window to encapsulate the majority of correlated events. This was judged to be 100 nanoseconds. A longer coincidence window would capture more of the correlated events but would also catch a greater fraction of uncorrelated events (Fig. 6.43).

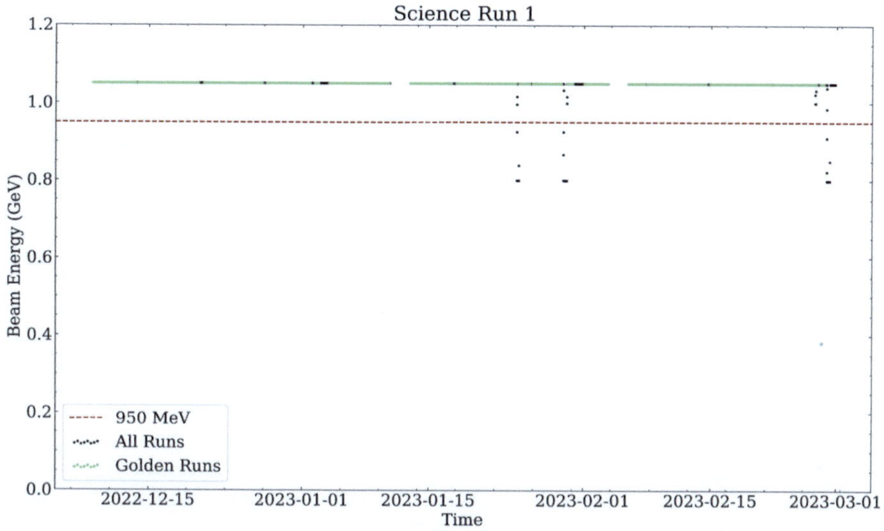

(a) Average beam energy for Science Run 1 data runs.

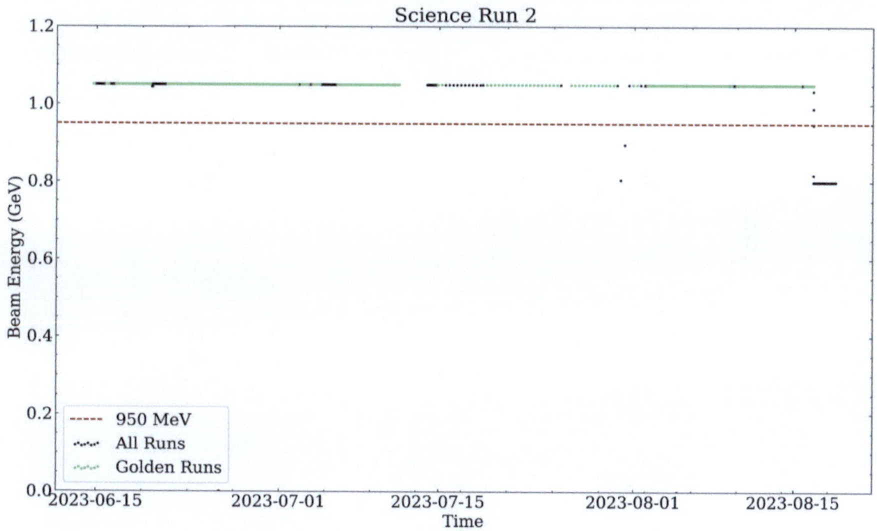

(b) Average beam energy for Science Run 2 data runs.

Fig. 6.39 The beam energy monitoring during both NuThor's Science Run 1 and 2. Green scatter points indicating which data runs passed the conditions for inclusion in the Golden run list

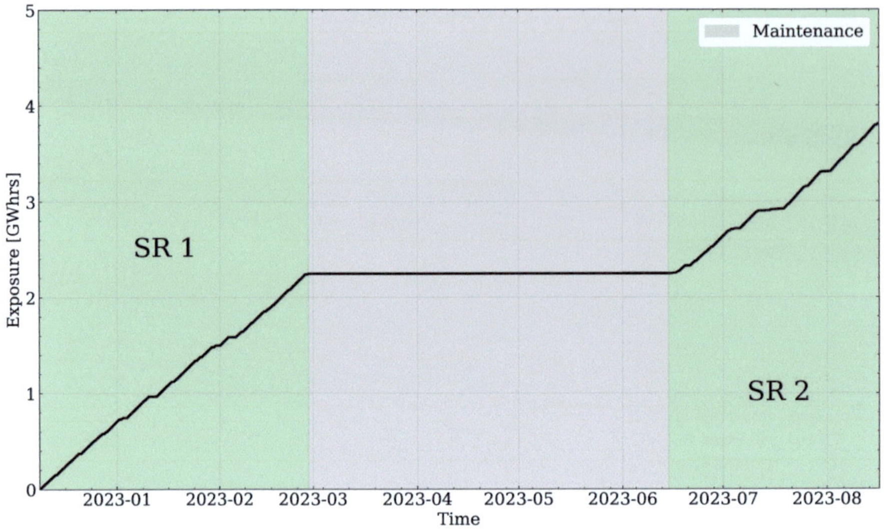

Fig. 6.40 The beam exposure attained through Science Runs 1 and 2. Only the data runs that pass the Golden run criteria are included in this integrated beam exposure

Table 6.4 NuThor exposure in the Golden Run list after applying the health checks. The neutrino flux for each respective beam exposure is estimated using the parametric function Eq. 3.6

Beam Period	Exposure [Days]	Exposure [GWhrs]	Estimated Neutrino Flux [cm^{-2}]
SR 1	64.80	2.25	9.25×10^{13}
SR 2	43.36	1.57	6.64×10^{13}
Total	108.16	3.81	1.59×10^{14}

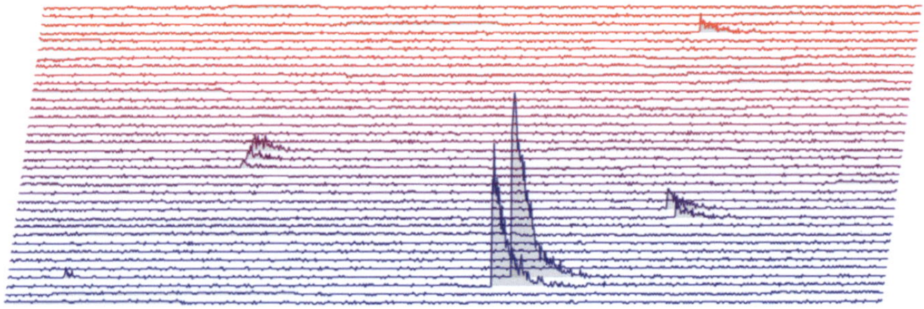

Fig. 6.41 Display of the NaI waveforms associated with a toy neutrino-induced nuclear fission event. This is a toy event generated from real noise data and the NaI signal waveform template meant simply for illustrative purposes

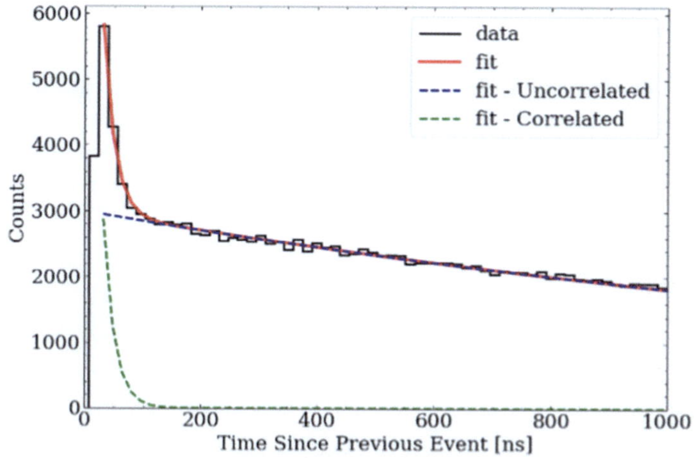

Fig. 6.42 Time since the previous event in the NuThor data. The previous event is restricted to being a different detector channel than the comparator

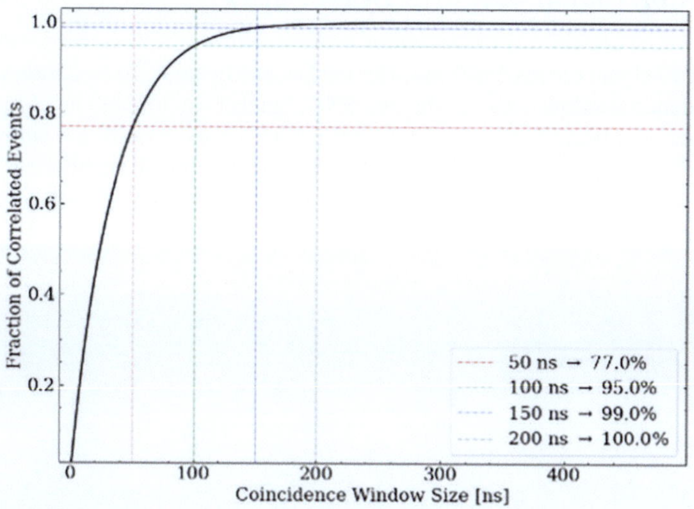

Fig. 6.43 N_{100} coincidence window selection determined by taking the cumulative sum of the exponential fit of the correlated events in Fig. 6.42

6.11.1.1 N_{100} Clusters

The clusters of NaI hits within the same coincidence window are dubbed N_{100} clusters. This naming convention is adopted from the Super-Kamiokande experiment [11]. Figure 6.44 illustrates how the hits across the NaI detector channels are clustered into neutron candidate events with the 100 ns sliding window.

6.11.1.2 Super Clusters

In addition to the N_{100} clusters, a wider sliding window of 15.360 μs is employed to create "Super clusters." (See Fig. 6.45) 15.360 μs is the maximum length of the waveform recorded in NuThor. These super clusters are all the events correlated to a single beam spill.

6.11.2 Correlation with SNS Beam Pulses

The Event 39 signal comes 2.667 ms after the Event 61 signal. To be conservative, a 3 ms coincidence window between the two SNS signals is required to label a particular SNS trigger as a beam ON event. All events with a corresponding beam pulse are labeled as such and are separated from the analysis pipeline to be blinded.

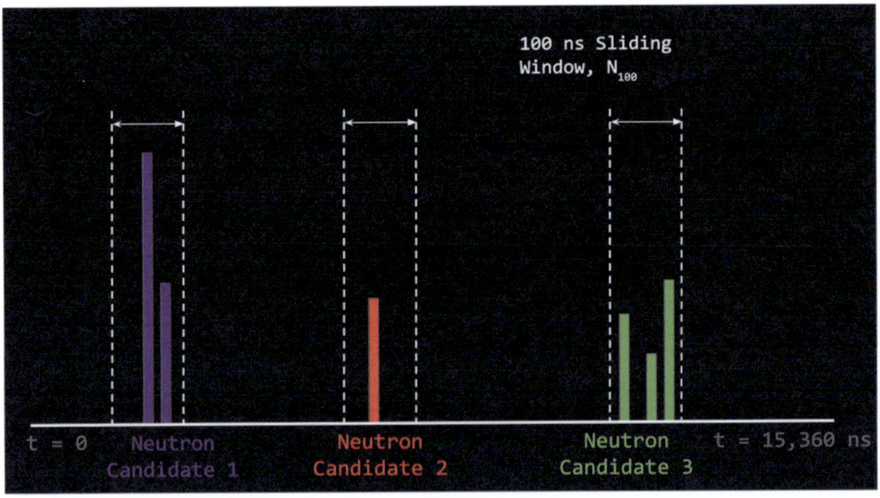

Fig. 6.44 N_{100} clustering to locate neutron candidate events across the 35 function NaI detectors. The horizontal position of the bars indicates their timing within the recorded acquisition window. The height of the bars indicates the energy of each deposition in arbitrary units

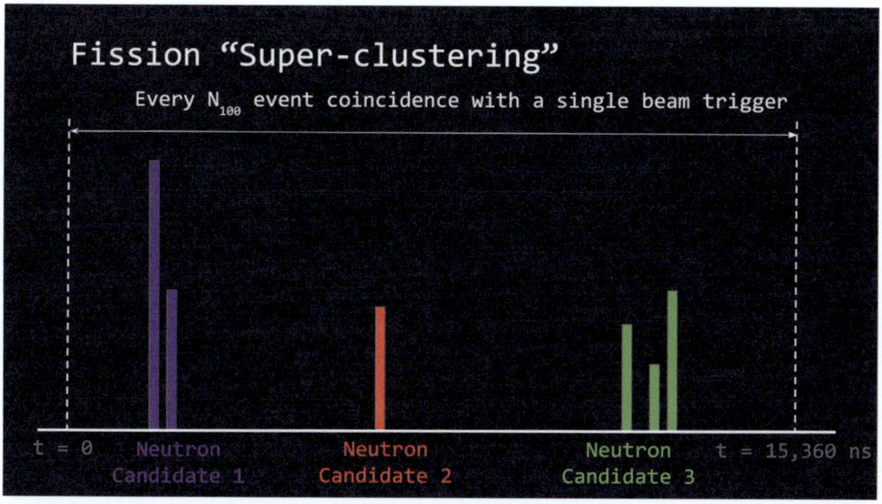

Fig. 6.45 Super-clustering events within a coincidence window of 15.360 μs–the maximum distance between any two events in a single external trigger in NuThor

References

1. Dalesio LR et al (1994) The experimental physics and industrial control system architecture: past, present, and future. Nucl Instrum Methods Phys Res Sect A: Accel Spectromet Detect Assoc Equip 352(1):179–184. ISSN: 0168-9002. https://doi.org/10.1016/0168-9002(94)91493-1. https://www.sciencedirect.com/science/article/pii/0168900294914931
2. Saethre RB, Plum MA (2017) SNS proton power upgrade requirements for magnet and kicker systems. In: IEEE 21st international conference on pulsed power (PPC), pp 1–6. https://doi.org/10.1109/PPC.2017.8291208
3. Brigham EO, Morrow RE (1967) The fast Fourier transform. IEEE Spectr 4(12):63–70. https://doi.org/10.1109/MSPEC.1967.5217220
4. Gerbier G et al (1999) Pulse shape discrimination and dark matter search with NaI(Tl) scintillator. Astropart Phys 11(3):287–302. ISSN: 0927-6505. https://doi.org/10.1016/S0927-6505(99)00004-3. https://www.sciencedirect.com/science/article/pii/S0927650599000043
5. Schafer RW (2011) What is a Savitzky-Golay gilter? [Lecture Notes]. IEEE Signal Proc Mag 28(4):111–117. https://doi.org/10.1109/MSP.2011.941097
6. Rich GC (2017) Measurement of low-energy nuclear-recoil quenching factors in CsI[Na] and statistical analysis of the first observation of coherent, elastic neutrino-nucleus scattering. Ph.D. Thesis, North Carolina U. https://doi.org/10.17615/972g-w286
7. Hedges S (2024) Measurement of quenching factors in NaI[Tl]. In: Low energy neutrino-nucleus interactions at the spallation neutron source. Springer Nature Switzerland, Cham, pp 45–65. ISBN: 978-3-031-68110-3. https://doi.org/10.1007/978-3-031-68110-3_4
8. Giannakis GB, Tsatsanis MK (1990) Signal detection and classification using matched filtering and higher order statistics. IEEE Trans Acoust Speech Signal Proc 38(7):1284–1296

9. Frigo M, Johnson SG (2005) The design and implementation of FFTW3. Proc IEEE 93(2):216–231. Special issue on "Program generation, optimization, and platform adaptation"
10. Storn R (1996) On the usage of differential evolution for function optimization. In: Proceedings of North American fuzzy information processing, pp 519–523.https://doi.org/10.1109/NAFIPS.1996.534789
11. Fukuda Y et al (1998) Evidence for oscillation of atmospheric neutrinos. Phys Rev Lett 81:1562–1567. https://doi.org/10.1103/PhysRevLett.81.1562

Chapter 7
Simulation of Signal and Backgrounds

Abstract The foundation of the simulation effort is the MCNP–Polimi simulation. However, a known issue with MCNP is its handling of gadolinium de-excitations. A package called GADGET was devised by the author of this work to account for the discrepancies in MCNP gadolinium by combining the trust worthy neutron propagation in MCNP with the data-driven, high-fidelity gamma-ray cascades of DICEBOX. This framework is used to estimate the NuThor detector's response to the signal neutrons emitted from the neutrino-induced nuclear fission events and both the steady-state and beam-correlated backgrounds expected to be faced in this measurement.

An idealized prediction of the signal products from a neutrino-induced nuclear fission event, as discussed in Sect. 4.4, is insufficient at describing the expected observables from the NuThor detector. The fission neutrons must be simulated within the NuThor detector geometry detailed in Sect. 5.2. This chapter details the use of particle transport simulation software, MCNP, to track neutrons through the apparatus until they result in a detectable interaction within a NaI detector volume or escape detection by being captured on a passive material or leaving the detector's boundaries. The simulation is rigorously evaluated and post-processed to replicate the output of the signal processing routine established in Sect. 6.2. The event clustering method from Sect. 6.11 is applied, and the reconstructed observables are scaled by the beam exposure shown in Sect. 6.10.

The Simulation

A series of simulations were leveraged to predict the expected nuFission signal in NuThor. The foundation of the simulation work done here is through MCNP-Polimi [1] (see the MCNP geometry in Fig. 7.1). Each NuThor Science Run is accompanied by eight MCNP-Polimi neutron simulations to examine the nuFission signal and the various background sources. The background sources simulated are from ^{232}Th NINs, ^{nat}Pb NINs, ^{127}I NINs, ^{56}Fe NINs, beam-related neutrons, thorium spontaneous fission, and cosmic muon-induced neutrons. The entire simulation work flow is displayed in Fig. 7.2. The remainder of this chapter is devoted to elaborating on the components of this scheme.

T. Johnson, *The First Indication of Neutrino-Induced Nuclear Fission*,
https://doi.org/10.1007/978-3-032-01936-3_7

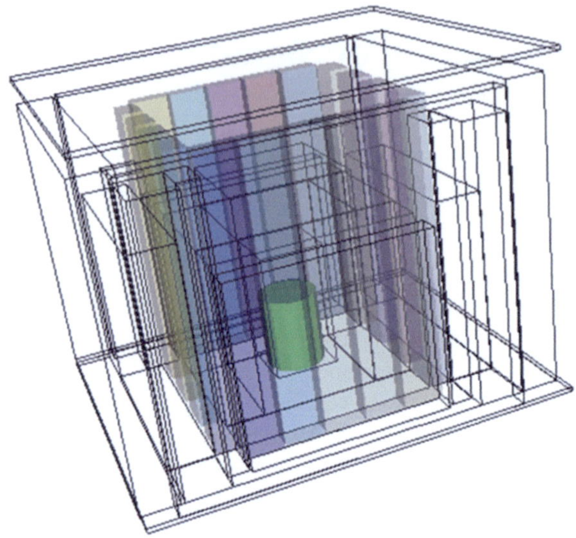

Fig. 7.1 The NuThor MCNP Geometry used for the simulations. The colorful vertical bars are the 36 NaI detectors. The green cylinder is the stack of thorium plates. The translucent boxes are all the other structures in the design including the lead in the inner core, the gadolinium loaded water bricks, borated polyethlyene shielding and steel plates

MCNP-Polimi was selected as the particle transport code for its robust, data-driven treatment of neutron scattering and absorption on a verity of materials. MCNP is also rigorously benchmarked against experimental measurements to assure maximal accuracy given its important function in nuclear security and weapons design [2]. However, there are a few shortcomings with MCNP-Polimi that must be overcome.

Both MCNP and GEANT4 have historically performed poorly when it comes to accurately describing the de-excitation of gadolinium after neutron captures. Several tools have been created to address this issue. Much effort went towards upgrading GEANT4 in particular, but a γ-ray cascade event generated called DICEBOX [3] has demonstrated a superior track-record for describing gadolinium cascade data collected by experiments [4].

There are many instruments in neutrino physics (and the broader physics community) that utilize and simulate gadolinium in their experiments that do not implement an advanced gadolinium emission code, though this largely depends on the specific design of the detector. For instance, experiments like Daya Bay [5] and RENO [6] are not as sensitive to the gadolinium-related shortcomings of physics simulator tools because they are large-scale calorimeters that contains the entirety of the gamma-ray cascade. The individual gamma rays don't necessarily matter, because these instruments reconstruct the entire Q-value of the reaction.

Conversely, LZ's outer detector requires a more sophisticated treatment of the gadolinium de-excitation, thus employ a DICEBOX driven post-processing routine [7]. LZ's approach is the inspiration for the work here [8]. In NuThor, many of the gadolinium neutron capture gamma rays will miss the detectors entirely and thus the efficiency of detected gamma rays of the proper energies must be modeled inside the geometry of NuThor. This is the motivation to produce the package called the Generation and Detection of Gadolinium Event Topologies (GADGET).

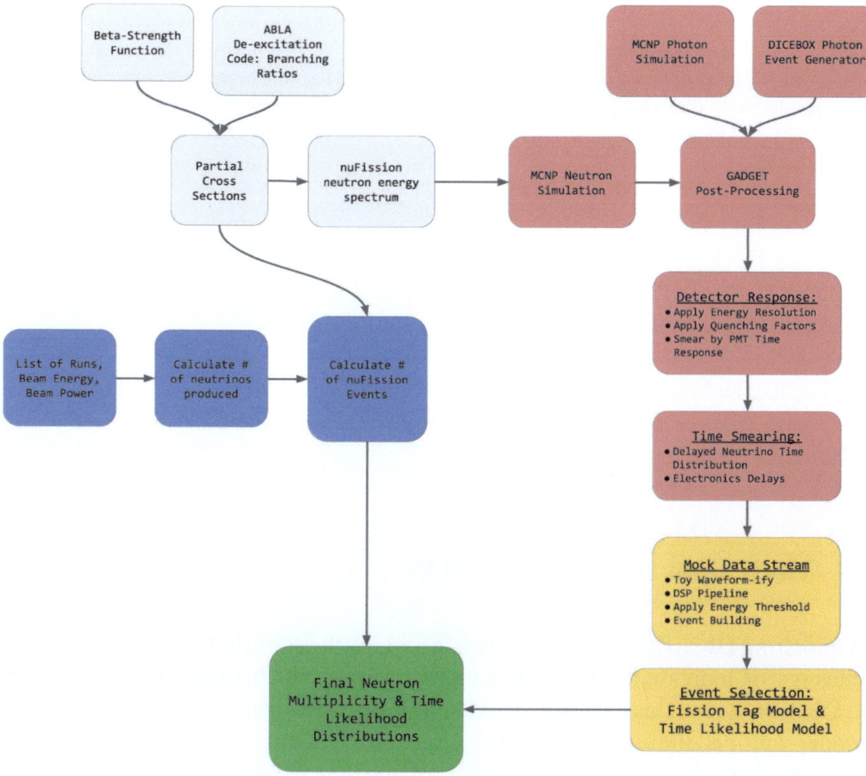

Fig. 7.2 The flowchart of the NuFission simulation and post-processing scheme. The colors indicate the various portions from the raw simulations in MCNP all the way to the final product

GADGET is the amalgamation of simulations produced by MCNP–Polimi and DICEBOX. GADGET only provides an advanced treatment of ^{157}Gd neutron captures (the most prominent capture agent) and not for ^{155}Gd (the second most prominent capture agent). MCNP simulations with the NuThor geometry indicate that approximately 81% of neutron captures on gadolinium occur on the ^{157}Gd isotope. Future initiatives can and should make progress towards a more complete GADGET implementation on both prominent gadolinium isotopes.

GADGET only intervening for ^{157}Gd is justified, in part, by the fact that the vast majority of gadolinium neutron captures occur on that isotope specifically. A more detailed analysis is provided in the following section. Neutrons simulated in MCNP are monitored, and their termination mechanisms are recorded. Focusing on only neutron capture events on gadolinium, Table 7.1 illustrates the odds of neutrons capturing on each individual isotope. When examining the overall termination of all simulated neutrons, it becomes evident that gadolinium captures neutrons with significantly higher efficiency, compared to any other element in the detector. As a brief aside, the neutrons lost to capture on hydrogen may still produce a visible

Table 7.1 The breakdown of gadolinium isotopes in the gadolinium nitrate hexahydrate within an MCNP—Polimi simulation of the NuThor detector geometry. The majority of the neutron captures occur on the ^{157}Gd isotope which is logical given the outsized thermal neutron capture cross section of ^{157}Gd relative to the other present isotopes. Naturally, the isotope with the next largest capture cross section, ^{155}Gd, has the next largest percentage of neutron captures within the NuThor geometry

Isotope	Percentage of neutron captures (%)
^{157}Gd	81.2
^{155}Gd	18.3
Other	0.5

signal in the NaI detectors because it results in the emission of a 2.2 MeV γ-ray. This is also true for neutron captures on ^{127}I which emits approximately 6.8 MeV after neutron capture (Figs. 7.3 and 7.4).

7.1 Motivation Behind DICEBOX

Incorporating DICEBOX was motivated, in part, by the study below, which compares the energy of individual photons in gamma-ray cascades between MCNP—Polimi and DICEBOX. For this exercise, the MCNP-Polimi geometry was a 100-meter cube of water with gadolinium nitrate hexahydrate evenly dispersed inside the volume at 2.5% (although the concentrate was not of too much relevance). Neutrons were started at thermal kinetic energies inside the volume of Gd-loaded water, to ensure that the only energy summed per history is due to the Q-value of the ^{157}Gd neutron capture reaction.

During post-processing, the energy of each photon is summed if the photon was generated from a neutron capture on ^{157}Gd.[1] As one be seen from the Fig. 7.5, the energy spectra have notable discrepancies that could have a significant effect on the event selection procedure outlined in Chap. 8. The energy depositions greater than 3.75 MeV shows significant tension between MCNP and the data-driven GADGET de-excitation cascade. This motivates the GADGET intervention.

It was also apparent that MCNP gadolinium de-excitations do not always abide by energy conservation. This is evidenced by Fig. 7.6, in which the total energy per history is expected to be exactly the Q-value of the reaction given by the mass difference

$$n +^{157}\text{Gd} \rightarrow^{158}\text{Gd}(Q = 7.94\ \text{MeV}) \tag{7.1}$$

It is worth noting that MCNP will propagate an event through a user-defined geometry, whereas DICEBOX simply determines the photons energies without deter-

[1] In MT codes, this is code = 102, interactionType = 0, and targetNucleus = 64157.

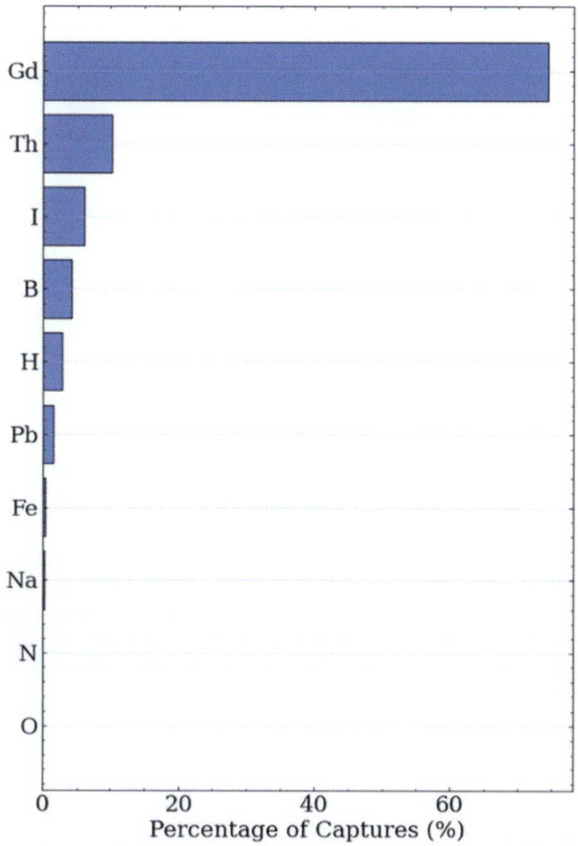

Fig. 7.3 The breakdown of nuclei that neutrons capture on inside the nuThor detector as evaluated through a MCNP-Polimi simulation in the NuThor detector geometry. The primary neutron capturing agent is the gadolinium dispersed throughout the water bricks in the neutron multiplicity meter. Notably, the send most common neutron capturing nucleus is the thorium itself. This is well within expectations of the design and simply constitutes an inherent inefficiency born from the decision to have a bulk volume thorium target mass. Neutron captures on iodine and hydrogen are still potentially observable experimentally because the iodine is inside the active detection material of the NaI scintillator and the 2.2 MeV gamma ray from hydrogen in the water bricks will sometimes hit a NaI detector

mining how those photons behave. Thus, the incorporation of DICEBOX needs to be accompanied with the propagation through the NuThor detector geometry.

Furthermore, gamma multiplicity of the de-excitation cascade using MCNP is also in tension with DICEBOX. Approximately one of the counted photons in the de-excitation cascade is from a 79.5 keV X-ray originating from the emission of an internal conversion electron (ICE). These are mostly likely not detectable by the hardware installed in NuThor, thus are of minimal importance. Even after eliminating

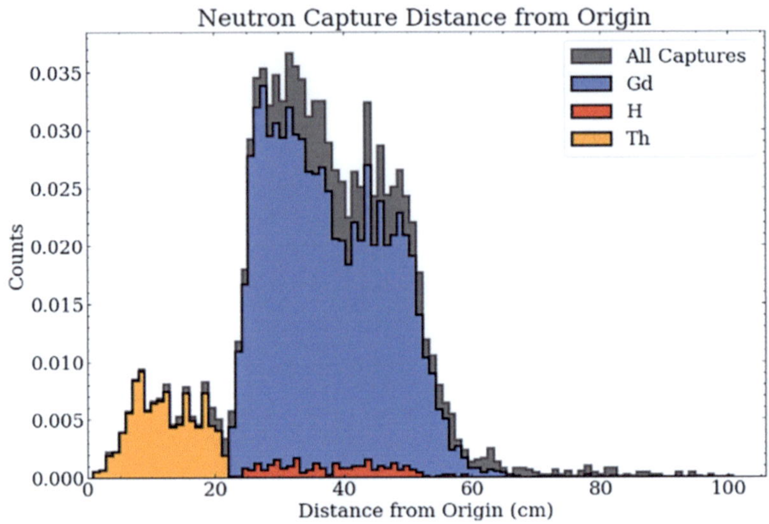

Fig. 7.4 The location of the neutron captures in the NuThor geometry according to an MCNP—Polimi simulation. All neutron captures are relative to the origin which is at the center of the thorium plates in the simulated geometry. This is a more vivid reflection of the above breakdown of neutron capture nuclei because one can see that obviously the neutron captures on gadolinium occur in the neutron multiplicity meter, but also that the hydrogen neutron captures occur in the water as opposed to the borated polyethylene thus potentially visible

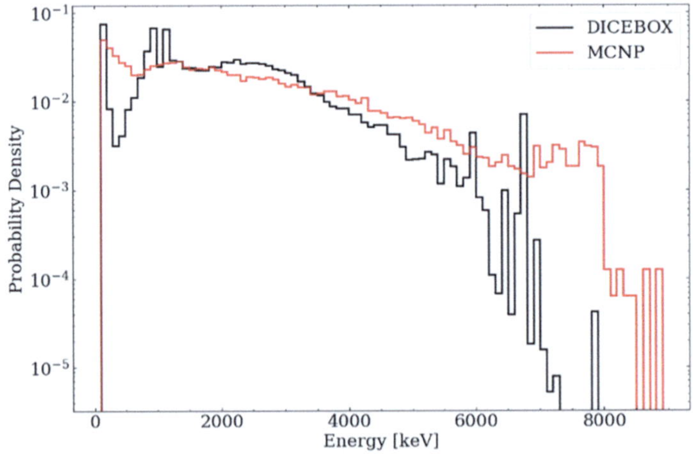

Fig. 7.5 The energy of each photon emitted after neutron-capture on ^{157}Gd nuclei using too simulation tools

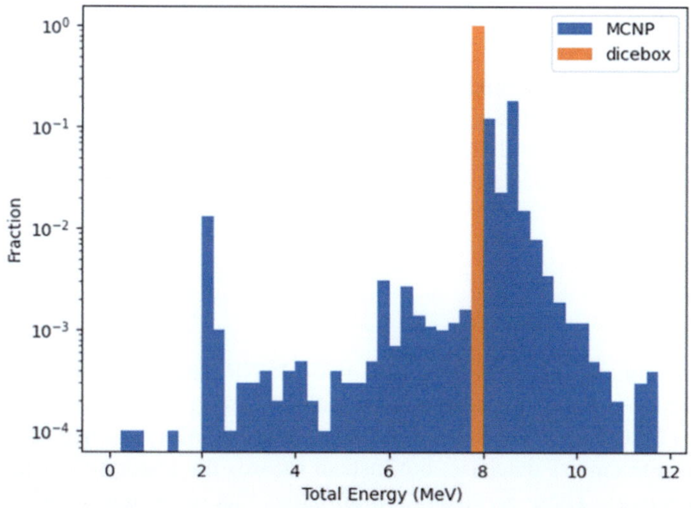

Fig. 7.6 The total simulated energy deposited per neutron capture of ^{157}Gd. DICEBOX shows a complete adherence to the Q-value of the reaction. MCNP does not adhere well to energy conservation as evidenced by the severe spread around 8 MeV

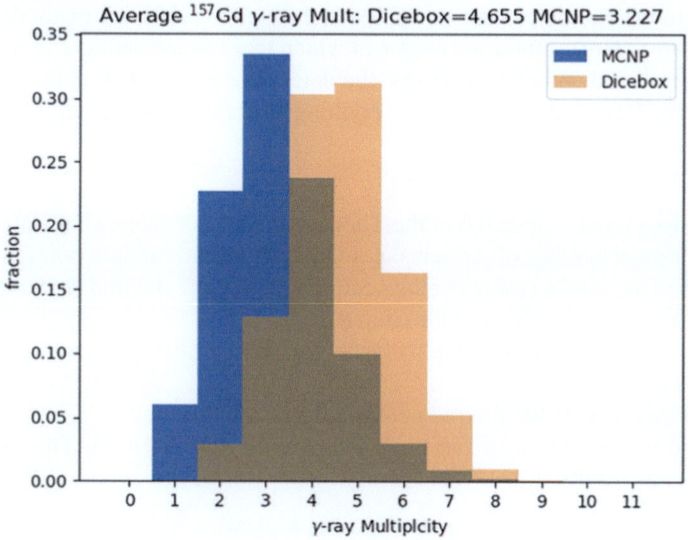

Fig. 7.7 The gamma-ray multiplicity distributions emitted from neutron capture on ^{157}Gd using MCNP and DICEBOX. There's a clear difference between the outcomes of the two simulation tools

that X-ray, there is a gap between the photon multiplicity estimates of DICEBOX and MCNP (Fig. 7.7).

The conclusion of these studies is that an augmented treatment of Gd-related γ-ray de-excitation cascade to the simulation pipeline was needed to increase confidence in the efficacy of the simulation. The energy conservation, γ-ray multiplicity and spectral shape deficiencies are all addressed with this alteration.

7.2 Gadget

Each of the simulations is divided into two MCNP databases—see Fig. 7.8. One database which contains the scatter information of every simulated history, labeled the "dumn1" database.[2] The second database contains information about the events recorded in the dumn1 database, but instead records the history's starting location and initial energy. These serve as look up tables that are strung together with the DICEBOX simulation output.

Only the neutron-captures on ^{157}Gd are replaced with a enhanced simulation output, so the remainder of the other possible channels for detecting events (e.g. n-capture on Hydrogen, n-capture on ^{155}Gd, n-capture on ^{127}I, and neutron elastic scattering in the NaI detector itself) proceed through the ordinary methods. Further, there are some (rare) simulation events in which there is detectable activity before a n-capture on ^{157}Gd. When this occurs, the detectable portion before the subsequent n-capture on ^{157}Gd is preserved and everything after the capture is replaced with the advanced simulation outlined here.

The overarching structure is as follows:

1. A random point is selected in the Thorium–or other volume (Fig. 7.9).
2. Determine a number of particles, n, to start from said random point.[3]
3. Voxelize the starting points of the neutron simulation[4] and find the voxel address of the random point selected in Step #1.
4. Query the neutron simulation database for events starting in sought after voxel address.
5. Randomly sample the filtered neutron events n times.
6. Record the end point of the n neutron events from Step #5. This could be a number of results:

    (a) ^{157}Gd capture (with a capture location logged)

[2] Based on the name of the MCNP-Polimi output file naming convention.

[3] This is a way to circumnavigate the built-in MCNP custom-source neutron multiplicity argument that is finicky and somewhat difficult to utilize.

[4] The voxels always have uniform dimensions, but that length is user-defined. For simulations with a sparse number of events, a larger voxel size may be appropriate to mitigate the likelihood of unpopulated voxels.

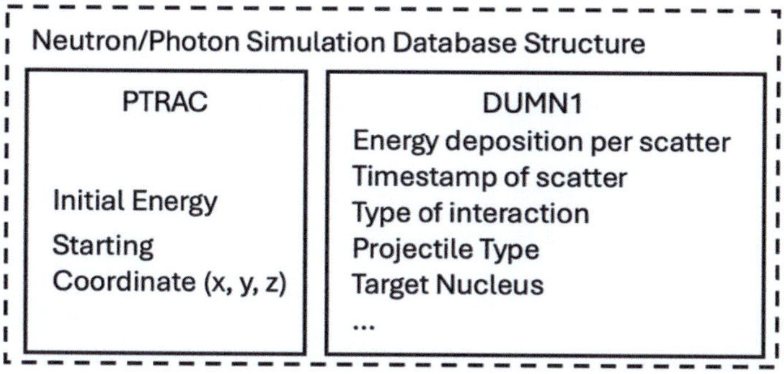

Fig. 7.8 Structure of the photon and neutron databases from MCNP-Polimi. There are two primary outputs from MCNP propagated forward: the DUMN1 and PTRAC file. Each contain different information about the same simulated events

(b) The neutron captures on a hydrogen nucleus, emitting a 2.2 MeV gamma ray which hits a NaI crystal (to a lesser extent capture on Fe nucleus with a 10 MeV gamma-ray).
(c) The neutron scatters or captures in the NaI depositing some non-zero energy
(d) Otherwise, any event that neither captures on ^{157}Gd nor produces a NaI energy deposition.

7. For ^{157}Gd n-captures, find the voxel address of the capture location
8. Randomly sample the DICEBOX gamma-ray cascade database for a list of photon energies to propagate through the detector geometry
9. Filter the photon simulation for events that match the voxel address and the energies identified in Step #8
10. Record the detectable hits in the NaI detectors from the propagated photon simulation. Note: not all events will have a detectable hit. These need to be accounted for in order to properly tabulate the efficiency of the simulated events.

7.2.1 MCNP Neutron Simulation

The foundation of the nuFission simulation are neutrons generated uniformly inside the thorium metal volume using MCNP-Polimi. The events' origin-points are voxelized with an address expressing the approximate coordinates within the thorium in order to pair multiple, separate simulated events together and form multi-neutron events.

The neutrons started in the thorium volume (Fig. 7.9) then propagate outwards through the thorium and lead in NuThor's inner core and into the neutron multiplicity meter where they quickly moderate and capture on gadolinium or hydrogen (Fig. 7.10). The GADGET framework high-jacks the simulation post-processing

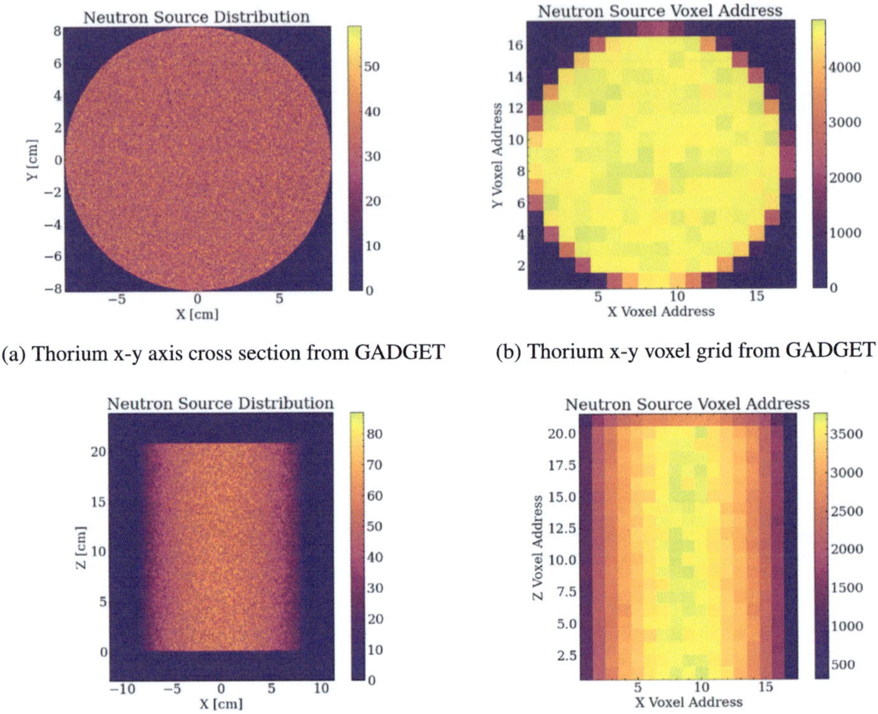

(a) Thorium x-y axis cross section from GADGET (b) Thorium x-y voxel grid from GADGET

(c) x-z cross section of Thorium in GADGET (d) x-y voxel grid of Thorium in GADGET

Fig. 7.9 GADGET neutron source initialization and voxelization of the thorium mass. The voxel size is user-defined at the outset of the code execution

steps if there is a neutron capture on a ^{157}Gd nucleus. All other events are handled within the base neutron simulation output.

7.2.2 *Photon Simulation*

A photon simulation is produced in MCNP-Polimi in which the photons are started uniformly throughout the gadolinium-water bricks with a uniform energy spectrum. This serves as a lookup table for propagating photons from a user-defined voxel and energy. Each of these photon events is then assessed for whether it engages with a NaI detector. If the photon deposits a non-zero amount of energy in one or more NaI detectors, its energy, time and detector channel are recorded. If the photon does not deposit any energy in a NaI detector channel, those are logged as misses. This pre-processing of the photon simulation allows for a speedy execution of the neutron post-processing.

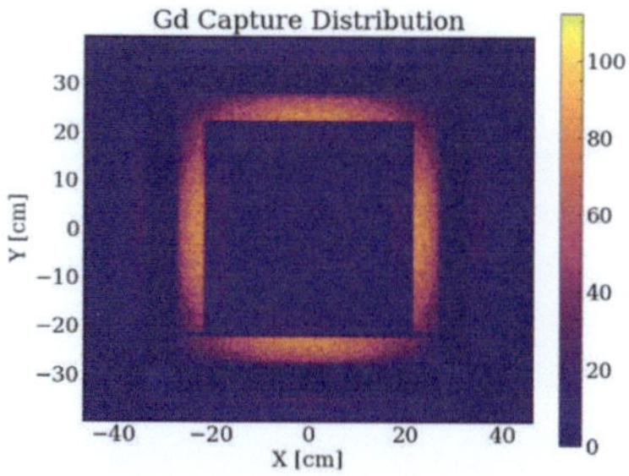

(a) Neutron Sim x-y cross section indicating the Gd capture location

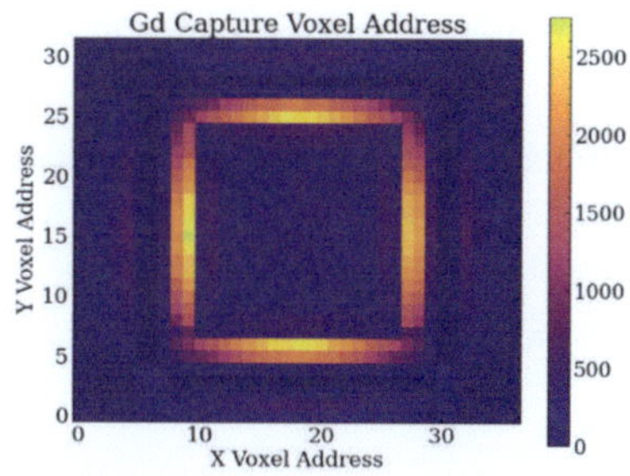

(b) Neutron Sim x-y voxel grid indicating the Gd capture location/voxel address

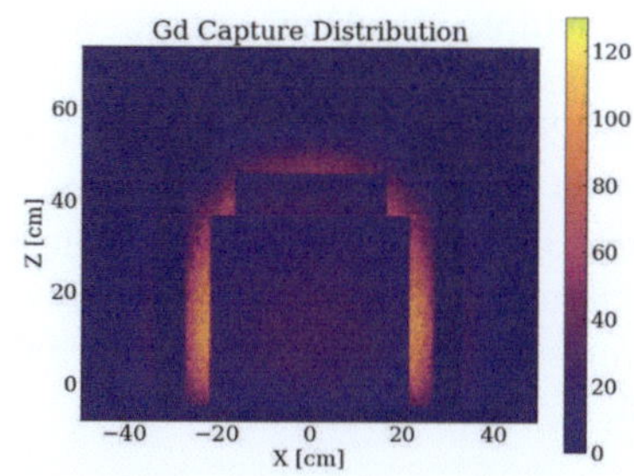

(c) Neutron Sim x-z cross section indicating the Gd capture location

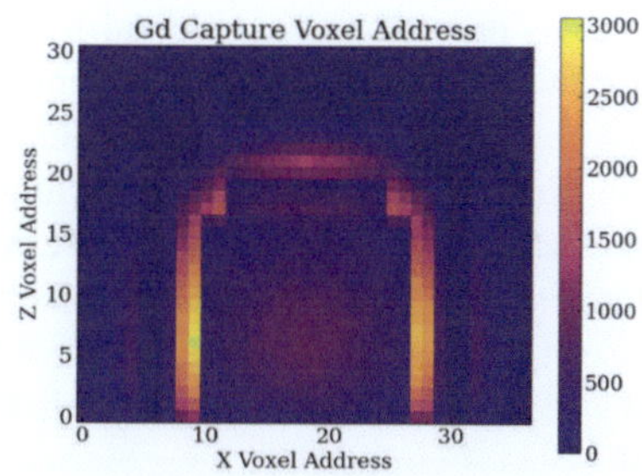

(d) Neutron Sim x-z voxel grid indicating the Gd Capture voxel address

Fig. 7.10 GADGET gadolinium capture location and voxelization of the gadolinium-loaded water volume in the neutron multiplicity meter

When a neutron from the neutron simulation is captured by a ^{157}Gd nucleus, the location of the capture is logged and converted into a voxel address which is displayed in Fig. 7.10. Then, a random ^{157}Gd de-excitation cascade is selected from the DICEBOX output. Each γ-ray cascade contains some number of photons each with an associated energy. The photon simulation output database is queried for events that fulfill both the voxel address and the energy condition within a user-defined tolerance. The bank of queried photons events is randomly sampled. The selected photon event has already been parsed for NaI interactions, so these are listed for each of the NaI channels for this event. These lists are dubbed "waveform seeds" because they will be inserted into a mock datastream, which includes a toy waveform evaluation.

7.3 Simulation Post-processing

The simulated MCNP output is considered “truth-level” information that does not incorporate detector effects such as energy resolution, nuclear recoil quenching factors, or software artifacts from the DSP routine. For this reason, the simulation output receives several levels of adjustment to better represent reality.

7.3.1 *The Truth Level*

The simulated MCNP output is organized into individual particle histories, or ‘throws,’ to prevent the accidental blending of multiple simulated events. Within each throw, events are further segmented by the specific NaI detector channel they interacted with, reflecting the fact that these are distinct detector volumes. Each detector channel records the energy and time of all particle interactions within its volume. At this stage, the outputs are not yet realistic, as they do not account for the finite energy and timing resolutions, thresholds, and timing structures of physics processes that caused the events. These factors must be incorporated to ensure the model’s accuracy for subsequent predictive work.

7.3.2 *Energy Smearing*

Energy Resolution of the NaI Detectors

The simulated energy in the MCNP truth level information is smeared in two steps. First, the energy resolution of the NaI crystals in NuThor is considered–see Fig. 7.11. The Gaussian energy resolution can be described with Eq. 7.2 [9]

$$\sigma = \frac{1}{2.355}\sqrt{a^2 + b^2 E^2} \tag{7.2}$$

where E is the energy deposited, a is the term describing the location-dependent light transmission and b is a noise term.

The computed energy resolution was used to form a Gaussian probability density function that was normalized and randomly sampled for every energy deposition in the active cells within the MCNP-Polimi geometry.

Nuclear Recoil Quenching Factors

Next, the energy deposited by neutrons elastically scattering in the NaI crystals is smeared by the following quenching factors for Sodium and Iodine, respectively (Fig. 7.12 [10]), where the curve labeled Joo is used in this work. The relationship between the true nuclear recoil energy, E_R, and its electron-equivalent energy, E_{ee}, is known as the quenching factor. Parameterization of the energy-depending quenching factor comes from the Lindhard theory [11] and is described with the formula:

$$f(E_R) = \frac{kg(\epsilon)}{1 + kg(\epsilon)} \quad (7.3)$$

where $\epsilon = 11.5Z^{-7.3}E_R$, $k = 0.133Z^{2/3}A^{1/2}$, Z is the proton number, A is the atomic number, and $g(\epsilon)$ (from [12]) is:

$$g(\epsilon) = 3\epsilon^{0.15} + 0.7\epsilon^{0.6} + \epsilon \quad (7.4)$$

Including the quenching factors has nearly a negligible effect on the post-processing output but is nevertheless included for completeness.

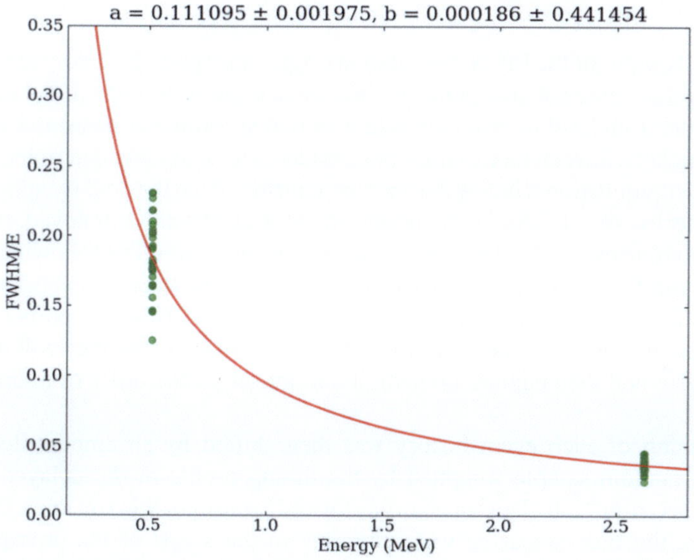

Fig. 7.11 Energy Resolution for the 35 functioning NuThor NaI scintillators. The photo-peaks at 511 and 2614 keV are used to evaluate the energy resolution

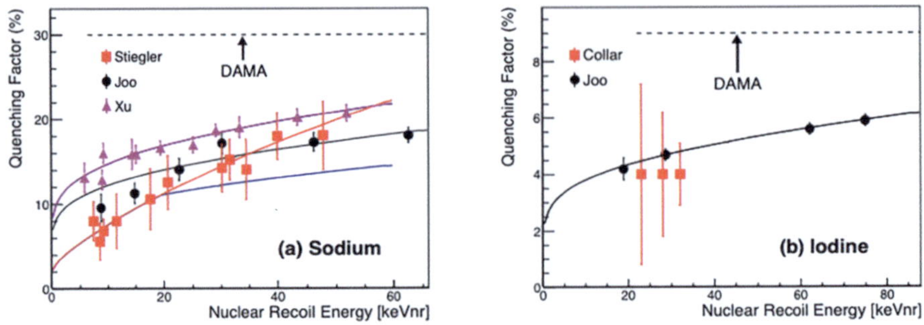

Fig. 7.12 Quenching Factors used for elastic scattering of neutrons on Sodium and Iodine [10]. The Lindhard model fit to the Joo et al data points is used for this work

7.3.3 Time Smearing

The default output of the MCNP simulation originates a particle history at t = 0, then everything that ensues at later points in time are in references to the initial time.[5] The timing of the truth level information is first shifted according to measured electronic and cable delays. Several factors must be considered to accurately determine where in the digitized acquisition window the onset of neutrinos from the SNS is expected. This is very detector specific due to the large variety of electronics employed, triggering mechanisms and even cable lengths used to relay information. The following exercise bears out the best estimate for where in the waveform trace to expect neutrinos relative to the start of the pre-trigger window for NuThor. This is important because neutrons are the signal for neutrino-induced nuclear fission, and they will need time to thermalize and then capture on gadolinium nuclei on the order of microseconds (Fig. 7.13).

The timing of each event history was then shifted by an amount determined through a random sample weighted by the timing profile of the delayed electron neutrino spectrum.[6] All electron neutrino-related events got this smearing treatment. For BRNs, the time smearing was according to the shape of the prompt pulse.[7] Any steady-state background such as spontaneous fission was varied by a uniform distribution throughout not only the available time window of the acquisition period, but also before it, to account for partially encapsulating events that start before the external trigger. These timing treatments are listed in the Table 7.2 (Fig. 7.14).

NuThor must concern itself with the timing of the neutrons emitted by various processes in and around the instrument. As mentioned in the previous section, the neutron simulations from MCNP all begin at a t = 0, thus we must impose a timing structure to reflect the neutrino beam pulses from the SNS. However, the MCNP sim-

[5] MCNP–Polimi time is expressed in shakes, where 1 shake = 10 ns.

[6] It is important to note that all particles in a multi-particle event must all receive the same shift in time from to the electron neutrino smearing random sample.

[7] The shape of the protons-on-target trace was used.

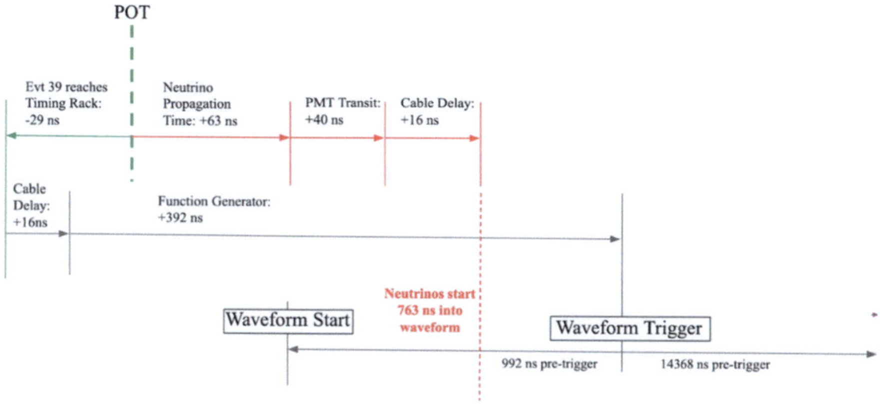

Fig. 7.13 Timing structure of the DAQ electronics. Cable delays are based on Nubes measurements and must be redone specifically for NuThor. With the current values, electron neutrinos begin 763 ns into the waveform trace triggered in the DAQ

Table 7.2 The method of smearing each individual source of neutrons. Because electron neutrinos are the only pertinent neutrino signal, only the delayed neutrino timing distribution is used. If the muon neutrinos are of interest in a future study, the prompt neutrino timing distribution should be used

Source	Smearing type
Neutrino-induced fission neutrons	Delayed neutrinos
Neutrino-induced neutrons	Delayed neutrinos
Beam-related neutrons	Protons-on-target trace
Spontaneous fission neutrons	Uniform
Cosmic Muon-induced neutrons	Uniform

ulations do intrinsically account for the moderation and capture time of the neutrons after their creation. So, the final timing structure of the signal neutron captures is a convolution of the exponential capture time and the time structure of the beam (or otherwise).

Other Time Smearing Approaches

The two-time shifts are illustrated in the Fig. 7.15. The red distribution is the fully time smeared MCNP output. Note that the NuThor DAQ is limited to taking waveforms that are 15,360 ns long. Approximately 75% of the electron neutrino spectrum occurs within this window for Science Run 2. Beginning with Science Run 3, this efficiency is reduced to 66% because the POT relative to the arrival Event 39 was delayed by 2 μs.

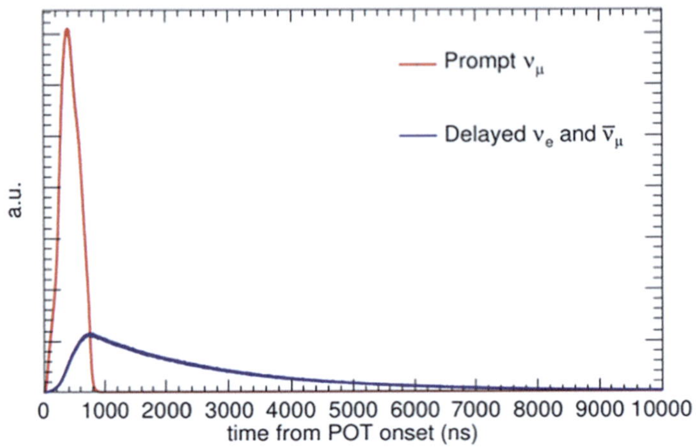

Fig. 7.14 Time profile of the prompt and delayed neutrino pulses from the SNS. These curves are normalized and used as PDFs for smearing simulated events in time

7.3.4 Mock Data Stream

After smearing the truth level information in time and energy, a toy waveform was produced (e.g. Fig. 7.16). The waveform noise was harvested directly from NuThor data. The simulated waveform signal seeded onto the noise was randomly sampled from the average shape of 10,000 NaI waveforms. This approach accounted for digital signal processing (DSP) quirks and efficiencies imposed by data-cleaning. A similar method was used by the PandaX-4T experiment for their recent ^{8}B Solar neutrino CEvNS result [13].

The DSP routine employed in the NuThor analysis was then applied to the above toy waveforms. This replicated the idiosyncrasies of the DSP in the simulation. After signals were identified through the DSP routine, pulses were integrated and a 50 keV (100 keV) threshold was applied for Science Run 2 (Science Run 1) to remove noise.

This approach precipitates the exact same extracted quantities as the actual implemented signal processing scheme DSPv2. These events were then clustered in time in two layers: (1) the neutron clustering in the 100-nanosecond, N_{100}, coincidence window, and (2) the fission super-clustering in the 15,360-nanosecond coincidence window.

7.4 NuFission Signal Expectations

The GADGET simulation framework was used to construct the signal expectations for neutrino-induced nuclear fission events in NuThor across a variety of parameters. This was achievable because multiple neutrons could be started from the same voxel

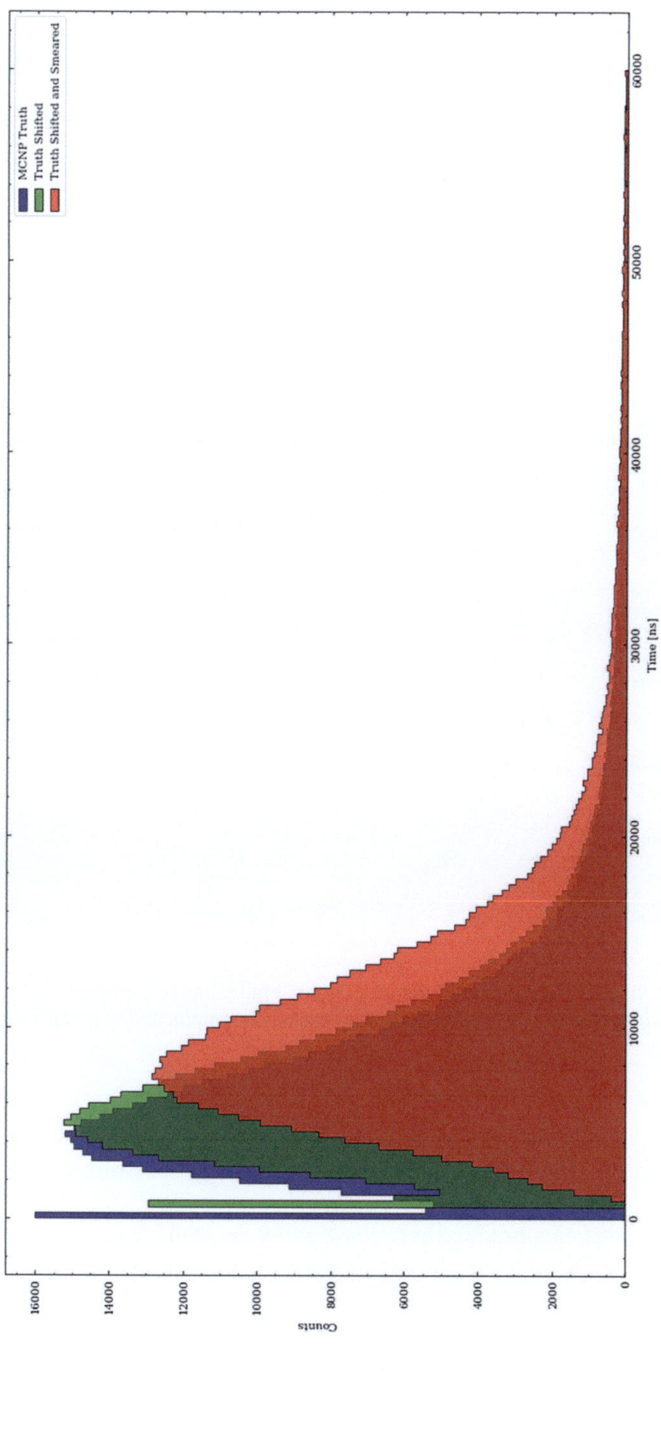

Fig. 7.15 Time Smearing Applied in two separate steps to the nuFission neutrons. The y-axis is in arbitrary units. The blue curve is the timing structure straight from the MCNP simulation with no intervention. The green distribution is the result from a 732 ns shift stemming from cable delayed and the physical hardware. The red distribution emerges from time smearing based on the delayed neutrino timing profile. Recall that the waveforms recorded by NuThor are limited to approximately 15 μs. Thus, events occurring past that maximum are lost without observation. That inefficiency is accounted by simulation

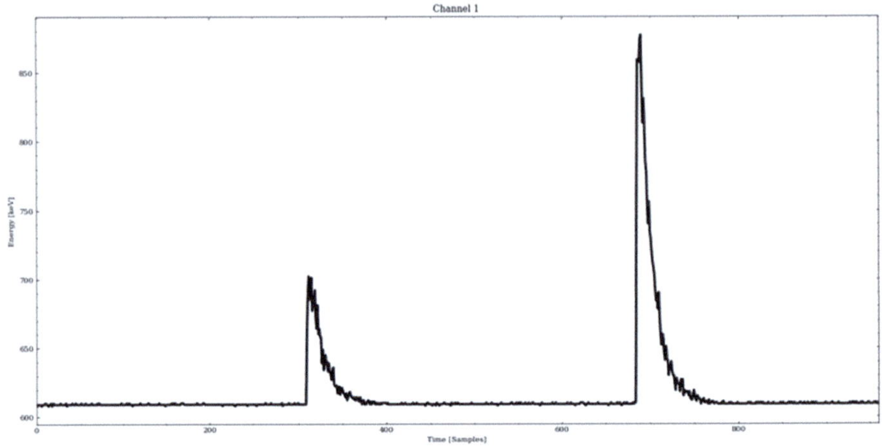

Fig. 7.16 An example toy waveform trace in which to signal events occur on the same channel within the acquisition window (i.e. pileup)

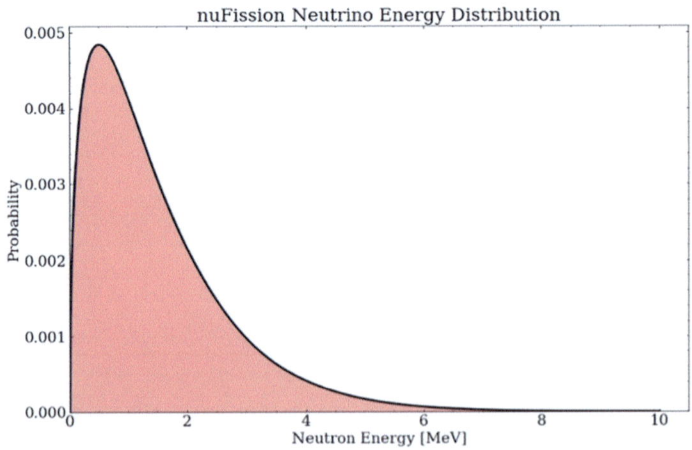

Fig. 7.17 The kinetic energy distribution of neutrons emitted after a neutrino-induced nuclear fission of a ^{232}Th nucleus at SNS neutrino energies and angular momentum expected from neutrino captures

address in the thorium target material. Further, the energies of the fission neutrons were sampled from the ABLA-produced kinetic energy spectrum shown below in Fig. 7.17.

The simulated events produce between 1 and 9 neutrons from a single voxel per iteration of the GADGET code, in roughly uniform proportions. However, in reality, the neutron multiplicity follows a distribution set by the nuclear physics of a stimulated a nucleus from a projectile above its fission barrier. This effect was translated to a neutrino-induced nuclear fission flux-average cross section as a function of the

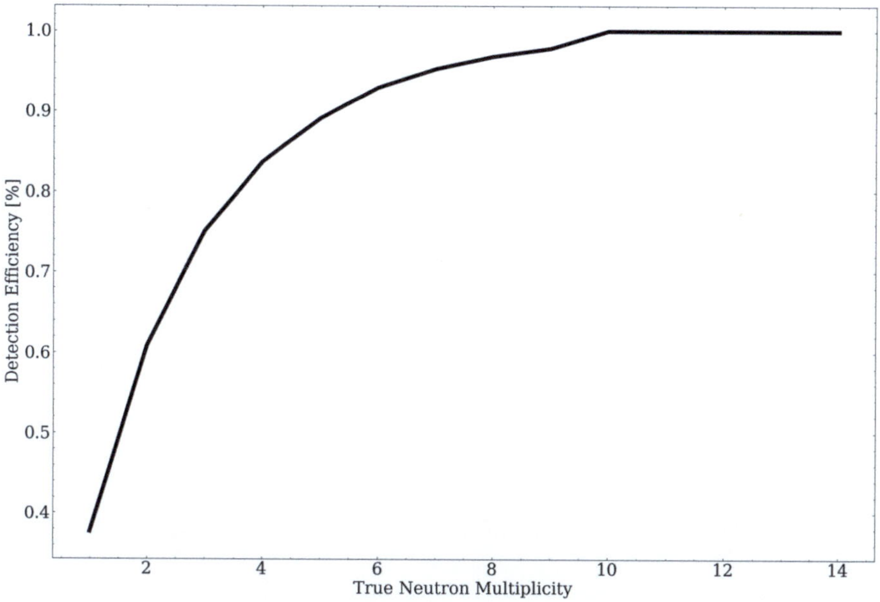

Fig. 7.18 The probability of NuThor detecting at least one neutron as a function of the true number of neutrons thrown per simulated event for NuThor Science Run 2

neutron multiplicity, which, in conjunction with the neutrino flux and target mass, was turned into predictions of observable parameters such as the event energy.

A key aspect of this exercise was the recording of all events that do not result in a detectable signal. These undetected events needed to be accounted for in the scaling of the probability density functions (PDFs). The probability of detecting any signal is found as a function of the true number of neutrons simulated (Fig. 7.18).

As the number of simulated neutrons per event increases, the likelihood of at least one neutron producing a detection increases and ever trends to nearly 100% efficiency at 10 neutrons. Unfortunately, a neutrino-induced nuclear fission event with that many neutrons is very unlikely according to the flux-averaged cross sections, simulation, and post-processing routines. The efficiency for detecting a single neutron is approximately 38%, which is in line with, or greater than, many other neutron multiplicity detectors. For instance, the supernova neutrino monitoring detector HALO has a neutron detection efficiency of 28% [14].

The true neutron multiplicity's reconstructed probability distributions were scaled by the fraction of expected counts that result in a successful detection in a particular data-taking campaign. The following figures (Figs. 7.19, 7.20, and 7.21) illustrate the most relevant reconstructed observables when scaled to the beam exposures of NuThor Science Run 1 and 2.

The nuFission energy spectrum tapers off precipitously after 15 MeV and dies off completely after 20 MeV. This informs where to place the boundary of the nuFission

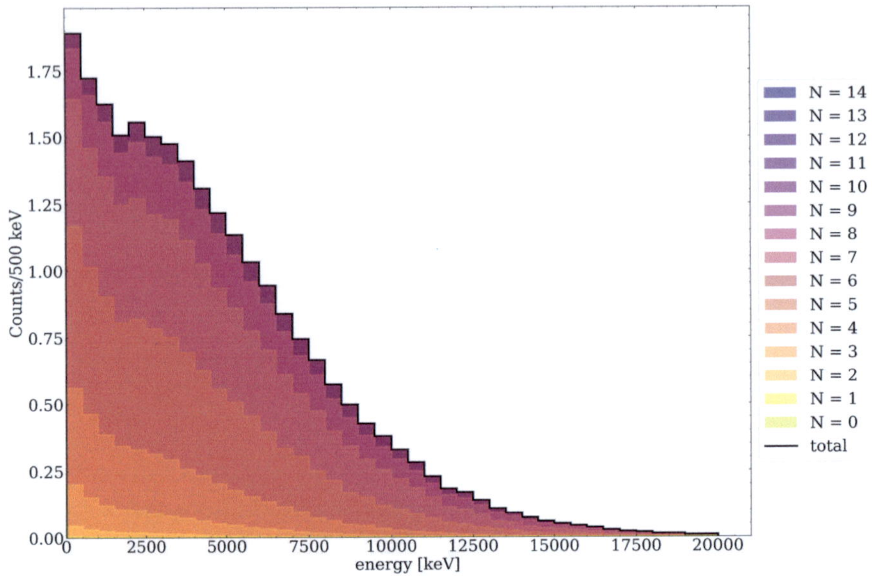

(a) Science Run 2

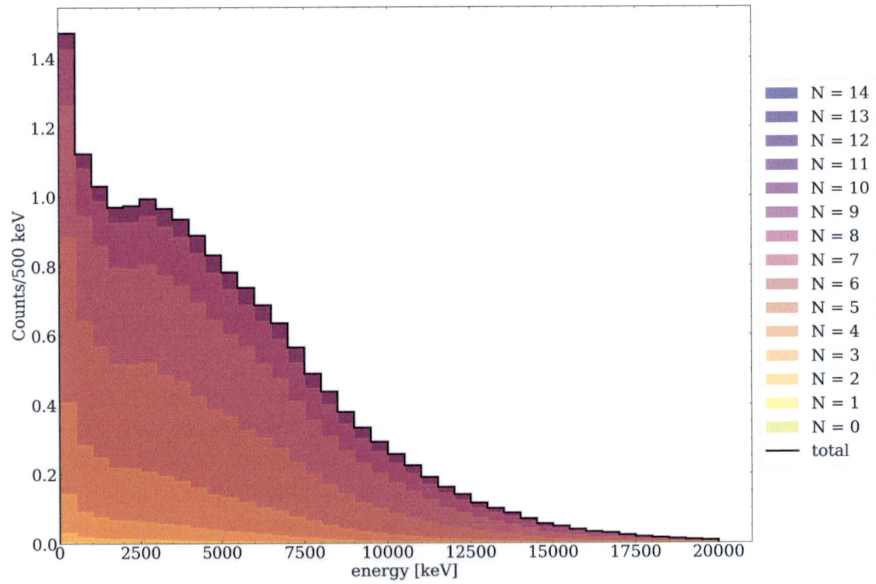

(b) Science Run 2

Fig. 7.19 Scaled energy distribution predicted according to the combination of the GADGET simulation output and the flux-averaged cross sections as a function of the neutron multiplicity

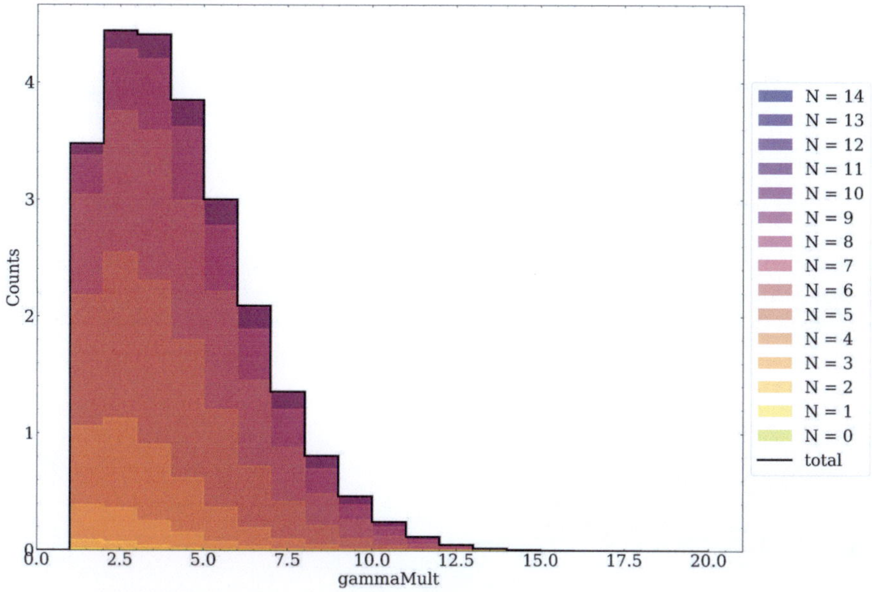

(a) Science Run 1

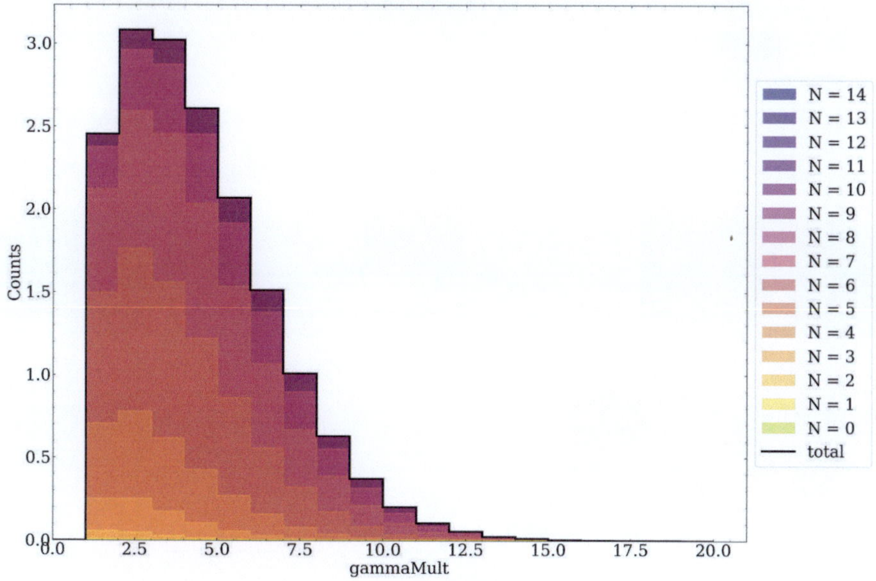

(b) Science Run 2

Fig. 7.20 Scaled gamma multiplicity distribution predicted according to the combination of the GADGET simulation output and the flux-averaged cross sections as a function of the neutron multiplicity

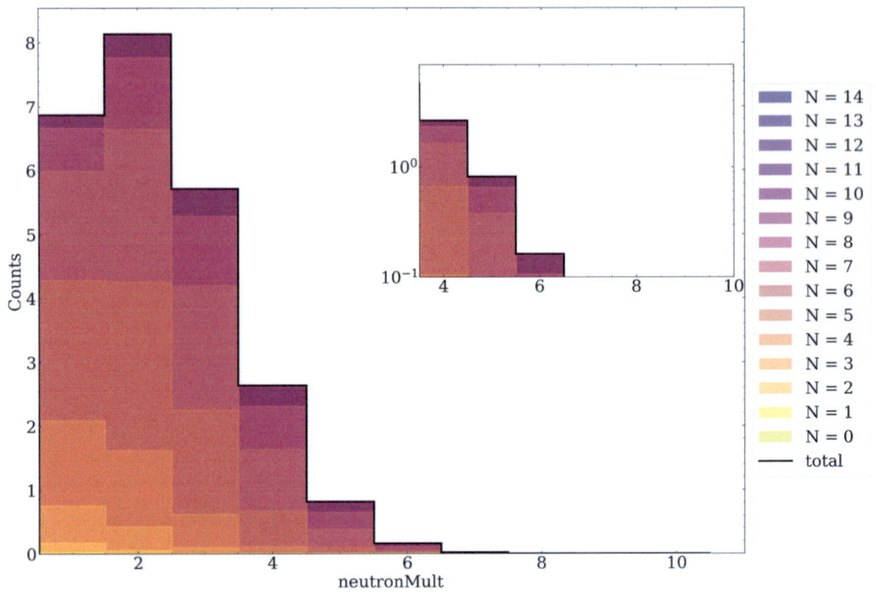

(a) Science Run 1

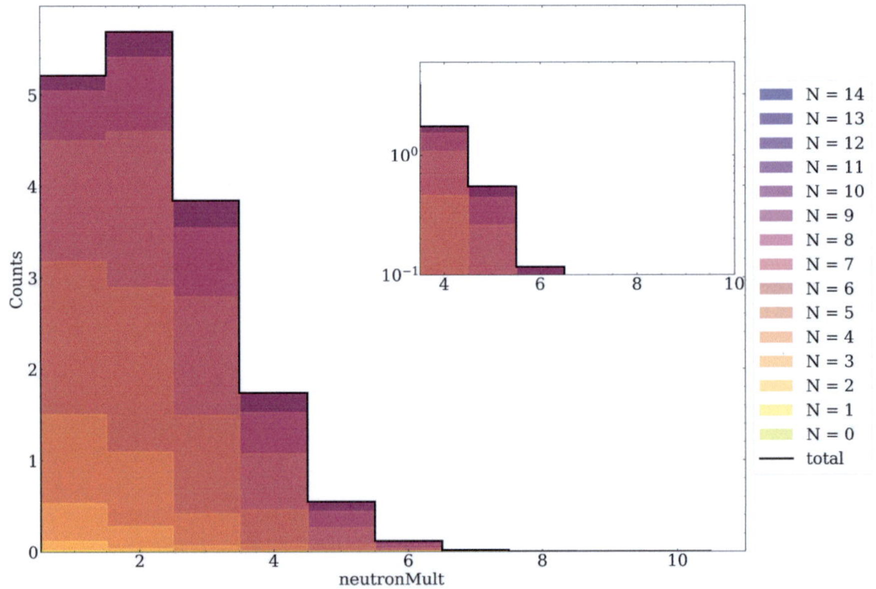

(b) Science Run 2

Fig. 7.21 Scaled neutron multiplicity distribution predicted according to the combination of the GADGET simulation output and the flux-averaged cross sections as a function of the neutron multiplicity region-of-interest and a high-energy sideband for testing the blinded dataset without adding in bias. The high-energy sideband dataset begins at 15 MeV–this will be explored in much greater detail in a later section.

Figure 7.21 indicates that the Science Runs 1 and 2 will likely be quite statistically limited. This work will only be reporting on the result from Science Run 2. As NuThor amasses more data, the higher neutron multiplicity bins will accrue counts–hence the sensitivity of the measurement will also increase.

7.5 Background Expectations

Several neutron background sources were simulated and post-processed through the GADGET framework. These can be categorized as either beam-related backgrounds or steady-state backgrounds.

7.5.1 Neutrino-Induced Neutrons on ^{232}Th

After neutrino-capture on thorium, the other likely de-excitation channel for the newly created protactinium nucleus is via neutron emission. The kinetic energy distribution of the thorium NINs was extracted from INCL/ABLA and passed to MCNP's source card. These particles originate in the thorium. Unlike neutrino-induced nuclear fission events, thorium NINs generally emit fewer neutrons (an average of 2). This results in a lower detector efficiency in the NuThor apparatus. Thus, many of these neutrons escape the detector altogether and the reconstructed neutron multiplicity is skewed lower. The distributions illustrated in Fig. 7.22 are normalized then scaled by the expected counts in the NuThor Science Run 2 dataset.

7.5.2 Neutrino-Induced Neutrons on ^{208}Pb

There are 877 kg of lead bricks stacked around NuThor's inner core. This is also a target for producing NINs. The background simulation for this source starts the neutrons in the lead geometry cells with a kinetic energy distribution calculated by MARLEY. MARLEY was used for ^{208}Pb because there is no need to compute the competition between de-excitation through nuclear fission and neutron emission (the fission barrier for lead is greater than the expected excitation energies) (Fig. 7.23).

7.5.3 Spontaneous Fission of ^{232}Th

The spontaneous fission background was simulated originating uniformly through the thorium cylinder's volume. MCNP's built-in ^{232}Th spontaneous fission source was

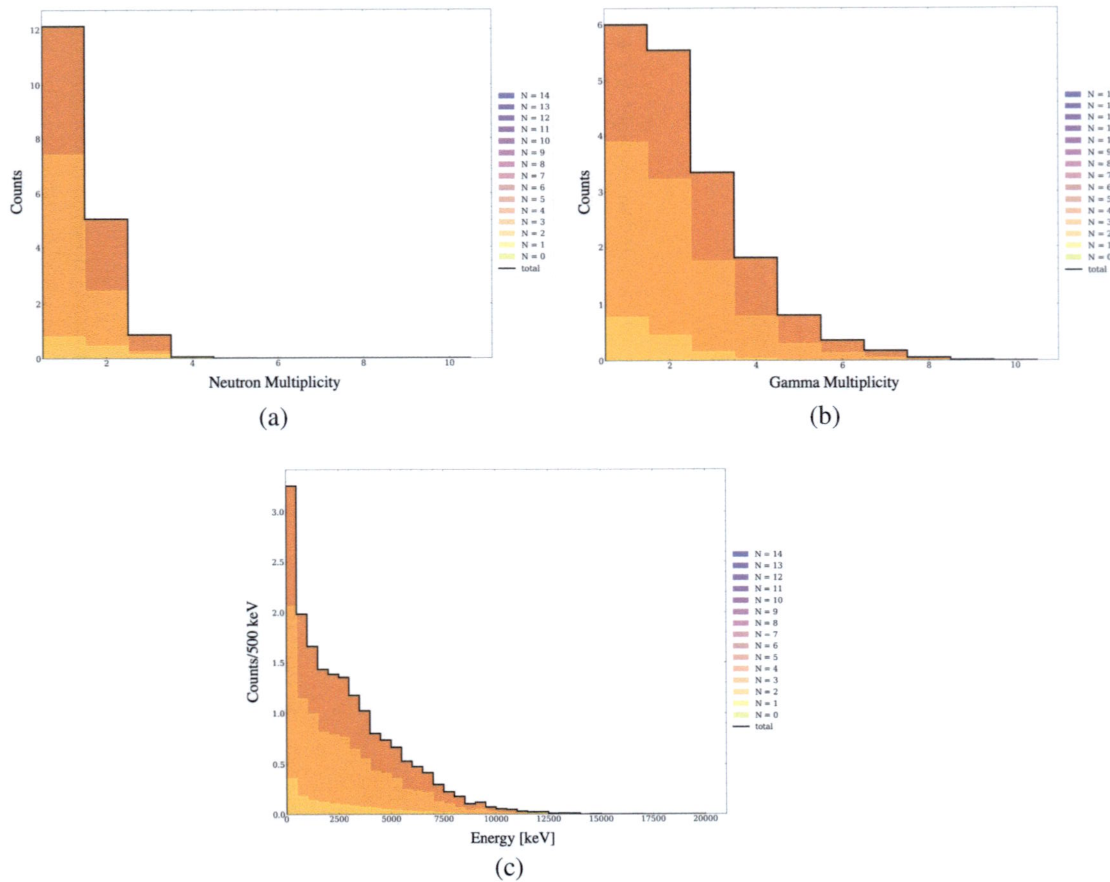

Fig. 7.22 Parameters from the Th NINs simulation. The distributions are scaled by the mass of thorium in the detector and the appropriate beam exposure for Science Run 2. The most pertinent parameter is the reconstructed neutron multiplicity shown in the upper left figure

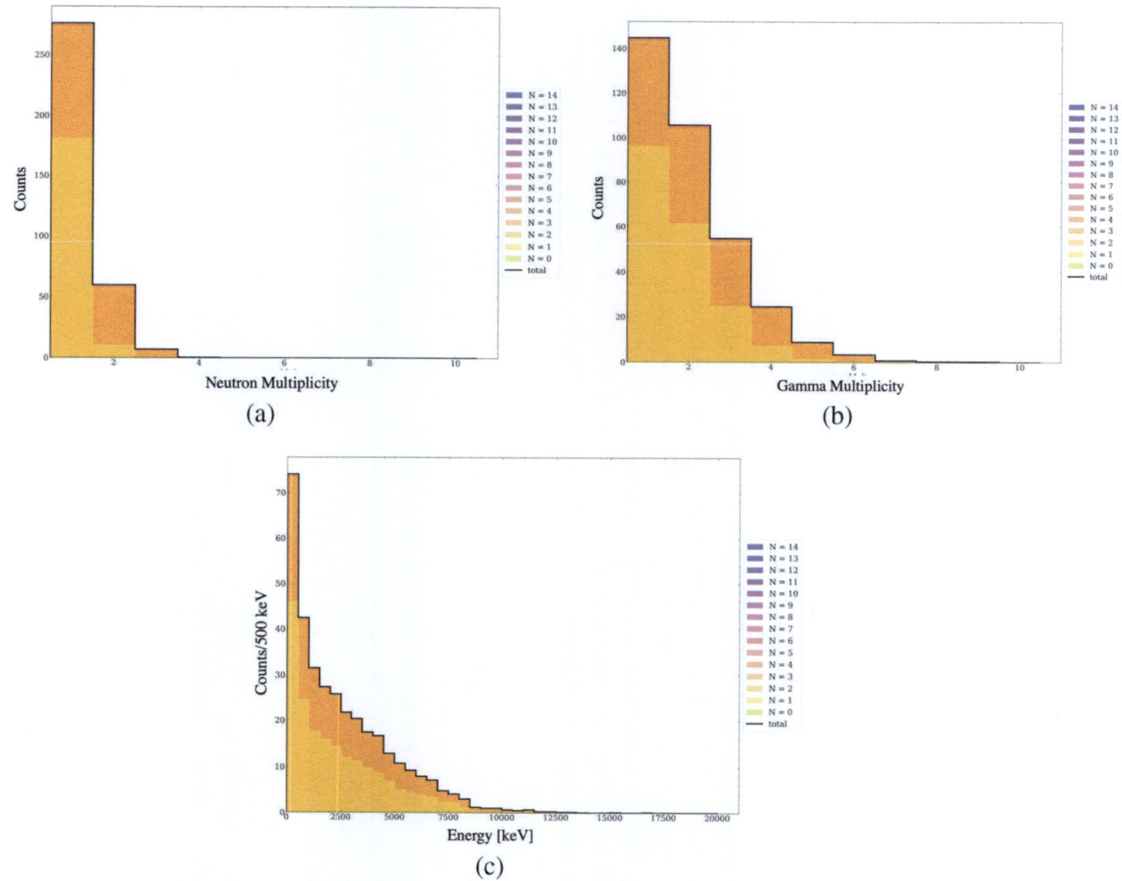

Fig. 7.23 Parameters from the Pb NINs simulation. The distributions are scaled by the amount of lead in the NuThor detector and the beam exposure for Science Run 2. The most pertinent parameter is the neutron multiplicity, which demonstrates a noticeably lower average value compared to the neutrino-induced nuclear fission distribution shown in Fig. 7.21

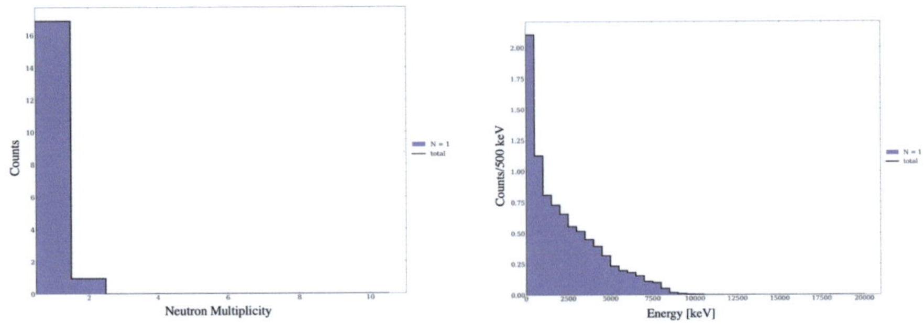

Fig. 7.24 Parameters from the thorium spontaneous fission simulation. The distributions are area normalized then scaled to the expected number of thorium spontaneous fissions expected within the the 15.360 μs acquisition window plus events that occur before the window and spill into it randomly. Very few events induce high neutron multiplicity observations

leveraged to generate these events. The output of the standard MCNP simulation was parsed with GADGET. The spontaneous fission background must be treated uniquely compared to the other simulated sources because it is steady-state. However, because the NuThor acquisition system is externally triggered and has a limited window that it is recording, spontaneous fission events must also be started in times before the external trigger in addition to during the acquisition window. The events that begin before the window can then leak into the recording period. A coincidence window of 42 μs was used to estimate the total number of expected spontaneous fission events and the half-life is pulled from nuclear data tables as 1.2×10^{21} years. The resulting reconstructed neutron multiplicity is shown in Fig. 7.24. The reconstructed neutron multiplicity distribution indicates that there are no high multiplicity events expected. Because of this, the spontaneous fission background is not specifically incorporated in the remainder of this analysis. Instead, the total steady-state background is measured and this source is a constituent of that population of events.

7.5.4 Beam-Related Neutrons

Neutrons emitted from the SNS target occasionally reach Neutrino Alley, and some of these beam-related neutrons (BRNs) can penetrate the NuThor detector shielding. The energy spectrum used to simulate these neutrons extends up to 400 MeV, meaning a non-zero fraction of BRNs exceed the 15 MeV maximum energy set for the region of interest. This simulation work was the foundation for a high-energy sideband analysis in a proceeding section, where events with energy greater than 15 MeV was examined to estimate the BRN flux in the region-of-interest (Figs. 7.25 and 7.26).

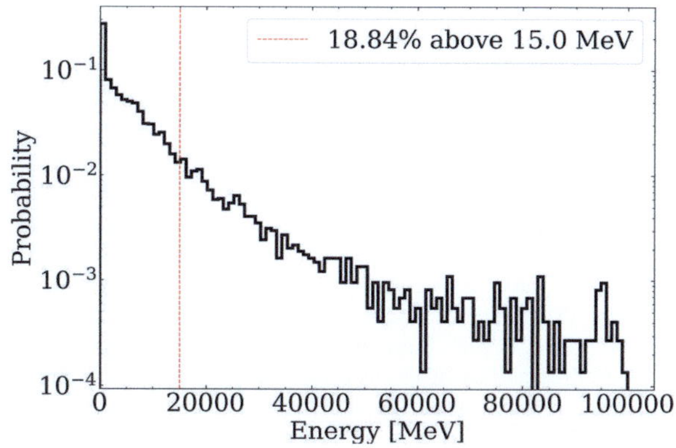

Fig. 7.25 GADGET Post-processed energy spectrum for Beam-related neutrons (BRNs) reconstructed in NuThor. Approximately 19.84% of the BRN events occurred in the high-energy region above 15 MeV

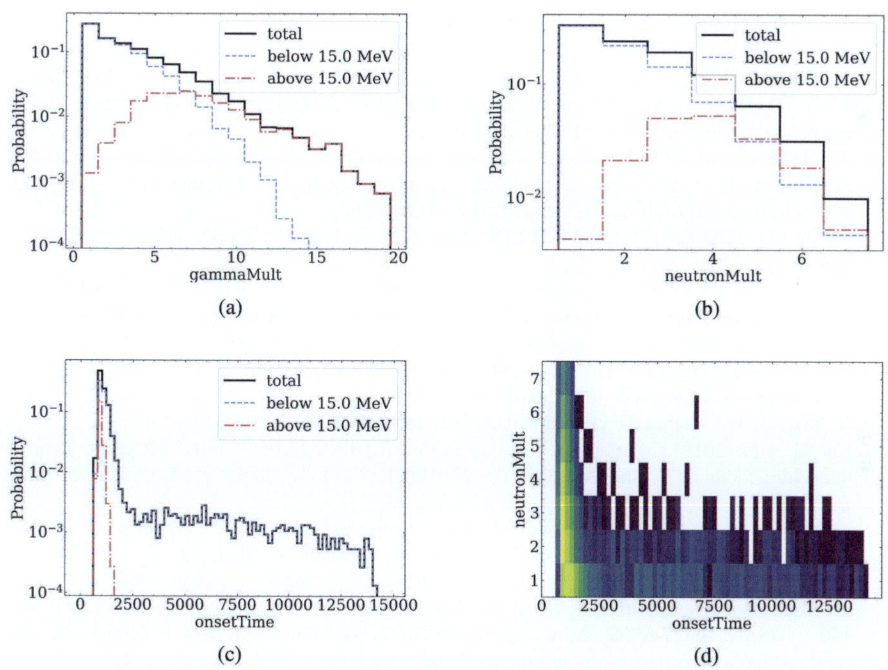

Fig. 7.26 Parameters from the GADGET post-processed BRNs simulation. The contributions from events with above or below 15 MeV total energy are demarcated

References

1. Pozzi SA, Padovani E, Marseguerra M (2003) MCNP-PoliMi: a Monte-Carlo code for correlation measurements. Nuclear Instr Methods Phys Res Sect A: Accel Spectrom Detec Assoc Equ 513(3):550–558. ISSN: 0168-9002. https://doi.org/10.1016/j.nima.2003.06.012. https://www.sciencedirect.com/science/article/pii/S0168900203023027
2. Clarke SD et al (2013) Verification and validation of the MCNPX-PoliMi code for simulations of neutron multiplicity counting systems. Nucl Instr Methods Phys Res Sect A: Accel Spectr Detec Assoc Equ 700:135–139. ISSN: 0168-9002. https://doi.org/10.1016/j.nima.2012.10.025. https://www.sciencedirect.com/science/article/pii/S0168900212011576
3. Bečvář F (1998) Simulation of cascades in complex nuclei with emphasis on assessment of uncertainties of cascade-related quantities. Nucl Instr Methods Phys Res Sect A: Accel Spectr Detec Assoc Equ 417(2):434–449. ISSN: 0168-9002. https://doi.org/10.1016/S0168-9002(98)00787-6. https://www.sciencedirect.com/science/article/pii/S0168900298007876
4. Chyzh A et al (2011) Measurement of the ^{157}Gd(n,γ) reaction with the DANCE γ calorimeter array. Phys Rev C 84:014306. https://doi.org/10.1103/PhysRevC.84.014306. https://link.aps.org/doi/10.1103/PhysRevC.84.014306
5. Cao J, Luk K-B (2015) An overview of the Daya Bay reactor neutrino experiment. Nucl Phys B 908:62–73. Neutrino Oscillations: Celebrating the Nobel Prize in Physics 2015. ISSN: 0550-3213. https://doi.org/10.1016/j.nuclphysb.2016.04.034. https://www.sciencedirect.com/science/article/pii/S0550321316300724
6. Ahn JK et al (2012) Observation of reactor electron antineutrinos disappearance in the RENO experiment. Phys Rev Lett 108:191802. https://doi.org/10.1103/PhysRevLett.108.191802. https://link.aps.org/doi/10.1103/PhysRevLett.108.191802
7. Mount BJ et al (2017) LUX-ZEPLIN (LZ) Technical design report. arXiv: 1703.09144 [physics.ins-det]
8. Akerib DS et al (2021) Simulations of events for the LUX-ZEPLIN (LZ) dark matter experiment. Astropart Phys 125:102480. . ISSN: 0927-6505. https://doi.org/10.1016/j.astropartphys.2020.102480. https://www.sciencedirect.com/science/article/pii/S0927650520300529
9. Dietze G, Klein H (1982) Gamma-calibration of NE 213 scintillation counters. Nucl Instr Meth Phys Res 193(3):549–556. ISSN: 0167-5087. https://doi.org/10.1016/0029-554X(82)90249-X. https://www.sciencedirect.com/science/article/pii/0029554X8290249X
10. Ko YJ et al (2019) Comparison between DAMA/LIBRA and COSINE-100 in the light of quenching factors. J Cosmol Astropart Phys 2019(11):008. https://doi.org/10.1088/1475-7516/2019/11/008
11. Lindhard J et al (1963) Integral equations governing radiation effects. Matematisk-Fysiske Meddelelser, Det Kongelige Danske Videnskabernes Selskab 33(10):1–42
12. Lewin JD, Smith PF (1996) Review of mathematics, numerical factors, and corrections for dark matter experiments based on elastic nuclear recoil. Astropart Phys 6(1):87–112. . ISSN: 0927-6505. https://doi.org/10.1016/S0927-6505(96)00047-3. https://www.sciencedirect.com/science/article/pii/S0927650596000473
13. PandaX Collaboration et al (2024) First measurement of solar ^{8}B neutrino flux through coherent elastic neutrino-nucleus scattering in PandaX-4T. arXiv: 2407.10892 [hep-ex]
14. Duba CA et al (2008) HALO - the helium and lead observatory for supernova neutrinos. J Phys: Conf Ser 136(4):042077. https://doi.org/10.1088/1742-6596/136/4/042077

Chapter 8
Event Selection

Abstract This chapter hones in on the Boosted Decision Tree (BDT) model used to classify signal events versus the plentiful steady-state backgrounds posed by the intrinsic radioactivity of the thorium target mass–the main background source. Two models are implemented: one for neutron classification and one for fission classification. A rigorous model validation is also pursued using cosmic muon-induced neutron events recorded in the NuThor dataset.

Neutrino-induced nuclear fission is a rare process and the vast majority of the events recorded and processed by NuThor are backgrounds from various sources. The spatial distribution of the 36 NaI detectors and the multi-particle emission following a neutrino-induced nuclear fission event create a distinctive topology of detector hits, characterized by the energies deposited, detector channels triggered, and the timing of those hits. The expected event topology, as predicted by simulations in Sect. 7.3, is seldom replicated by spurious backgrounds. To capitalize on this distinctive signature, a multivariate analysis pipeline was developed, which is detailed in this chapter.

Multivariate Analysis

Once the events are correlated with the beam pulse and grouped into super-clusters, each is assigned a score indicating the likelihood of it being a signal event using a multivariate analysis (MVA) approach [1]. The process of classifying events with a high likelihood of being signal is called event selection. The central technique employed for event selection is a set of Boosted Decision Tree models [2, 3].[1] Accompanying calculations are made that leverage other MVA quantities, such as the Fisher linear discriminant [6], to further the goal of background discrimination. Hereafter the results shown will be the Science Run 2 dataset only.

The overarching goal of the event selection is to maximally isolate neutrino-induced nuclear fission events from the many sources of background. The algorithms

[1] There were a number of other MVA techniques including Random Forest [4], Multilayer Perceptron [5], etc. The Gradient Boosted Decision Tree and multilayer perceptron approaches resulted in the best performances, but due to the significantly shorter training time, the XGBoost implementation of the Gradient Boosted Decision Tree model was chosen over the multilayer perceptrons.

T. Johnson, *The First Indication of Neutrino-Induced Nuclear Fission*,
https://doi.org/10.1007/978-3-032-01936-3_8

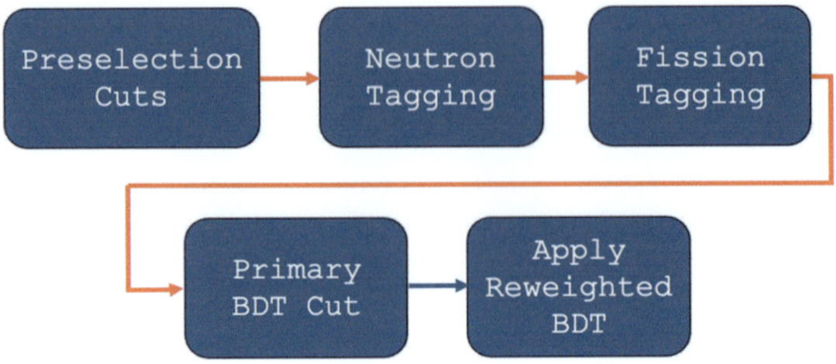

Fig. 8.1 Event selection workflow used to isolate a small population of signal events within an onslaught of background events

are built upon the comparison of simulated neutrino-induced nuclear fission signals and measured steady-state backgrounds collected in the beam anti-coincident dataset. The two quantities that guide the optimization of the event selection are efficiency and purity. Where the efficiency is the fraction of the true neutrino-induced nuclear fission signal events that are selected with the algorithm.

$$\epsilon = \frac{\text{Number of True Signal Events Selected}}{\text{Number of Total Signal Events Generated}} \tag{8.1}$$

The purity is the fraction of the selected events that are signal.

$$p = \frac{\text{Number of True Signal Events Selected}}{\text{Total Number of Events Selected}} \tag{8.2}$$

The event selection workflow is illustrated in Fig. 8.1. The remainder of this chapter is devoted to establishing each of these enumerated steps.

8.1 Preselection Cut

There are two cuts performed on the dataset to improve the overall performance of the BDT and increase the signal purity of the sample.

Beam-Related Neutron Time Cut

The first cut considered focused on the Beam-Related Neutrons (BRNs) interacting with the detector. The utility of cutting out the BRNs was both to diminish the number of background counts at the higher reconstructed neutron multiplicities and also to mitigate any doubt related to the modeling of high-energy neutrons scattering in the detector (MCNP cross section libraries are limited above 20 MeV). Where possible, this work adopts an extreme aversion to analysis cuts unless in situations where the underlying events can not be reliably reconstructed.

The BRN cut removed events that had NaI hits in the first 2 microseconds of the waveform trace (1 microsecond after the external trigger). Both simulations in MCNP[7]/GEANT4[8] and previously obtained NaIvE-185 data [9] demonstrated that BRNs largely arrive in sync with the protons-on-target (albeit with added delay due to propagation time to Neutrino Alley). This sudden influx of BRNs on the NaI detectors within the prompt region of the beam spill provided an efficient handle for removing BRN events. The time profile of the neutrino-induced nuclear fission signal is a convolution of the protons-on-target trace, the muon decay, and the neutron thermalization curve inside NuThor. As a result, it arrives later than the BRNs and is largely unaffected by this cut.

In Table 8.1, the acceptance efficiency of the neutrino-induced nuclear fission signal at higher neutron multiplicities becomes noticeably reduced. As the rate of random events (neutron multiplicity) increases, the timestamp of the earliest event tends to occur earlier, in accordance with Poisson statistics. This means that events with neutron multiplicity of 7, for example, had an earlier arriving first neutron capture on average, which also made it more susceptible to being cut by the BRN time cut. Given the exposure of Science Run 2, so few N = 7 events were expected that this is deemed a safe loss. Conversely, this time cut appears to be very efficient at rejecting BRN background events—See Table 8.2.

Table 8.1 Neutrino-induced nuclear fission signal acceptance efficiency after BRN time cut by neutron multiplicity

Neutron multiplicity	Time cut efficiency (%)
2	93.43
3	87.78
4	83.35
5	76.92
6	72.43
7	67.34

Table 8.2 Steady-state background rejection efficiencies by neutron multiplicity after the BRN time cut

Neutron multiplicity	BRN rejection efficiency (%)
2	93.67
3	96.83
4	98.91
5	99.58
6	99.50
7	96.97

8.1.1 Maximum Energy Cut

The second preselection cut that was applied imposed a maximum total energy for an event of 15 MeV. The purpose of this cut was three-fold. First, this removed a large portion of ^{127}I Neutrino-Induced Neutron emission (NIN) events. Second, it cleared the way for a high-energy sideband to study the high-energy BRNs and benchmark simulations. Third, the NaIvE-185 detector demonstrated the need for a high-energy calibration with Michel electron for a stable energy reconstruction beyond 10–15 MeV using these same NaI crystals. This would be a challenging endeavor in the background-rich context of the NuThor detector (though worth further investigation). Lacking the high-energy calibration in this work, this regime is excluded from the analysis.

Recall that the ^{127}I nuclei in the NaI detectors are a likely target for the electron neutrinos to capture on, due to their large neutron excess and ample mass in the 277.2 kg of NaI crystals. According to the MARLEY [10] neutrino event generator, each neutrino capture on ^{127}I has a 85.34% chance of emitting at least one NIN. The maximum energy cut's effect on the ^{127}I electromagnetic emission was also evaluated with MARLEY and is represented in Fig. 8.2. The 15 MeV cut is overlaid on the electromagnetic energy emitted following neutrino captures that emit *at least one* NIN in the final state. The background rejection efficiency of ^{127}I NIN events using this cut is 94.94%. Note that this is without consideration for detector response such as energy resolution and the quenching of nuclear recoils on Na and I.

Using the NaIvE-185 result, which reported an exclusive cross section of $\sigma(\geq 1n) = 2.2^{+3.5}_{-2.2} \times 10^{-40}\text{cm}^2$, coupled with the BRN time cut, the number of true ^{127}I events that passed these criteria was $0.48^{+0.77}_{-0.48}$ for the Science Run 2 exposure. Given the low estimated rate, ^{127}I will not be considered further in the analysis.

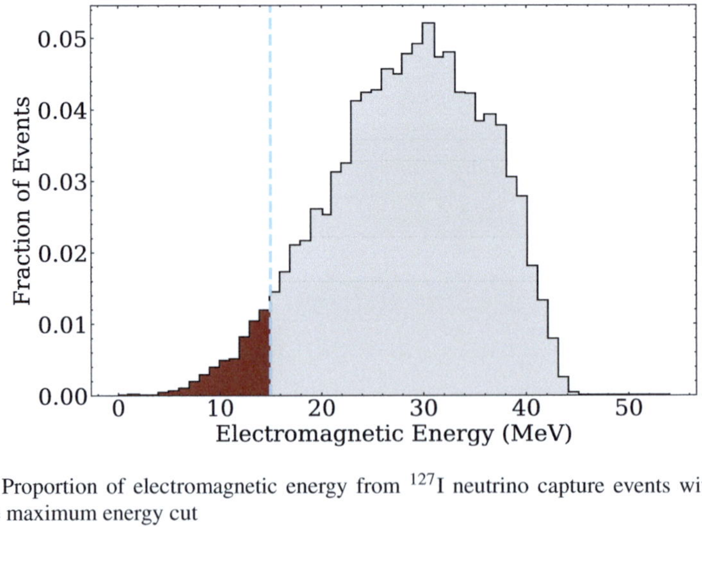

Fig. 8.2 Proportion of electromagnetic energy from ^{127}I neutrino capture events with 1+ NIN below the maximum energy cut

Table 8.3 Neutrino-induced nuclear fission signal acceptance efficiency after maximum energy cut by neutron multiplicity

Neutron multiplicity	Max energy cut efficiency (%)
2	99.89
3	98.63
4	93.71
5	84.00
6	71.78
7	56.61

The neutrino-induced nuclear fission signal acceptance following the maximum energy cut is summarized in Table 8.3. Similar to the BRN Time Cut, the maximum energy cut has a greater effect on the high neutron multiplicities due to the nature of stacking multiple neutron captures together in the same super cluster. Because so few counts are expected in these bins, this loss was deemed acceptable.

8.2 Building the BDT Model

A boosted decision tree is a type of supervised machine learning algorithm commonly used for classification and regression tasks. At its core, a boosted decision tree builds upon a basic decision tree. A decision tree is a structure in which data is split at each node based on specific criteria, continuing this process through several branches until a final classification is made, such as distinguishing between signal and background (Fig. 8.3).

Boosting refers to the process of creating multiple decision trees in sequence, with each tree aimed at improving the classification performance of the previous ones. In this work, the method used is the Extreme Graduate Boost (XGBoost) binary classifier [11].

8.2.1 Training and Testing Samples

The boosted decision tree model was constructed using an independent training dataset and a testing dataset to evaluate the quality of the model. Both datasets are pulled from the same sources. The signal dataset was generated from MCNP neutron simulations that are fortified with the GADGET gadolinium post processing. The background dataset was harvested directly from beam anti-coincident data in NuThor to incorporate all species of steady-state backgrounds. Because one class was from simulation and the other was from data, it was imperative that the simulation models the system with high fidelity.

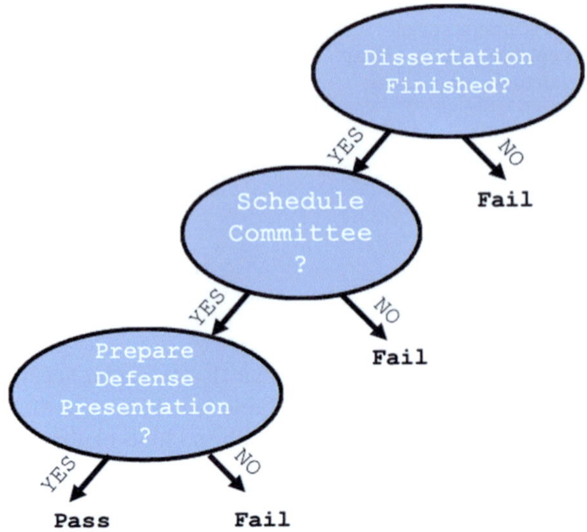

Fig. 8.3 An example of a decision tree illustrating the concept behind the architecture of the boosted decision tree (BDT)

8.2.2 Neutron Tagging

The first application of a BDT model was the creation of a neutron tag score for every N_{100} cluster found in the super-clusters. Each of the N_{100} clusters receives a neutron-tag in the form of a BDT score. No cut is made based solely on this neutron BDT score. The BDT scores of all N_{100} clusters within a super-clusters are used to evaluate an event as fission-like (more on this later).

Each neutron capture-induced γ-ray cascade can interact with anywhere between 0 and 35 NaI detectors, deposit a range of energies. The spatial relationship between the activated NaI channels and the deposited energies is used to construct a set of features. These features help the BDT model classify N_{100} events as either true neutron captures or random stochastic γ-rays from single or multiple thorium decays.

8.2.2.1 Feature Space

A total of 16 features are used to build the BDT model. These features are:

1. Gamma Multiplicity: the number of NaI hits, N_{100}
2. Total Energy: the sum of the energies for all NaI hits.
3. Energy Variance: the standard deviation between each of the NaI energy depositions.
4. Average X: using the central coordinate in (x, y, z) of each NaI detector, the average X.
5. Average Y: using the central coordinate in (x, y, z) of each NaI detector, the average Y.

6. Average Z: using the central coordinate in (x, y, z) of each NaI detector, the average Z.
7. Maximum Distance: the greatest distance between NaI hits in the N_{100} cluster. If the gamma multiplicity is 1, this value is 0.
8. Minimum Distance: the smallest distance between NaI hits in the N_{100} cluster. If the gamma multiplicity is 1, this value is 0.
9. Center of Energy—X: the average x-coordinate of the NaI hits weighted by the energy of each hit.
10. Center of Energy—Y: the average y-coordinate of the NaI hits weighted by the energy of each hit.
11. Center of Energy—Z: the average z-coordinate of the NaI hits weighted by the energy of each hit.
12. E_1: the energy of the first NaI hit in the N_{100} cluster.
13. E_2: the energy of the second NaI hit in the N_{100} cluster. If the gamma multiplicity is 1, this value is 0.
14. Acceptance Probability, P_{acc}: a likelihood score computed with the probability distribution of NuThor channels hit.
15. $P_{\text{Posterior}, S/B}$: a likelihood of signal/background based solely on the channel hit with an energy spectrum prior.
16. Fisher Discriminant: a common tool in multivariate analysis used to project a n-dimensional set of parameters into 1-dimension while simultaneously minimizing spread within the signal/background classes and maximizing separation between the two.

A more detailed description of the features is explored below, while illustrating how each parameter differs between the signal and background classes.

Total Energy

The total energy of the NaI hits within the N_{100} cluster.

$$E_{\text{total}} = \sum_{i=1}^{N_{100}} E_i \tag{8.3}$$

The gadolinium γ-ray cascade after neutron capture leads to events up to 8 MeV, while the majority of the spurious radioactivity that composes the background is constrained below the 2614 keV peak from the ^{208}Tl β-decay. However, there are still background events that leak into the 3+ MeV range from some sources.

Among these, cosmic muons can induce activity in this energy region if they pass through a corner of the detectors. Also, accidental coincidences between lower energy γ-rays in the same 100 nanosecond window can mimic a higher energy signal (Fig. 8.4).

Gamma Multiplicity

The number of NaI hits within a cluster, N_{100}. The main mechanism for neutron detection is through the capture on gadolinium nuclei. After capture, the gadolinium

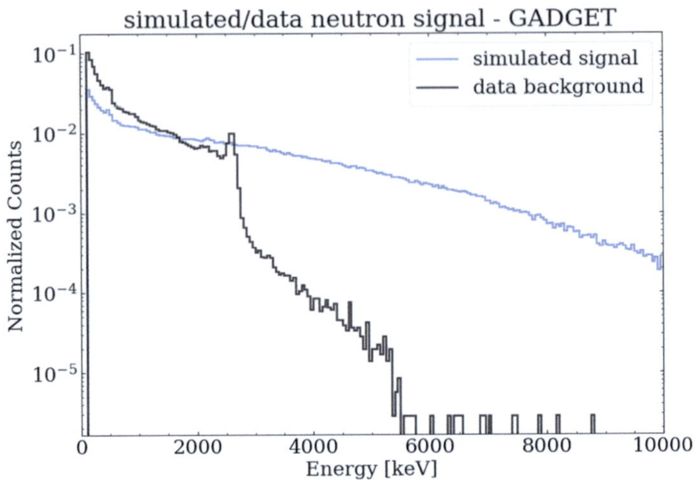

(a) Science Run 1

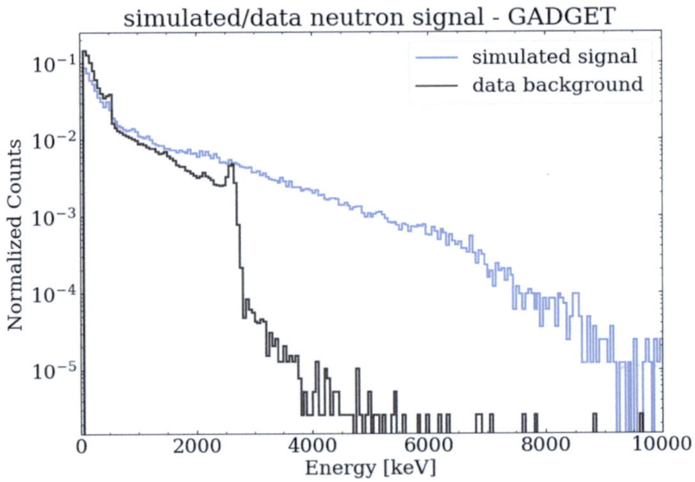

(b) Science Run 2

Fig. 8.4 Total energy of N_{100} clusters for NuThor Science Runs 1 and 2

compound nucleus will de-excite through the emission of multiple γ-rays that can hit multiple NaI detectors. If two γ-rays hit the same NaI within the same 100-nanosecond window, it is counted as a single hit.

The Science Run 1 dataset exhibits a higher average neutron multiplicity thanks to the increased capture efficiency and larger total volume of Gd-water capture agent (Fig. 8.5).

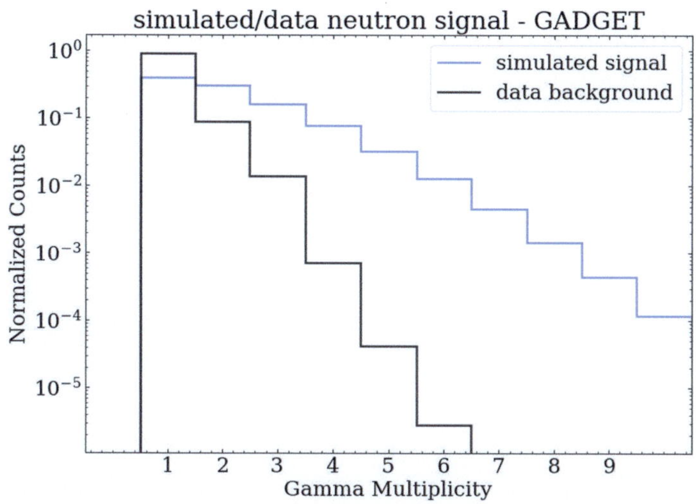

(a) Science Run 1

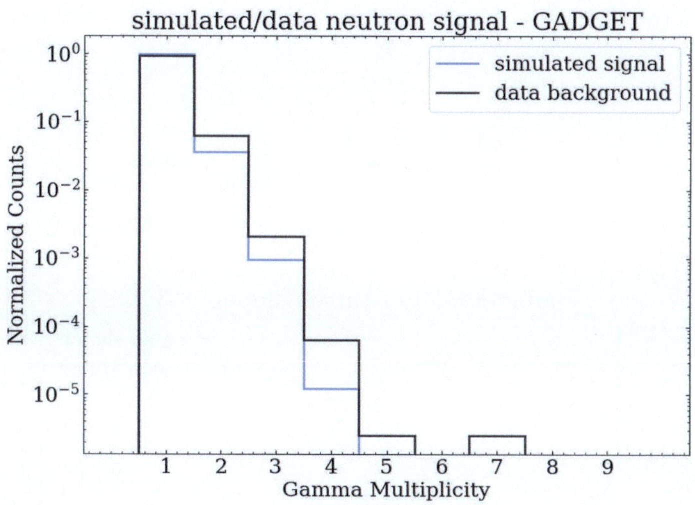

(b) Science Run 2

Fig. 8.5 Gamma multiplicity of N_{100} clusters

Energy Variance

The standard deviation of the energies in each of the NaI hits inside a N_{100} cluster (Fig. 8.6).

$$E_{var} = \frac{\sqrt{(E - E_{avg})^2}}{N_{100}} \tag{8.4}$$

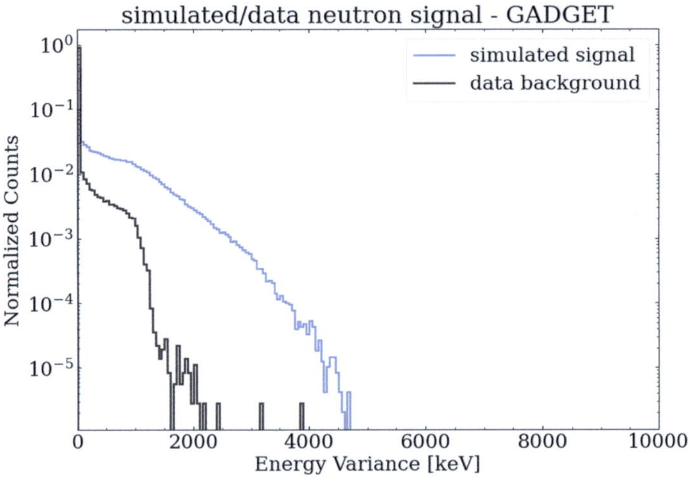

(a) Science Run 1

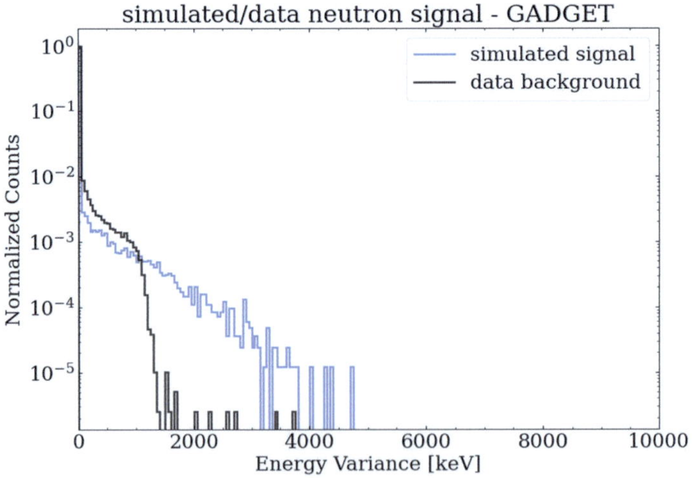

(b) Science Run 2

Fig. 8.6 Energy variance of N_{100} clusters

E_l

The energy of the earliest NaI hit in time inside the N_{100} cluster. Note that although this feature does relate the time it does not bias the selection of neutrons to any portion of the acquisition window because the timing is evaluated at such a local level constrained to a mere 100 nanoseconds (Fig. 8.7).

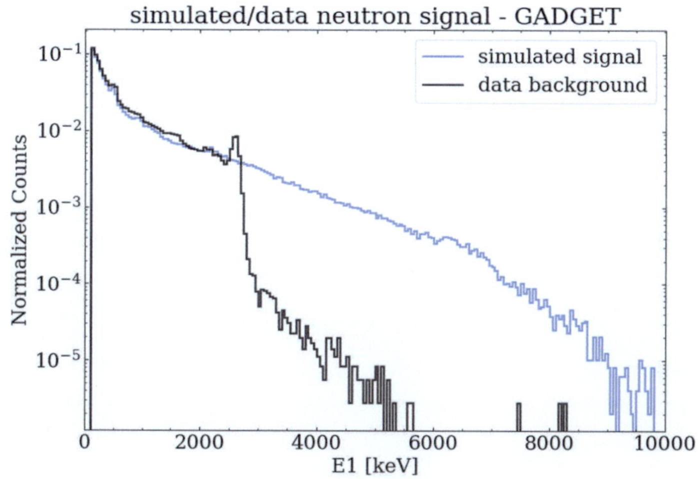

(a) Science Run 1

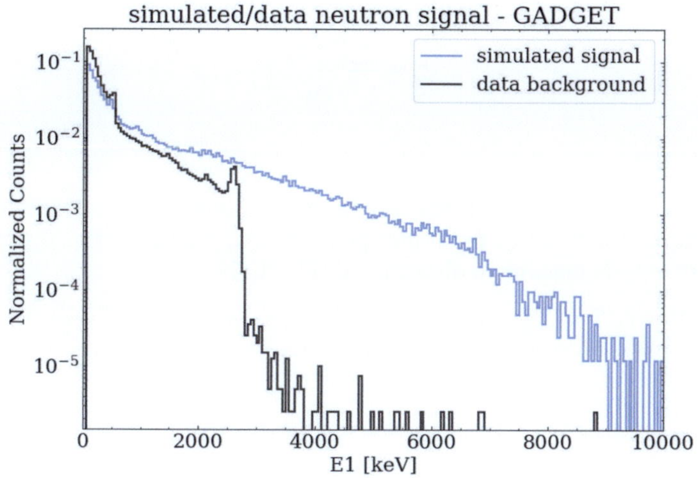

(b) Science Run 2

Fig. 8.7 E_l comparison of N_{100} clusters

E_2

The energy of the second earliest NaI hit in time inside the N_{100} cluster (Fig. 8.8).

X

The average X-value of the NaI's hits in N_{100}. The location of the NaI detectors are taken to be the center point in x, y, z coordinates and extracted from the detector geometry used for the MCNP simulation (Fig. 8.9).

$$x = \frac{\sum_{i=1}^{N_{100}} x_i}{N_{100}} \tag{8.5}$$

Y

The average Y-value of the NaI's hits in N_{100}. The location of the NaI detectors are taken to be the center point in x, y, z coordinates and extracted from the detector geometry used for the MCNP simulation (Fig. 8.10).

$$y = \frac{\sum_{i=1}^{N_{100}} y_i}{N_{100}} \tag{8.6}$$

Z

The average Z-value of the NaI's hits in N_{100}. The location of the NaI detectors are taken to be the center point in x, y, z coordinates and extracted from the detector geometry used for the MCNP simulation (Fig. 8.11).

$$z = \frac{\sum_{i=1}^{N_{100}} z_i}{N_{100}} \tag{8.7}$$

MaxDist

The greatest distance between NaI hits in the N_{100} cluster. The position of each detector channel is taken as the center point in x, y, z coordinates. If the gamma multiplicity = 1, the maximum distance = 0 (Fig. 8.12).

$$\text{MaxDist} = \max\left\{\sqrt{(x_i - x_j)^2 + (y_i - y_j)^2 + (z_i - z_j)^2}\right\} \tag{8.8}$$

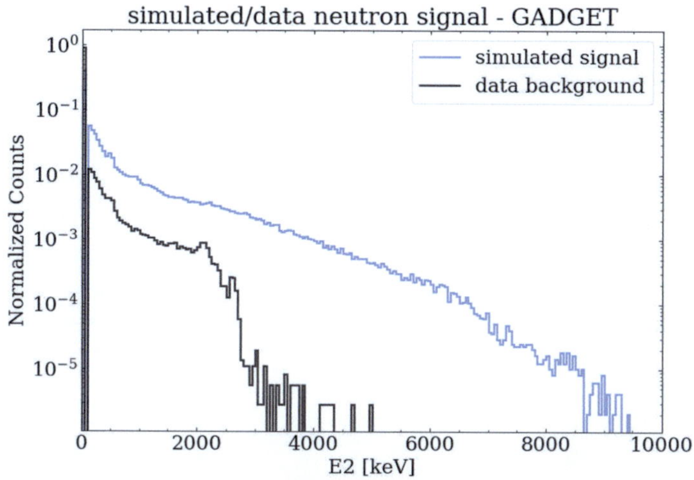

(a) Science Run 1

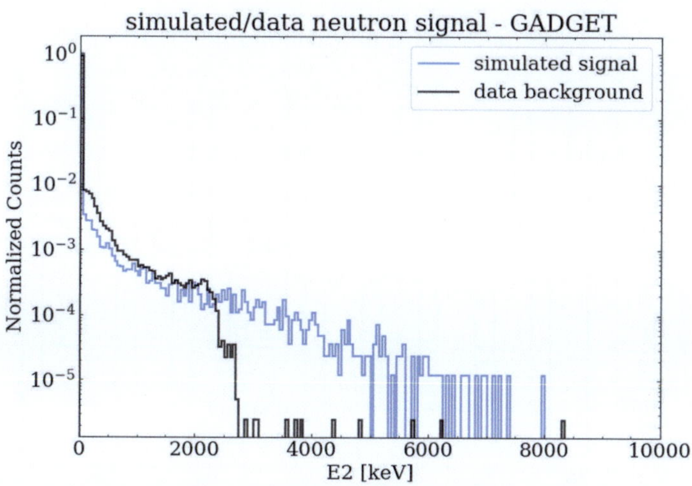

(b) Science Run 2

Fig. 8.8 E_2 comparison of N_{100} clusters

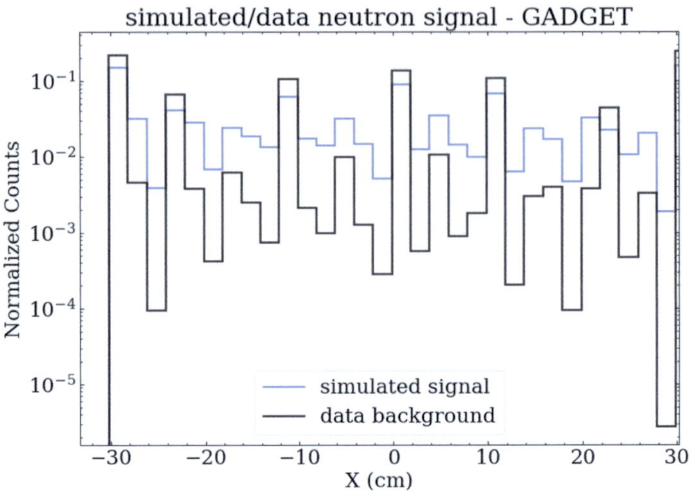

(a) Science Run 1

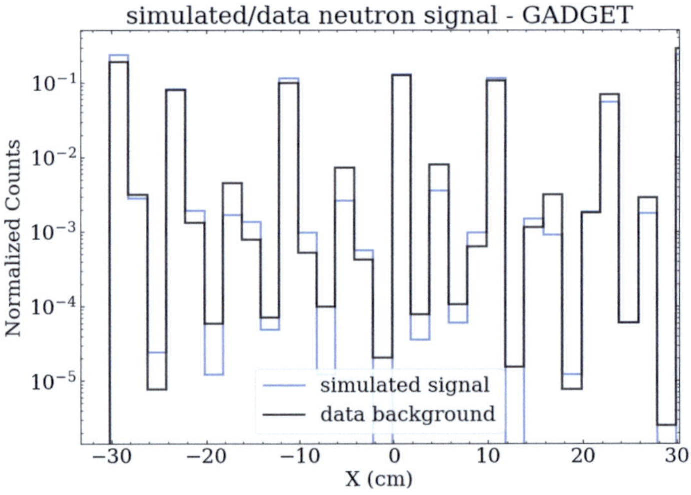

(b) Science Run 2

Fig. 8.9 Average X of N_{100} clusters

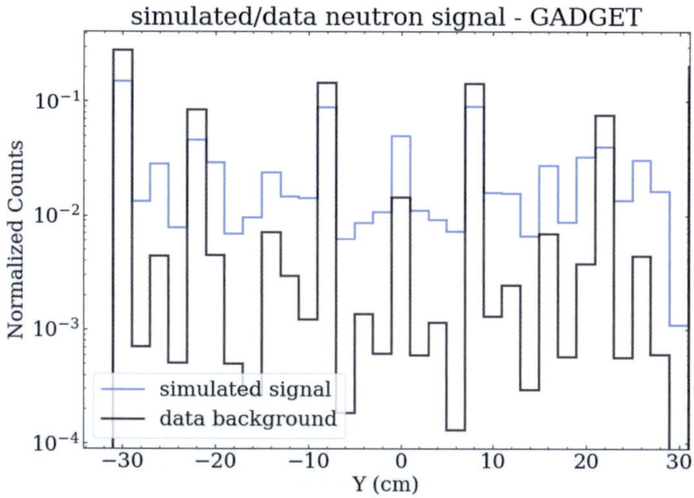

(a) Science Run 1

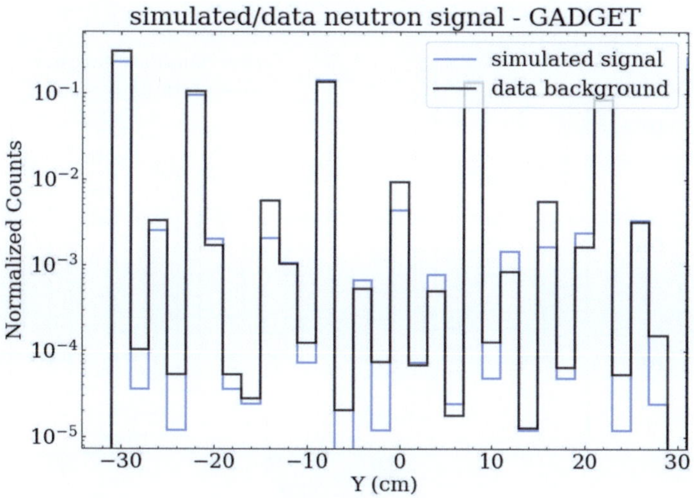

(b) Science Run 2

Fig. 8.10 Average Y of N_{100} clusters

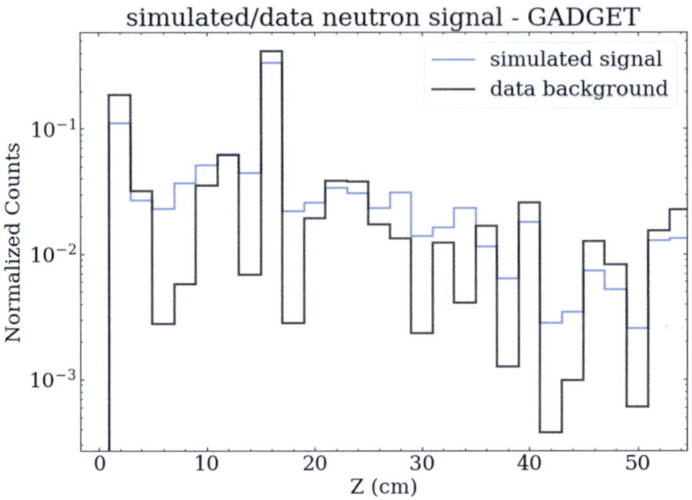

(a) Science Run 1

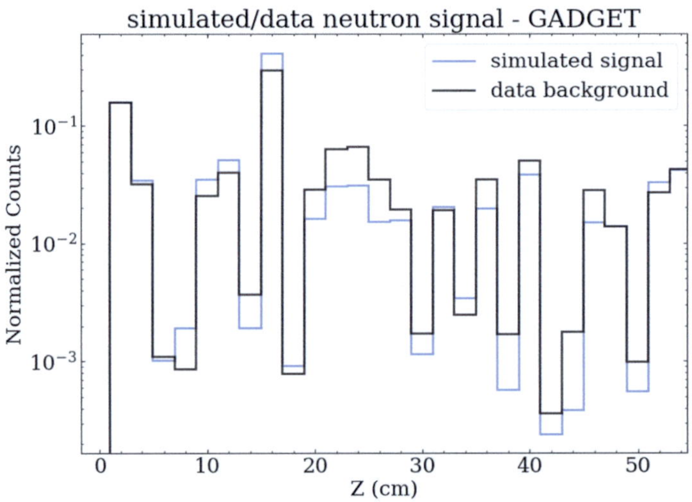

(b) Science Run 2

Fig. 8.11 Average Z of N_{100} clusters

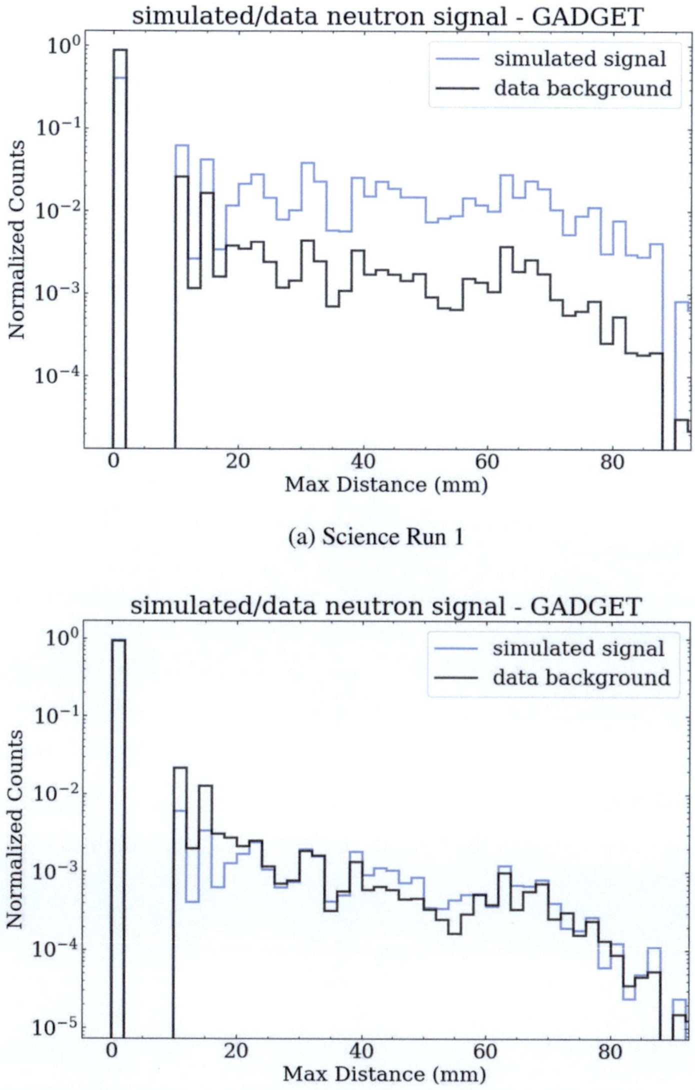

(a) Science Run 1

(b) Science Run 2

Fig. 8.12 Maximum distance between NaI hits

MinDist

The shortest distance between NaI hits in the N_{100} cluster. The position of each detector channel is taken as the center point in x, y, z coordinates. If the gamma multiplicity = 1, the maximum distance = 0 (Fig. 8.13).

$$\text{MinDist} = \min\left\{\sqrt{(x_i - x_j)^2 + (y_i - y_j)^2 + (z_i - z_j)^2}\right\} \tag{8.9}$$

Center of Energy—X

The average x-coordinate of the NaI hits in the N_{100} cluster weighted by the energy of each hit (Fig. 8.14).

$$x_E = \frac{\sum x_i E_i}{\sum E_i} \tag{8.10}$$

Center of Energy—Y

The average y-coordinate of the NaI hits in the N_{100} cluster weighted by the energy of each hit (Fig. 8.15).

$$y_E = \frac{\sum y_i E_i}{\sum E_i} \tag{8.11}$$

Center of Energy—Z

The average z-coordinate of the NaI hits in the N_{100} cluster weighted by the energy of each hit (Fig. 8.16).

$$z_E = \frac{\sum z_i E_i}{\sum E_i} \tag{8.12}$$

Acceptance Probability

The $P_{\text{Acceptance}}$ is an assessment of the likelihood that a given set of detector channels would be engaged from an actual neutron capture event. The basis of this is a log of detectors hit in both beam anti-coincident background data and MCNP + GADGET-based neutron simulations. First a probability distribution is created for signal and background by logging which detector is engaged, then normalizing by the total number of events. This is done for both the signal simulation and the background data. Then, a ratio is taken for each of the channels between the signal and background probability distributions.

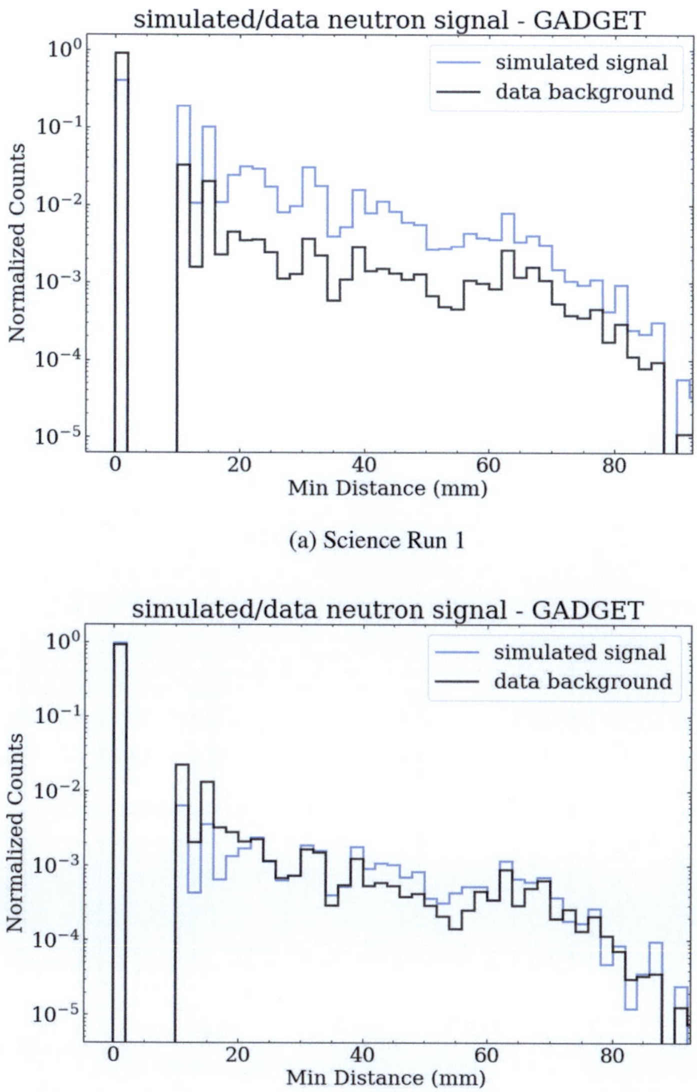

(a) Science Run 1

(b) Science Run 2

Fig. 8.13 Minimum distance between NaI hits

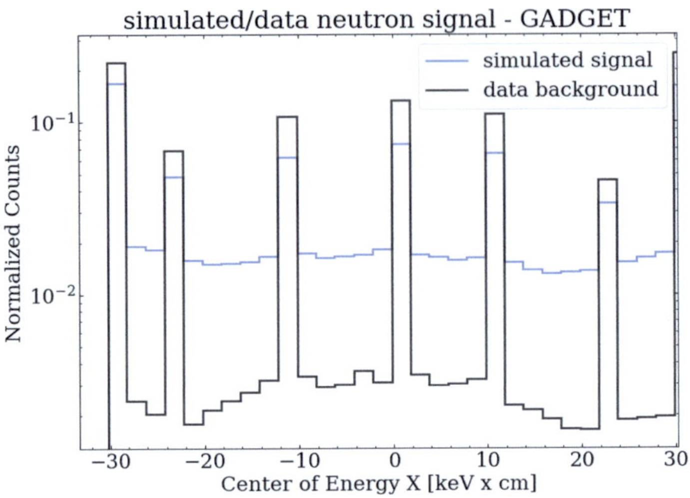

(a) Science Run 1

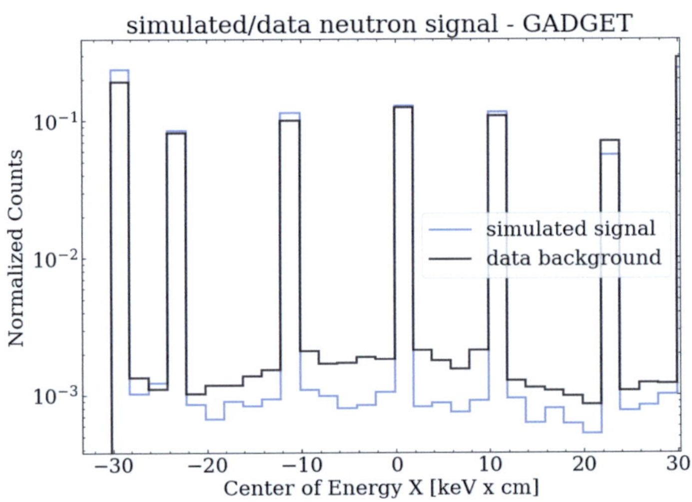

(b) Science Run 2

Fig. 8.14 Center of Energy—X

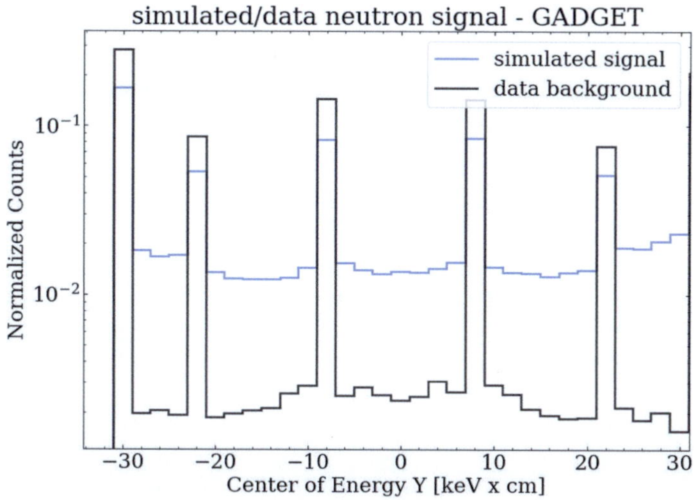

(a) Science Run 1

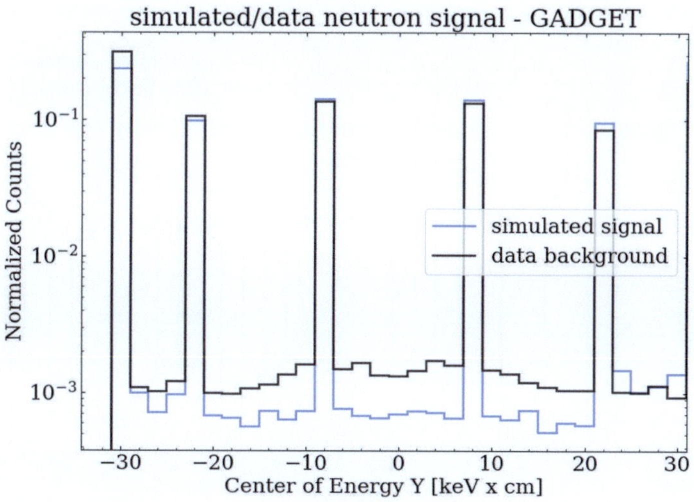

(b) Science Run 2

Fig. 8.15 Center of Energy—Y

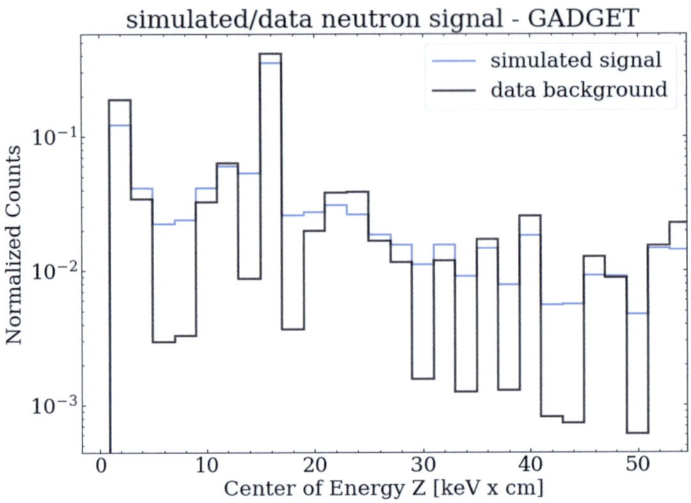

(a) Science Run 1

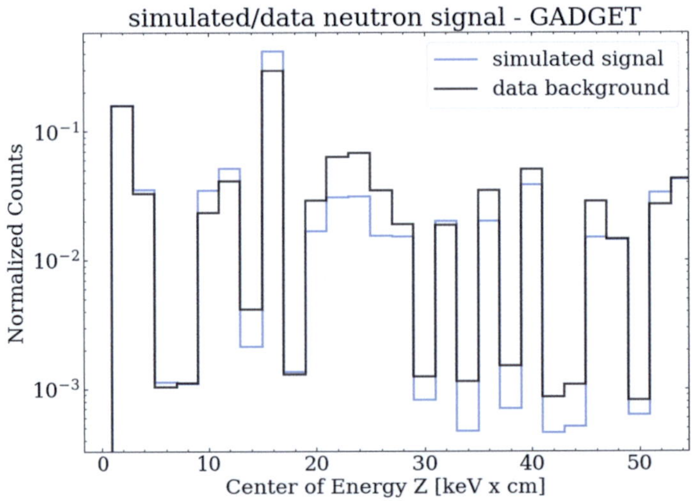

(b) Science Run 2

Fig. 8.16 Center of Energy—Z

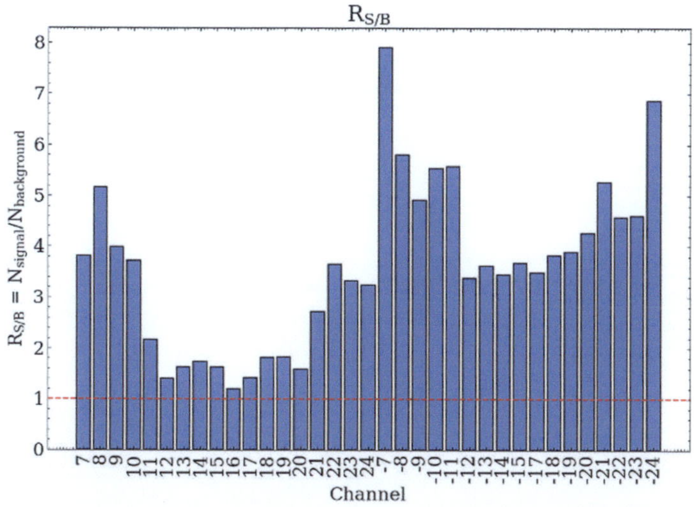

Fig. 8.17 The ratio of probability densities per detector channel between simulated signal neutrons and beam anti-coincident background data

The values tabulated in Fig. 8.17 are then combined with the following expression:

$$P_{Acceptance} = \frac{\log\left(\prod_{i=1}^{N_{100}} P_i\right)}{N_{100}} \tag{8.13}$$

The distributions are then (Fig. 8.18).

Bayesian Inference Parameters

A Bayesian estimate of the likelihood of an N_{100} cluster given a posterior distribution defined as a probability density function in energy on each NaI detector channel (Fig. 8.19).

$$\text{Posterior}_{S,B} = \frac{P_{S,B}(E) \times P_{S,B}(\text{Ch})}{P_S(E) \times P_S(\text{Ch}) + P_B(E) \times P_B(\text{Ch})} \tag{8.14}$$

Fisher Discriminant

The fisher discriminant is its own multivariate analysis technique used in a number of particle physics experiments. The fisher discriminant is a handy method of projecting multiple dimensions down into a single axis. Further, this projection is performed in such a way that the two classes of events being projected are optimized to (1) maximally space the two class's means from each other and (2) minimize the spread

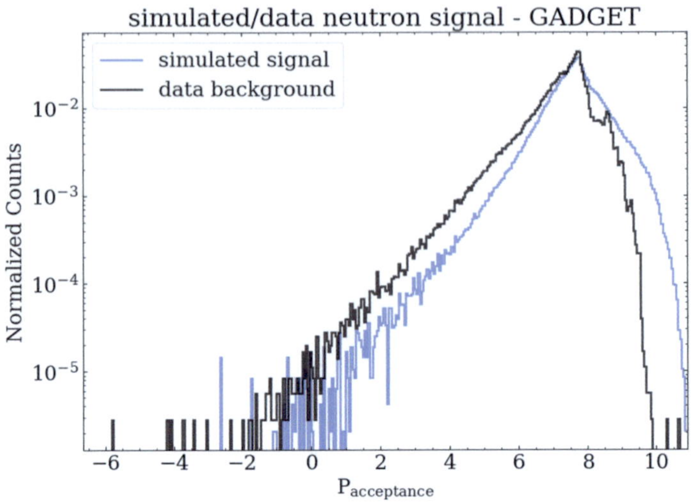

(a) Science Run 1

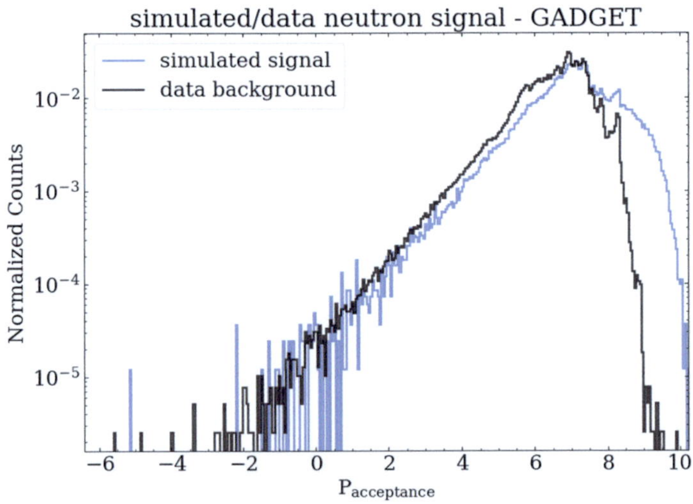

(b) Science Run 2

Fig. 8.18 Acceptance probability of N_{100} clusters

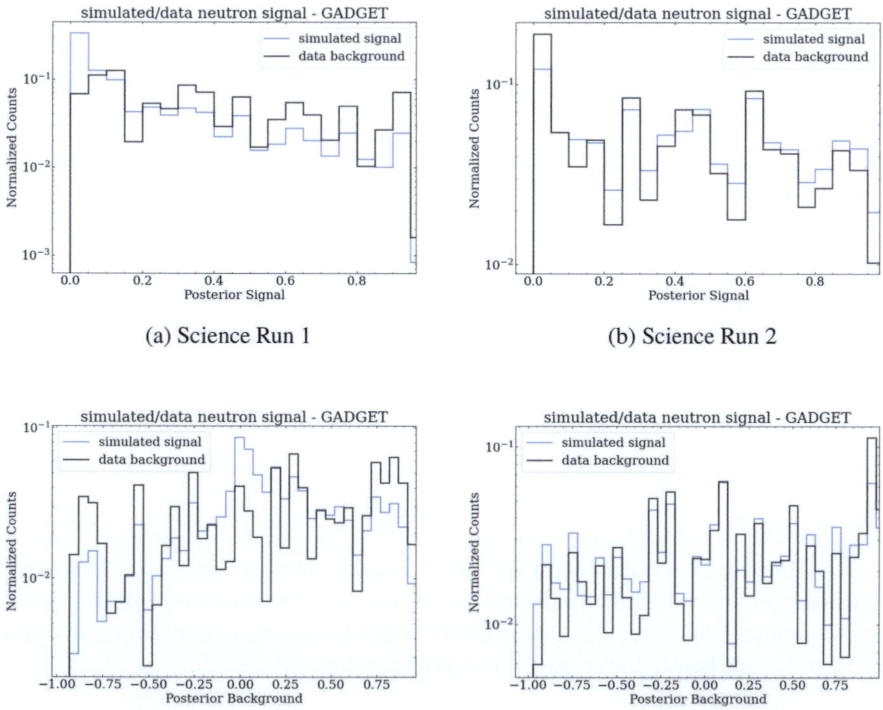

(a) Science Run 1 (b) Science Run 2

(c) Science Run 1 (d) Science Run 2

Fig. 8.19 Posterior parameters of N_{100} clusters

of events inside each event class. The process works to combine the values of each of the features as a linear combination with coefficients (Eq. 8.15) computed with the above criteria in mind.

$$t(\mathbf{x}) = a_0 + \sum_{i=1}^{n} a_i x_i \tag{8.15}$$

where $t(\mathbf{x})$ is the fisher linear discriminant test statistic and a_0 is the bias term which acts as an offset. The aforementioned objectives of maximum separation between classes and minimum spread within classes is achieved via the maximization of Eq. 8.18.

$$J(\mathbf{a}) = \frac{(\tau_0 - \tau_1)^2}{\Sigma_0^2 - \Sigma_1^2}, \tag{8.16}$$

$$(\tau_0 - \tau_1)^2 = \sum_{i,j=1}^{n} a_i a_j (\mu_0 - \mu_1)_i (\mu_0 - \mu_1)_j, \tag{8.17}$$

$$\Sigma_0^2 - \Sigma_1^2 = \sum_{i,j=1}^{n} a_i a_j (V_0 + V_1)_{ij} \tag{8.18}$$

where $(\mu_k)_i$ is a vector of mean values and $(V_k)_{ij}$ is the covariance matrix for the kth hypothesis class (signal or background). One then needs to find the optimum of 8.18 by taking the derivatives of $J(a)$ with respect to the a_i and setting it equal to 0. This yields a set of coefficients with maximum separation given by Eq. 8.20.

$$\mathbf{a} \propto W^{-1}(\mu_0 - \mu_1), \tag{8.19}$$

$$W_{ij} = (V_0 + V_1)_{ij} \tag{8.20}$$

The values of each variable's mean and covariance matrix is harvested from Monte Carlo simulation and used to train and test the model. The fisher discriminant model here is trained and evaluated using ROOT's TMVA and the variables receive a few transformations before being utilized in the projection (Fig. 8.20).

- Decorrelation
- Gaussianization
- Principle Component Decomposition

The coefficients calculated by TMVA for the two Fisher linear discriminant models are shown in Table 8.4.

8.2.2.2 Hyperparameter Optimization

The BDT model receives user-defined specifications on how it should be built. These are called hyperparameters. The hyperparameters have a variety of different roles. Some hyperparameters are meant to make the BDT model more resilient to over-training. One means of preventing over-training is regularization, where the model becomes less sensitive to individual data points. Other hyperparameters inform the training process by limiting the complexity of the decision trees.

A set of hyperparameters was varied simultaneously to determine what the most effective combination was. This was implemented with Scikit-Learn's built-in grid search tool which boasts considerable speed through multi-processing. The following is a list of the hyperparameters that were used in the optimization:

1. **ETA**: The ETA (also called the "learning rate") is a measure of the "shrinkage" or the extent that each successive boosting round is constrained to influence the training as to not accidentally over fit to the training dataset. Lower values of

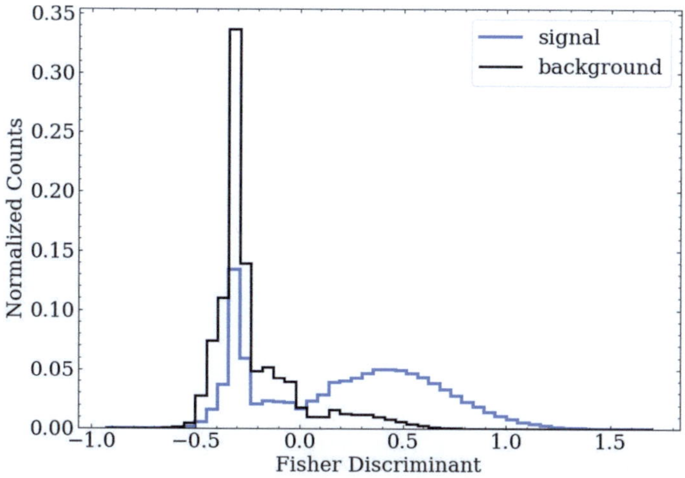

(a) Science Run 1

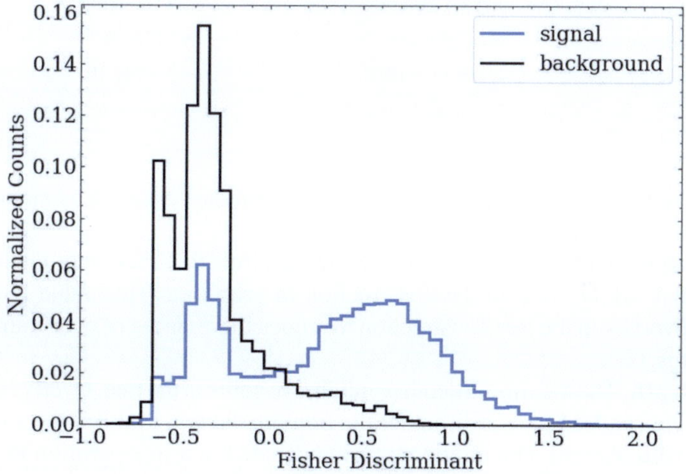

(b) Science Run 2

Fig. 8.20 Fisher linear discriminant distributions for the signal and background classes. There is a clear difference that is exploitable to distinguish the signal

Table 8.4 The coefficients of the variables incorporated into the trained Fisher discriminant model

Feature name	Science Run 2 coefficient	Science Run 1 coefficient
Bias term	1.66×10^{-3}	4.40×10^{-3}
Total energy	8.44×10^{-5}	-2.81×10^{-1}
Gamma multiplicity	4.39×10^{-5}	2.79×10^{-5}
Max. distance	1.05×10^{-1}	1.12×10^{-1}
Min. distance	-2.56×10^{-3}	-2.36×10^{-3}
X	1.23×10^{-2}	7.52×10^{-3}
Y	-5.10×10^{-4}	-8.85×10^{-5}
Z	3.48×10^{-3}	2.14×10^{-3}
Energy variance	-8.37×10^{-3}	-5.17×10^{-3}
$P_{Acceptance}$	-4.76×10^{-5}	1.62×10^{-5}
Posterior signal	4.78×10^{-2}	5.31×10^{-2}
Posterior background	-1.49	-1.13
E_1	-7.64×10^{-1}	-5.23×10^{-1}
E_2	1.26×10^{-4}	7.56×10^{-5}
Center of Energy—X	3.56×10^{-5}	3.32×10^{-5}
Center of Energy—Y	-5.53×10^{-5}	5.31×10^{-6}
Center of Energy—Z	-3.73×10^{-4}	-1.14×10^{-3}

the ETA mean that each boosting round contributes a smaller amount model prediction. The default is 0.3.

2. **Colsample_bytree**: The colsample_bytree controls the fraction of the input features that can be used in the construction of each tree. This value is meant to force diversity in the tree construction to reduce the chances of over-training. The default is 1.
3. **Max Depth**: The maximum depth or number of splits in the tree. Overly increasing this value can lead to over-fitting of the training dataset. The default is 6.
4. **Min Child Weight** The minimum weight needed in a tree partition to continue to branch out into a new leaf. The default is 1.
5. **Sub-Sample**: The sub-sample or sub-sample ratio is the fraction of the training dataset that is sampled to construct an individual decision for that boosting round. The default is 1.
6. **N Estimators**: Number of boosting rounds or the number of trees that will be grown.

The metric used to quantify each grid points effectiveness is the accuracy. The accuracy is the percentage of data points that are correctly classified. Conversely, one can optimize based on a value called the Area Under the Curve (AUC). The AUC is derived from the Receiver Operating Characteristic curve (ROC). The ROC curve is the a graphical representation of a binary classifier's ability to correctly label data points. The curve itself is the true positive rate plotted over the false positive rate (See

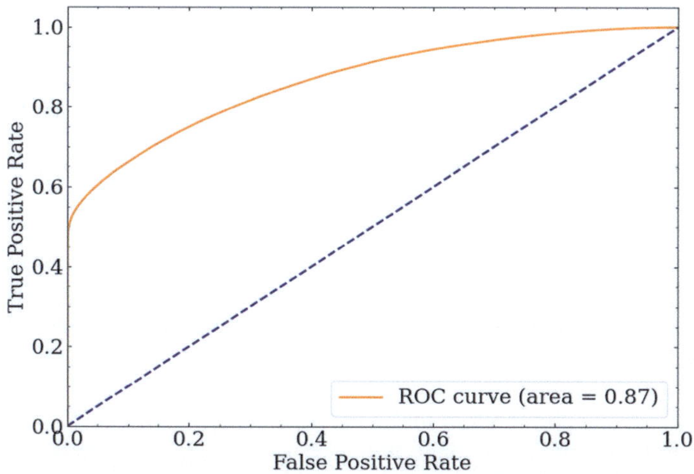

Fig. 8.21 The Receiver operating characteristic curve for the trained XGBoost Classifier to tag the simulated neutron events among background harvested from data for Science Run 2

Fig. 8.21). As the name would suggest, the AUC is the area under the ROC curve. A perfect classifier would have an area of 1. A classifier that selects a label completely at random would have an AUC of 0.5.

Several of the enumerated hyperparameters have the potential to cause the BDT model to be over-trained. This is generally the risk of a boosted decision tree as opposed to a decision tree model like Random Forest which is more resilient to over-training. Despite the hazard, there are methods for examining the trained BDT model to ensure over-training is being avoided.

The approach used in this work was to first optimize the hyperparameters in Table 8.5 through a grid search. With these optimized hyperparameters in hand, the hyperparameters with the highest risk of inducing over-training were selected. These are the number of estimators (N_estimators) and the maximum depth of the trees (max_depth).

Table 8.5 The optimized hyperparameters selected to train the XGBoost classifier model

Parameter name	Value
ETA	0.1
Colsample by tree	1
Max depth	15
Min child weight	1
Sub-sample	0.9
N Estimators	100

8.2.2.3 Over-Training Checks

Number of Trees

Beginning with the N_estimators parameter. All of the hyperparameters besides N_estimators are taken to be the values output by the grid search optimization. A new value for N_estimators are selected e.g. N_estimators = 10. The model was retrained on the training dataset with this new value of N_estimators and applied to the testing dataset. If the model had been over-trained, one would expect the model to perform worse for the testing set compared to the training set. Two quantities were used to measure the performance of the trained model: the Area Under the Curve (AUC) and the Area Under the Precision-Recall Curve (AUCPR). The AUCPR is a measure of the trade off between the precision (the signal purity of the sample) and the recall (the signal efficiency of the sample).

Some over-performance on the training set is fine, but it is important to control for when the differential between the two start to diverge or plateau. A second new value for the N_estimators was selected and the above steps were repeated. This was done incrementally from N_estimators = 10 to 250. After 250 training becomes too computationally expense to warrant further increase.

Figures 8.22 and 8.23 indicate that the testing dataset plateaus after around 100 trees while the train dataset continues to climb. There was even a slight decrease in performance after 100 trees. This supports the notion that continuing to increase the number of trees would cause the these quantities to continue to diverge to greater degree. Thus, 100 was the chosen number of boosting rounds/number of trees.

Maximum Depth of Trees

The maximum tree depth receives the same evaluation as the number of trees. Figures 8.24 and 8.25 illustrate the diminished returns associated with further increase of the Max_depth parameter after around 15. Therefore, 15 was the chosen Max_depth parameter value.

8.2.3 Neutron Model Performance

The aforementioned features were all leveraged into the BDT score. The BDT score is the log-odds transformation (the "logit") [12] of the classification probability found by the model, which takes the form of Eq. 8.21.

$$\text{BDT Score} = \log\left(\frac{p}{1-p}\right) \tag{8.21}$$

After training the XGBoost model on the simulated signal dataset and the beam anti-coincident background dataset, the following distributions of BDT scores for the two classes of events were produced by applying the model to a testing dataset (See Fig. 8.26b).

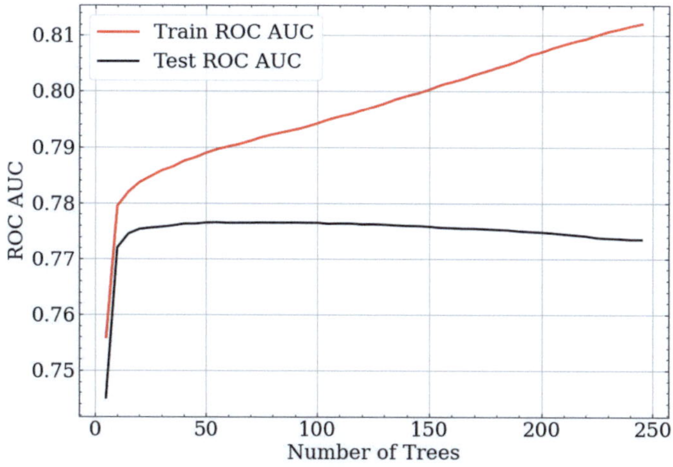

(a) Science Run 1

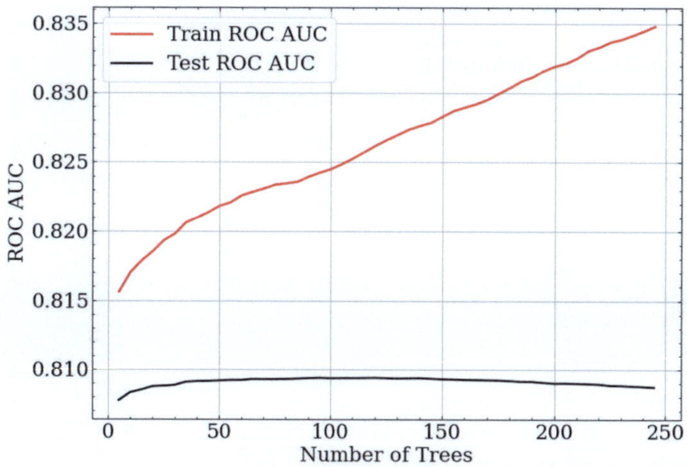

(b) Science Run 2

Fig. 8.22 Over-training checks for the number of trees hyperparameter with the AUC. There are diminishing returns after 100 trees so that was the value used going forward. The training dataset continues to increase in accuracy while the testing dataset does not (even slightly decreasing), which insinuates that the model begins to be over-trained

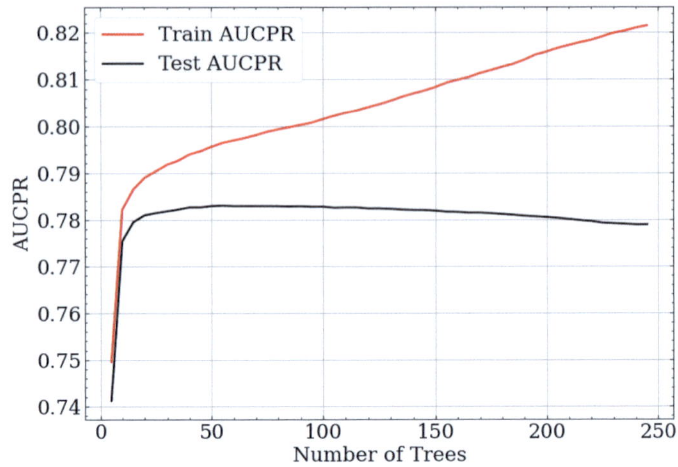

(a) Science Run 1

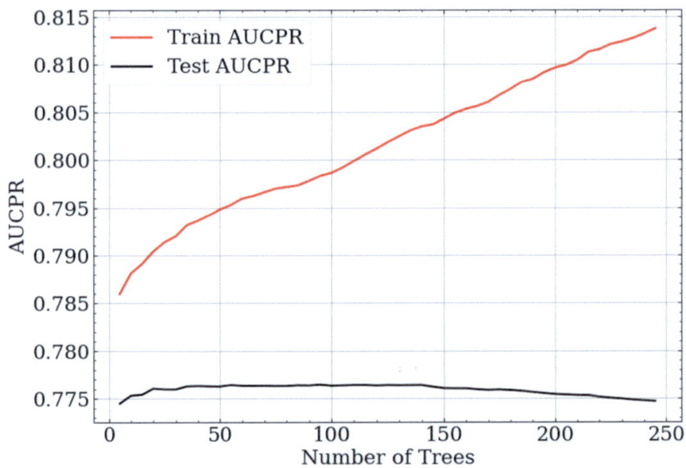

(b) Science Run 2

Fig. 8.23 Over-training checks for the number of trees hyperparameter with the AUCPR. There are diminishing returns after 100 trees, so that was the value used going forward. The training dataset continues to increase in accuracy while the testing dataset does not (even slightly decreasing), which insinuates that the model begins to be over-trained

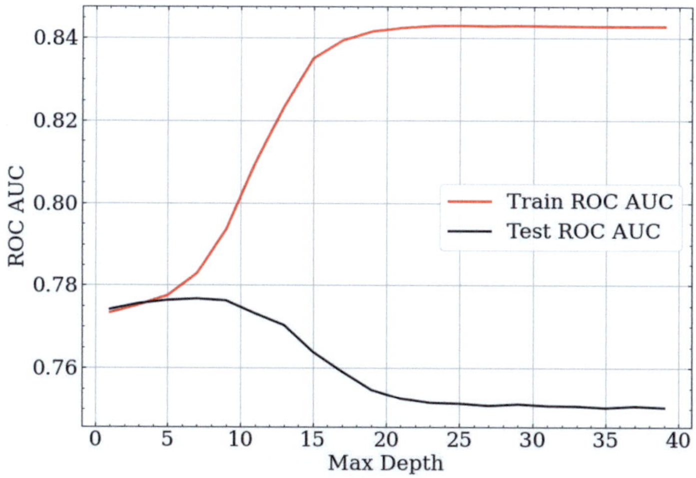

(a) Science Run 1

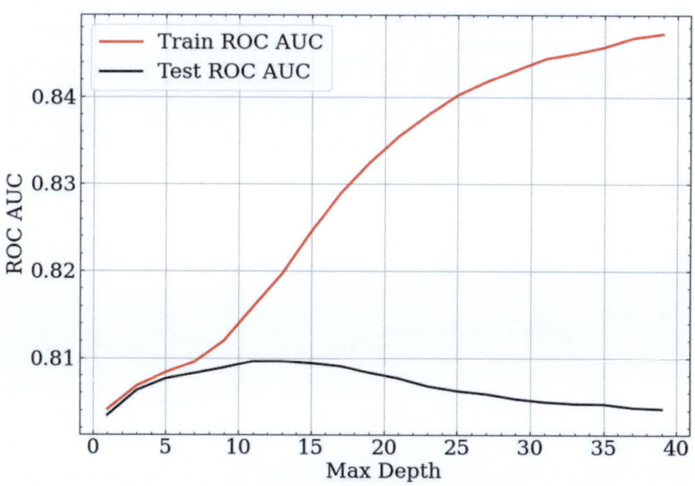

(b) Science Run 2

Fig. 8.24 Over-training checks for the max depth hyperparameter using the AUC

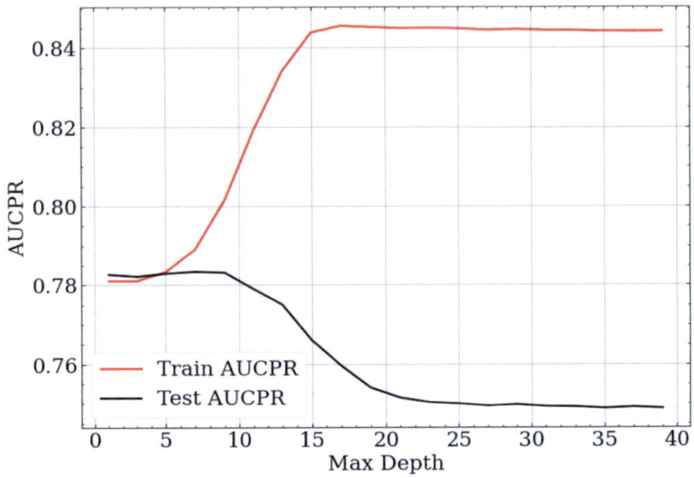

(a) Science Run 1

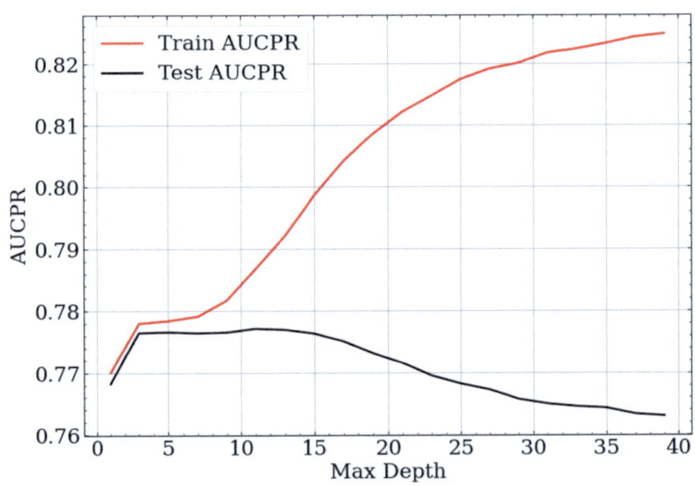

(b) Science Run 2

Fig. 8.25 Over-training checks for the max depth hyperparameter using the AUCPR

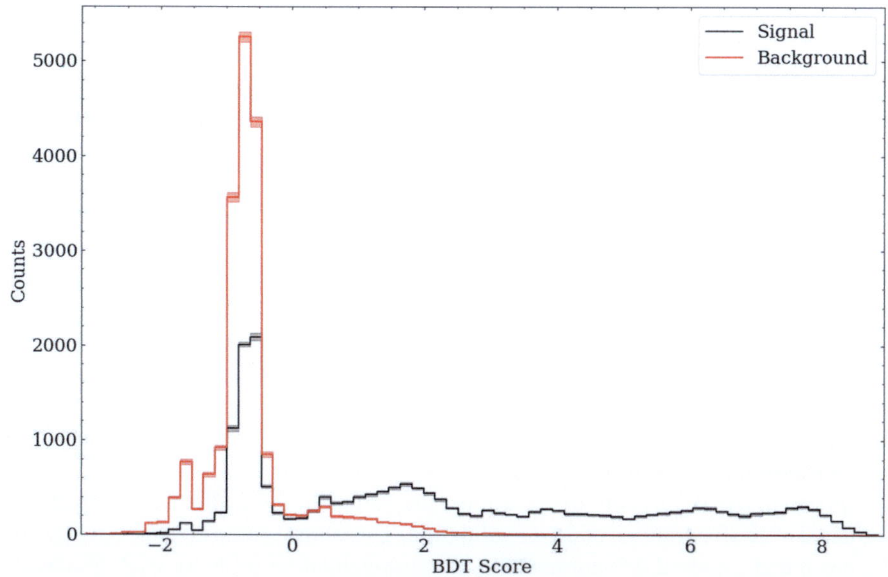

(a) Science Run 1

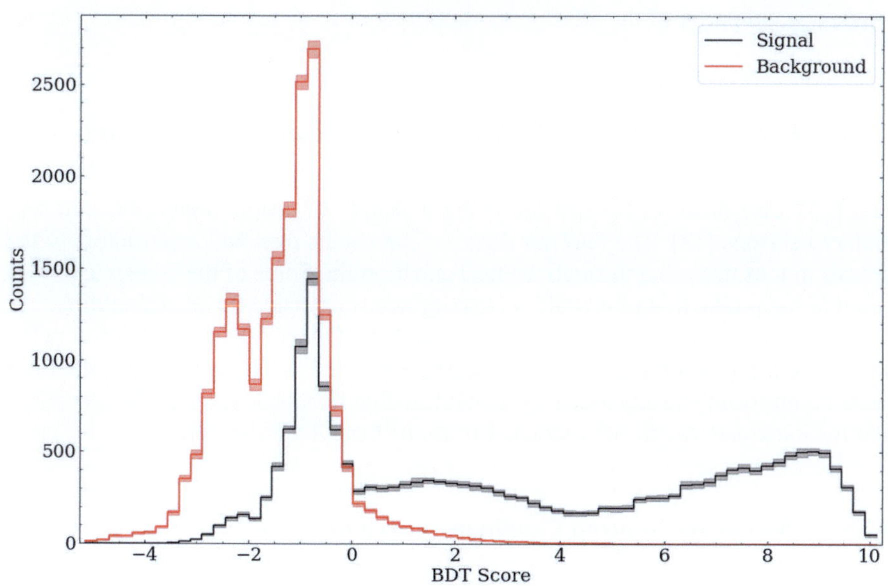

(b) Science Run 2

Fig. 8.26 BDT Score distributions for the NuThor Science Run 1 and 2 models

The central value and error bands in each distribution were generated by randomly sampling the testing dataset for equal number of signal and background then applying the neutron tag BDT model. This was done 30 times to cross validate the model's performance on a variety of different datapoints. For each BDT score bin, the mean and the standard deviation were found and used as the central value and the uncertainty band, respectively.

The performance of the neutron tag BDT model was summarized using a confusion matrix, which showed the number of true positives, true negatives, false positives, and false negatives. A prediction threshold of 0.5 was used for the matrix, meaning any BDT output above 0.5 was classified as signal. The BDT's prediction and the true label of each data point determine its placement in the confusion matrix. The inclusion of confusion matrices for Science Runs 1 and 2 in this work serves to illustrate the quality of discrimination based on the neutron tag score alone. No cut was applied using the neutron tag score, because the neutron tag score is used for a broader fission classification.

Similar to the BDT score distributions, the values in the confusion matrices are tabulated over 30 iterations of cross-validation. The central value and uncertainty are the mean and standard deviation of the 30 cross-validation folds for each quadrant, respectively (Fig. 8.27).

Another angle to weigh the effectiveness of the classification model is to view the efficiency and purity of the BDT score distributions as shown in Fig. 8.28.

8.2.4 BDT Validation with Cosmic Muon-Induced Neutrons

The BDT classification is validated using a sample of cosmic muon-induced secondary neutrons [13]. The NuThor detector can use its own NaI scintillators to tag cosmic muons that cross through the lead and thorium. Some of these cosmic muons result in the production of cosmic secondary neutrons inside the lead shielding.

The secondary neutrons are modeled through the GADGET framework following a muon-induced neutron energy spectrum (See Fig. 8.29). The simulation generates neutrons uniformly throughout the lead with no directional preference. The secondary neutron candidate events are extracted from the beam OFF dataset.

8.2.4.1 Secondary Neutron Candidate Search Criteria

The search used the NaI crystals to tag muon events in which high energy depositions occur simultaneously on opposite sides of the detector. This event topology insinuated that a muon passed through the detector with a path that likely passes through the inner lead shielding. In the background data, one of the only possible methods to depositing greater than 10 MeV in the detector is through cosmic muons. Some of

Fig. 8.27 The confusion matrix for the neutron tagging XGBoost Classifier after training with the optimized parameters and over-training checks performed. This describes the performance of a cut on BDT score equal to 0.5, which was not actually how this model will be implemented in this analysis, but was a useful touchstone. The signal and background classes are in equal proportion. 30 subsets of the total testing data bank were randomly selected and evaluated with the trained model. The mean and standard deviation of those 30 evaluations is represented in the elements of the matrix

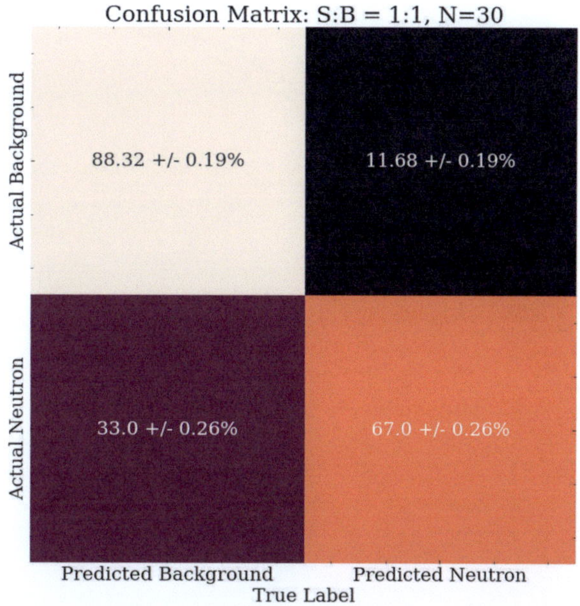

(a) Science Run 1

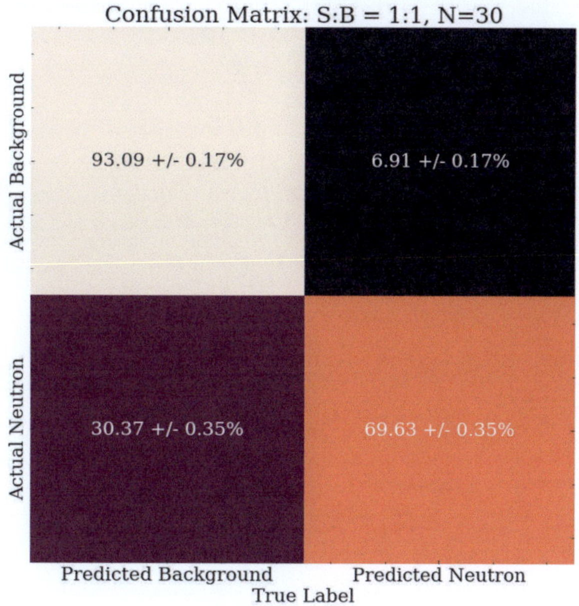

(b) Science Run 2

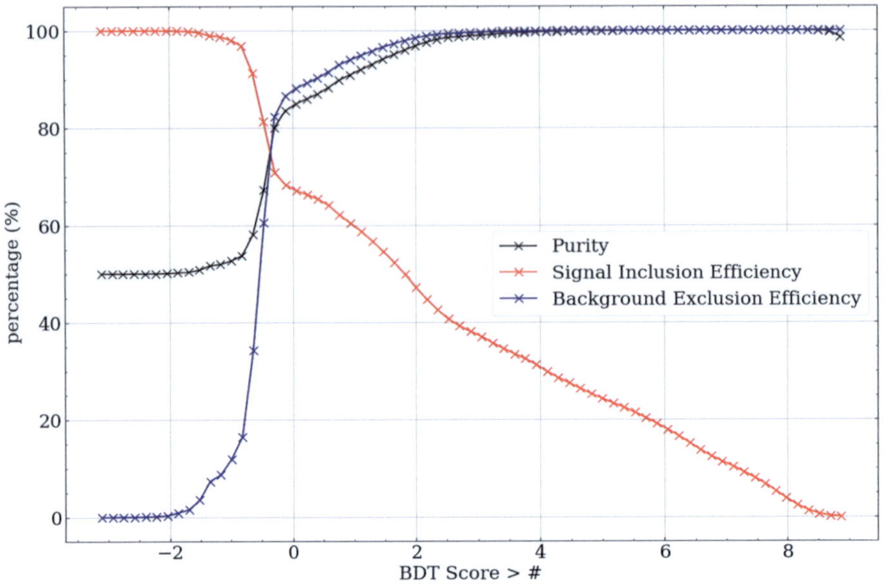

(a) Science Run 1

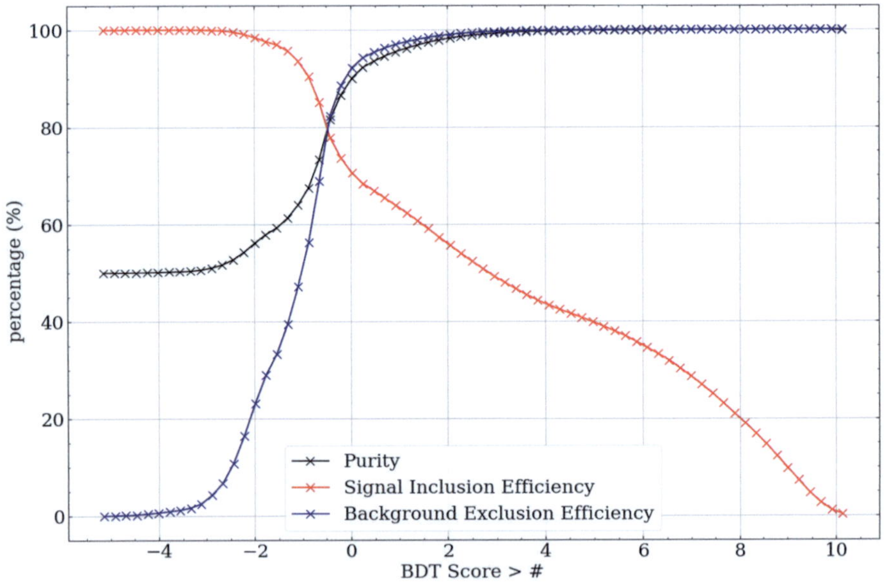

(b) Science Run 2

Fig. 8.28 The efficiency and purity of the signal and background BDT score distributions when iterating through a range of thresholds

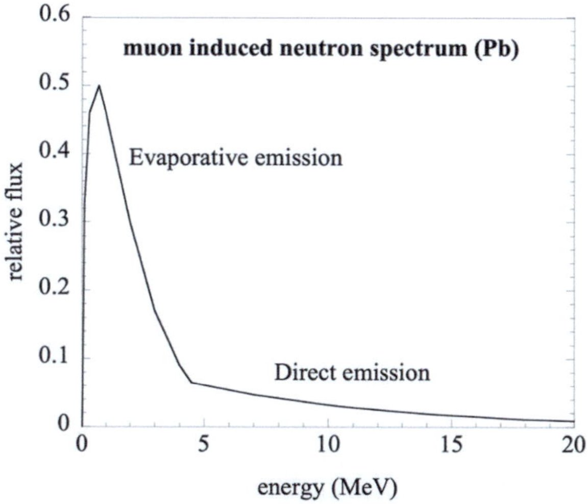

Fig. 8.29 Muon-induced neutron energy spectrum emitted from Pb with 20 m.w.e. [14]

these muons passing through the lead will produce secondary neutrons that propagate outwards to the active region of the neutron multiplicity meter for observation.

- Muon candidate: total Energy must exceed 10 MeV
- Muon candidate: detector channels hit must be present on at least two opposing sides of the detector.
- Neutron candidate: the timing must be within the same 15.360 microsecond acquisition window, and only the next nearest event is considered.
- Neutron candidate: the neutron event must have a total energy greater than 3 MeV to avoid excessive contamination by random radioactivity.

The muon-induced neutrons are passed through the GADGET simulation package to better model the neutron capture γ-rays emitted from the gadolinium nuclei. For this simulation the neutrons are started in the lead inner shielding with an energy spectrum modeled with the spectrum shown in Fig. 8.30. Figures 8.31 and 8.32 demonstrate the output of a GADGET simulation post-processing.

After identifying a population of secondary neutron candidates, a very similar XGBoost classifier model is trained on the simulated cosmic-induced neutrons for the signal class and background class events harvested from calibrated production run 2 nuThor data.

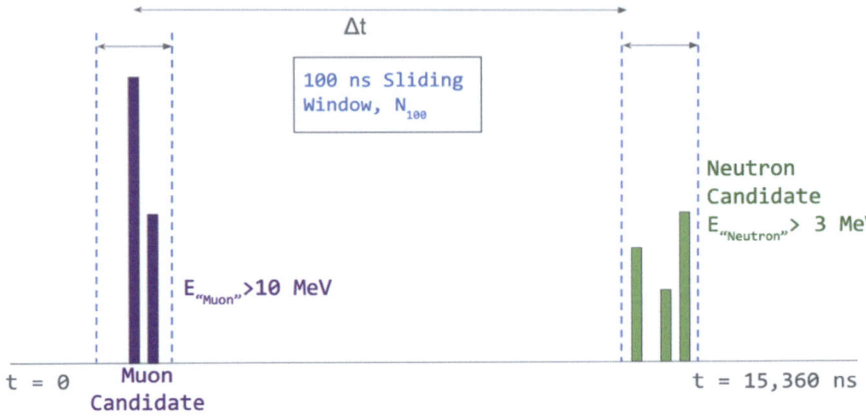

Fig. 8.30 Diagram of the Cosmic muon-induced neutron search criteria in NuThor. The muon candidate event must have a total energy in its N_{100} cluster greater than 10 MeV. The neutron candidate must be the event immediately following the muon candidate and have a total energy greater than 3 MeV to largely avoid background from radioactivity

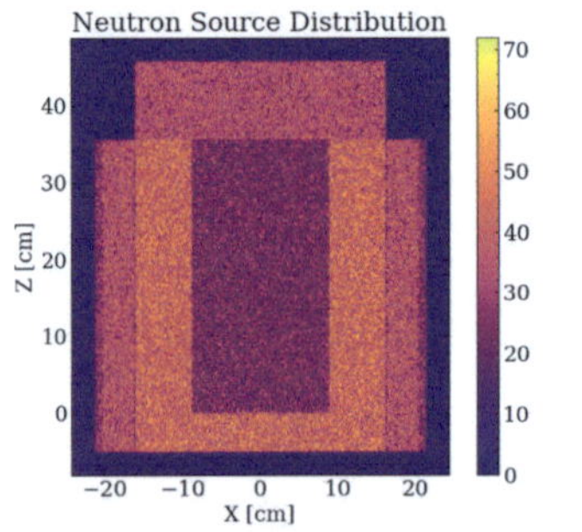

(a) muon-induced Neutron Sim neutron starting position in lead x-z

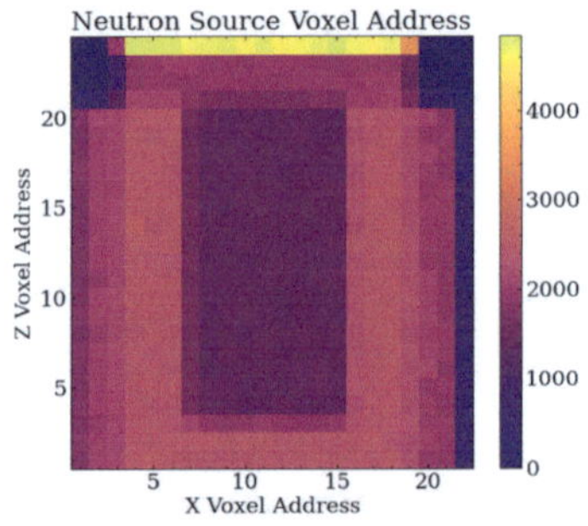

(b) muon-induced Neutron Sim neutron starting position voxel x-z

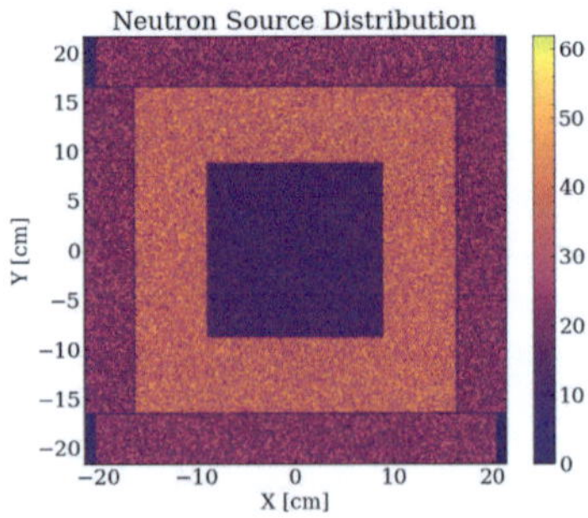

(c) muon-induced Neutron Sim neutron starting point in lead in x-y

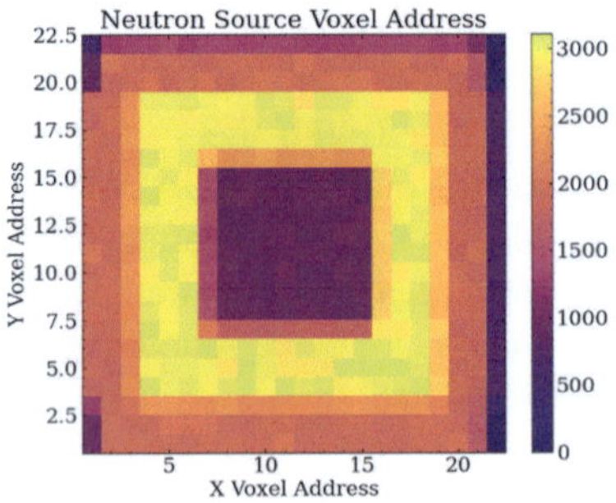

(d) muon-induced Neutron Sim neutron starting point in lead vox address

Fig. 8.31 GADGET muon-induced neutron simulation indicating the starting point of the neutrons inside the lead inner shielding

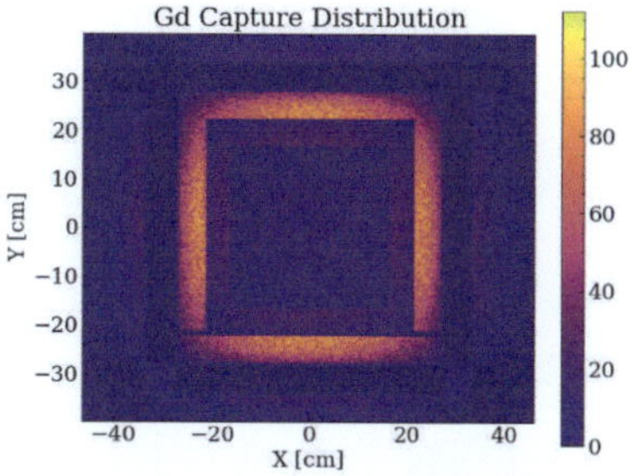

(a) muon-induced Neutron Sim Gd capture location in x-y

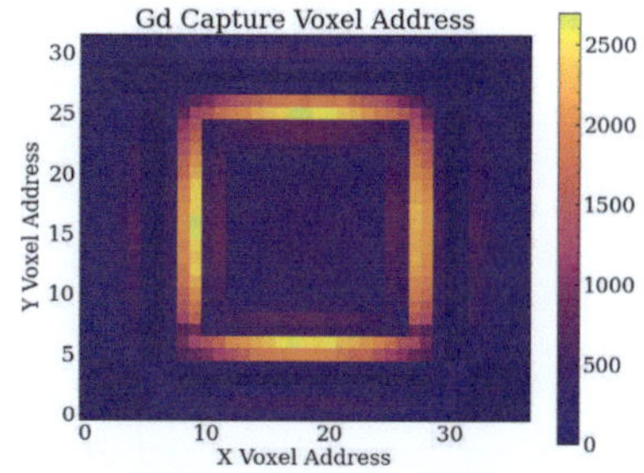

(b) muon-induced Neutron Sim Gd capture location voxel

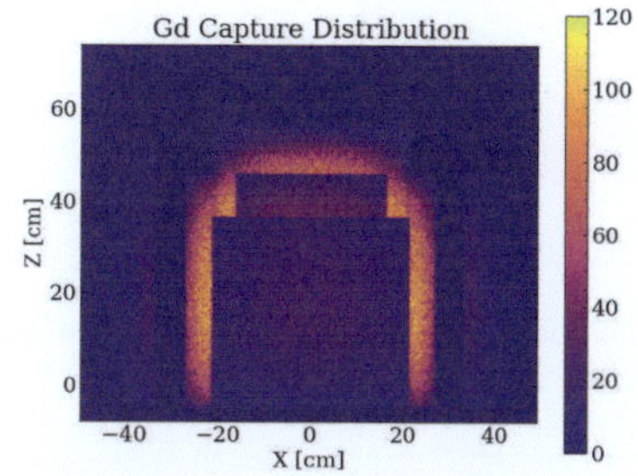

(c) muon-induced Neutron Sim Gd capture location in x-z

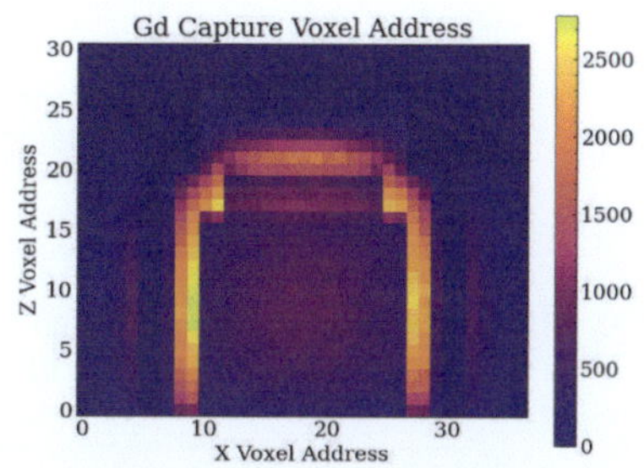

(d) muon-induced Neutron Sim Gd Capture voxel in x-z

Fig. 8.32 GADGET muon-induced neutron simulation indicating the location of the gadolinium neutron captures

The BDT model was applied to the secondary neutron candidate dataset and was used to predict whether each individual event was a neutron or background candidate. The total predicted neutrons and background candidates were then used to scale the normalized distributions of the model feature parameters. The sum of the two scaled distributions was then shown along with the secondary neutron candidate data points for each respective feature. It is important to note that this is not a fit of the collections of data points in the secondary neutron candidate dataset. It is successive evaluations of the model on each individual data point without paying heed to the rest of the dataset. Therefore, agreement between the data points and the scaled feature distributions should instill a degree of confidence in the classification's performance (Figs. 8.33, 8.34, 8.35, 8.36, 8.37, 8.38, 8.39, 8.40, 8.41, 8.42, 8.43, 8.44, 8.45, 8.46, 8.47, 8.48 and 8.49).

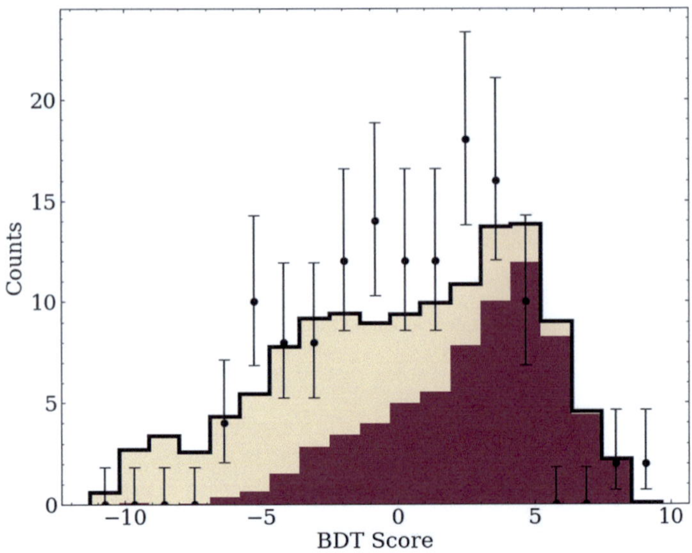

Fig. 8.33 BDT score distributions for signal neutrons and background scaled by the outcome of applying the trained classification model to the cosmic neutron candidate dataset

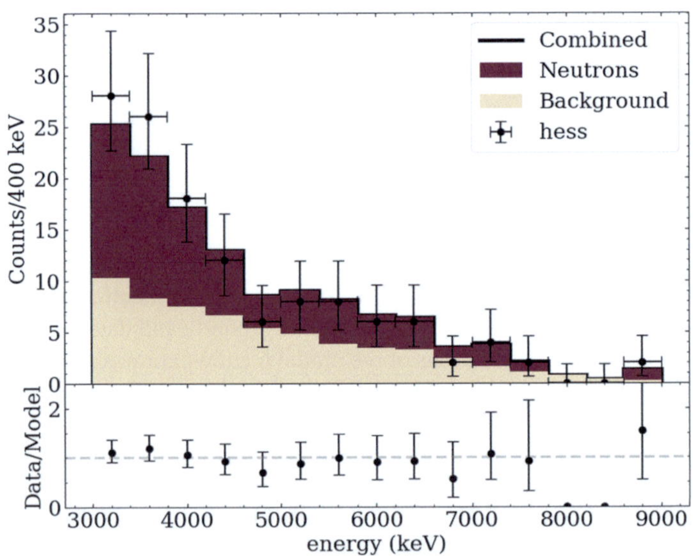

Fig. 8.34 Total energy distributions for cosmic neutron candidates classified by the trained BDT model compared to the data

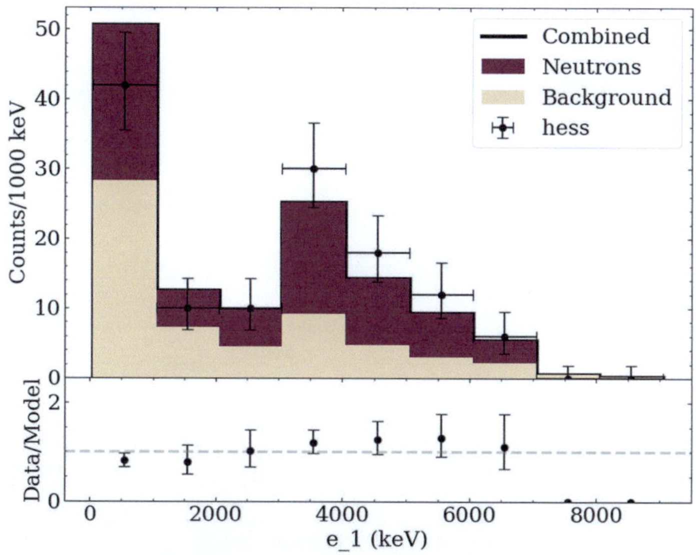

Fig. 8.35 Energy of the first hit in N_{100} cluster (E1) distributions for cosmic neutron candidates classified by the trained BDT model compared to the data

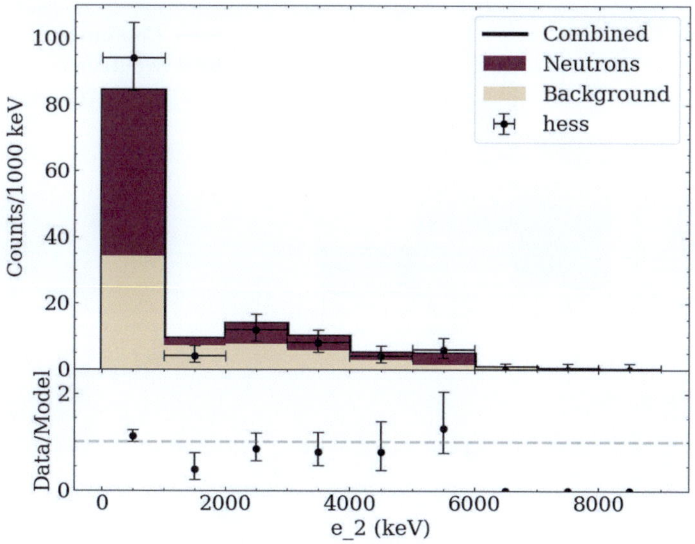

Fig. 8.36 Energy of the second hit in N_{100} cluster (E2) distributions for cosmic neutron candidates classified by the trained BDT model compared to the data

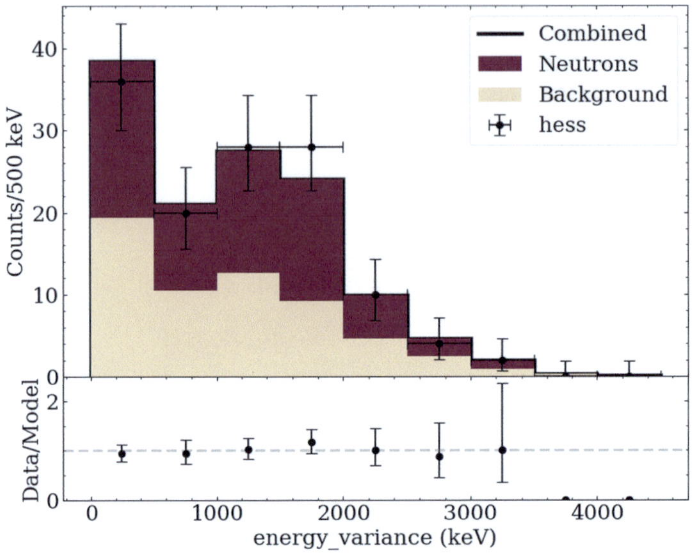

Fig. 8.37 Energy variance distributions for cosmic neutron candidates classified by the trained BDT model compared to the data

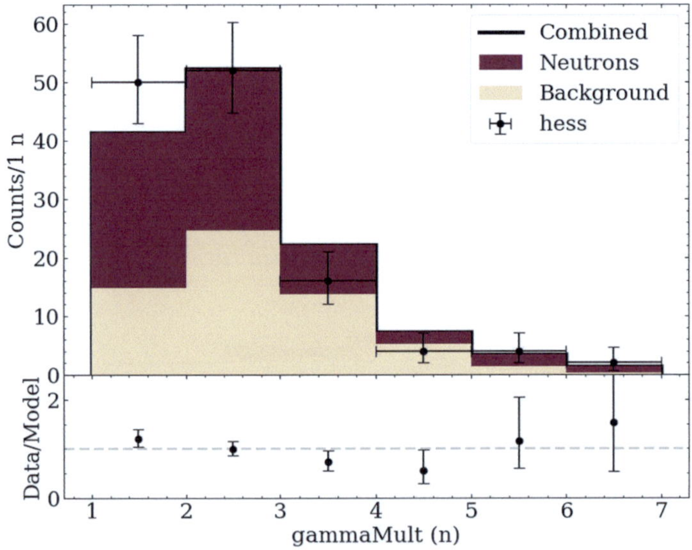

Fig. 8.38 Gamma multiplicity distributions for cosmic neutron candidates classified by the trained BDT model compared to the data

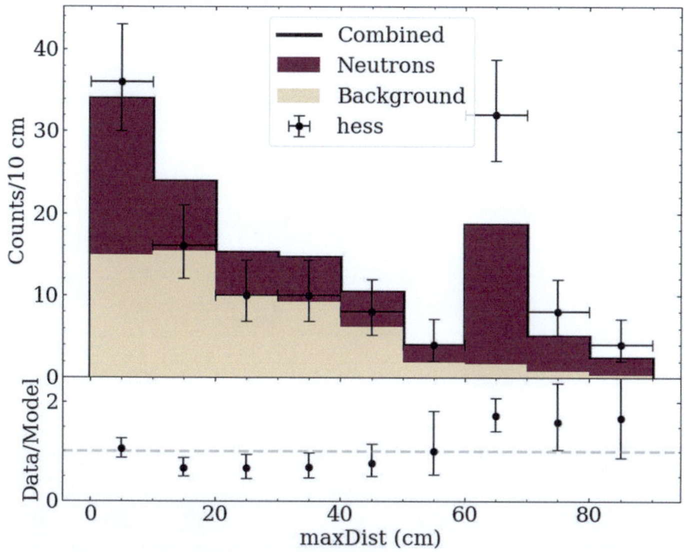

Fig. 8.39 Maximum distance distributions for cosmic neutron candidates classified by the trained BDT model compared to the data

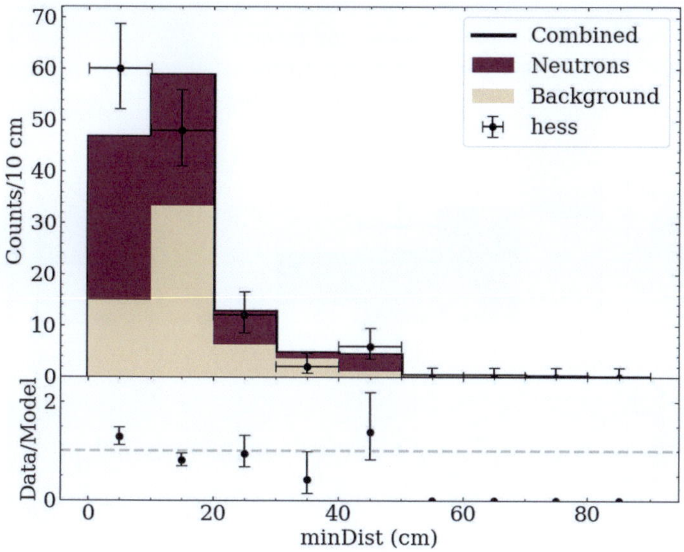

Fig. 8.40 Minimum distance distributions for cosmic neutron candidates classified by the trained BDT model compared to the data

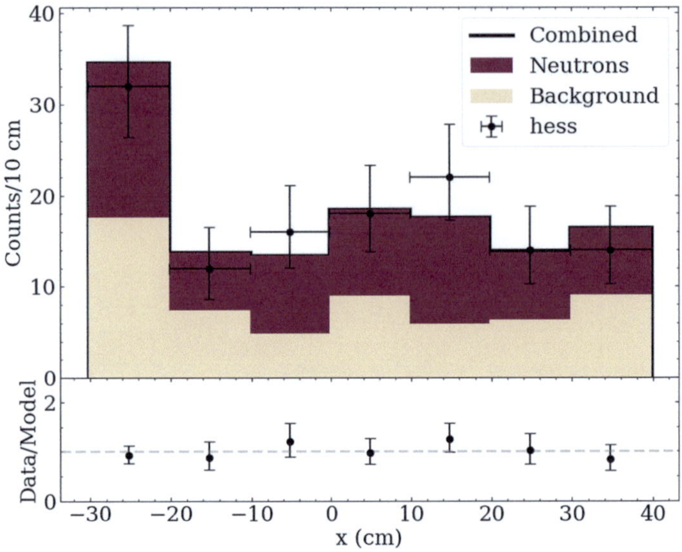

Fig. 8.41 Average X distributions for cosmic neutron candidates classified by the trained BDT model compared to the data

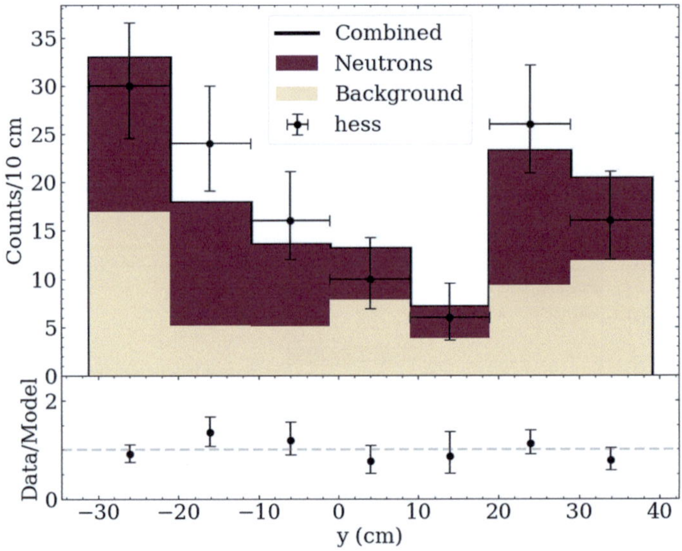

Fig. 8.42 Average Y distributions for cosmic neutron candidates classified by the trained BDT model compared to the data

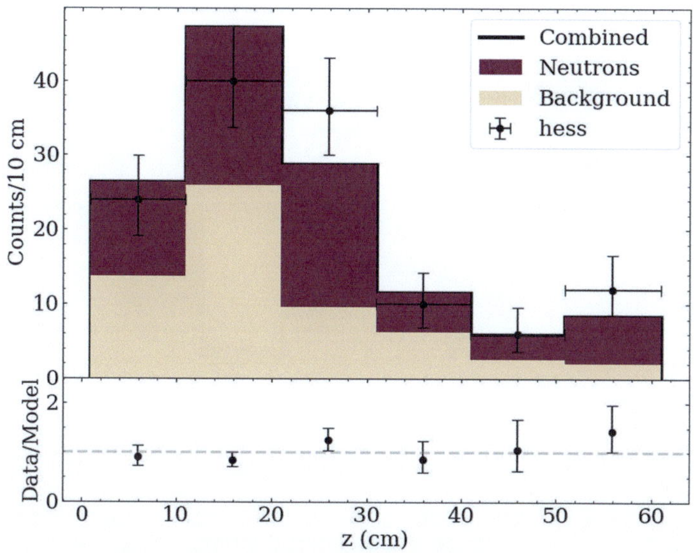

Fig. 8.43 Average Z distributions for cosmic neutron candidates classified by the trained BDT model compared to the data

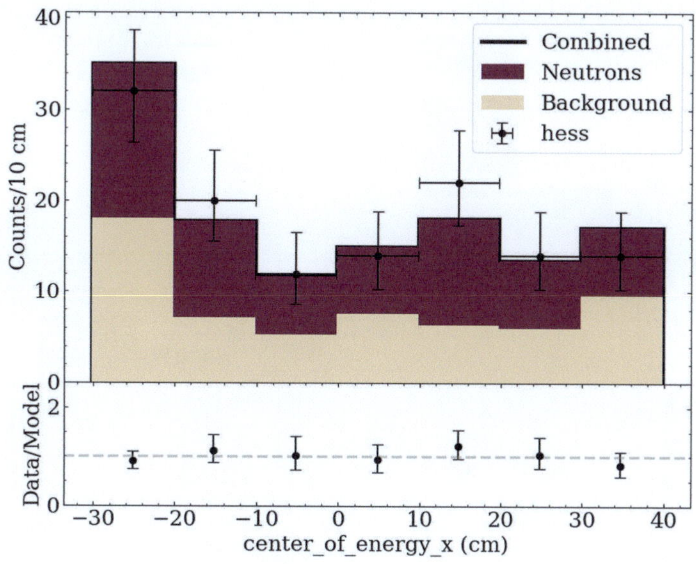

Fig. 8.44 Center of Energy in the X-axis distributions for cosmic neutron candidates classified by the trained BDT model compared to the data

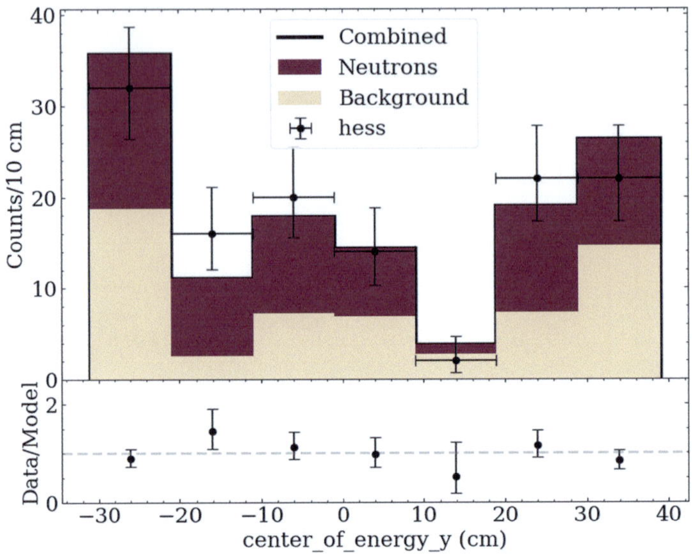

Fig. 8.45 Center of Energy in the Y-axis distributions for cosmic neutron candidates classified by the trained BDT model compared to the data

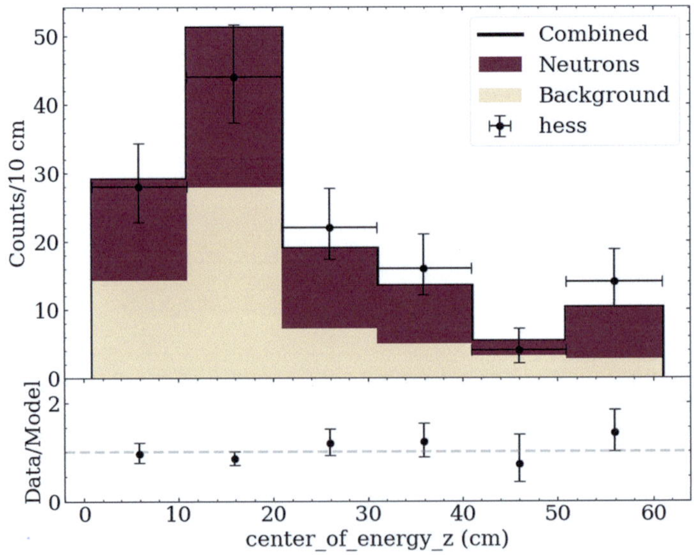

Fig. 8.46 Center of Energy in the Z-axis distributions for cosmic neutron candidates classified by the trained BDT model compared to the data

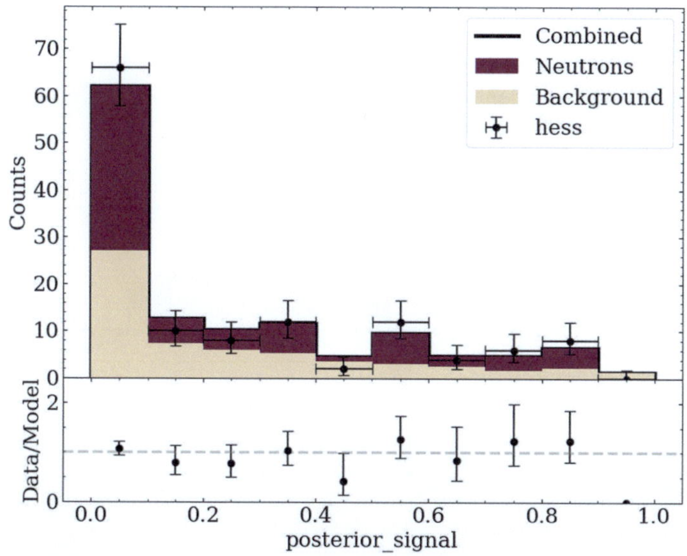

Fig. 8.47 Posterior signal distributions for cosmic neutron candidates classified by the trained BDT model compared to the data

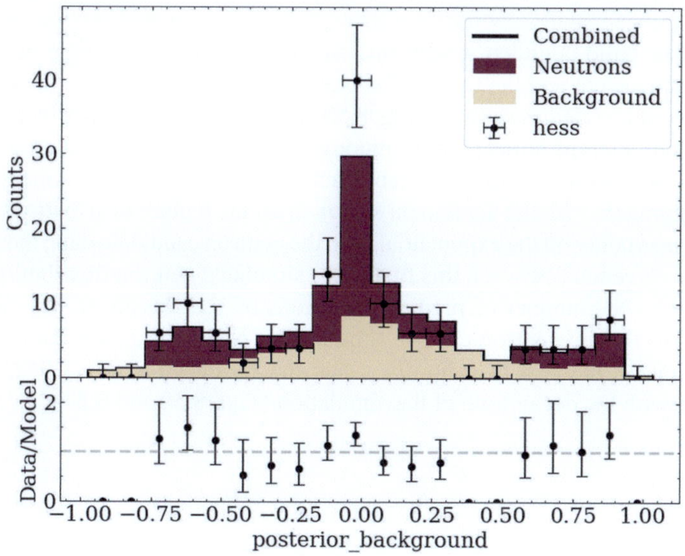

Fig. 8.48 Posterior background distributions for cosmic neutron candidates classified by the trained BDT model compared to the data

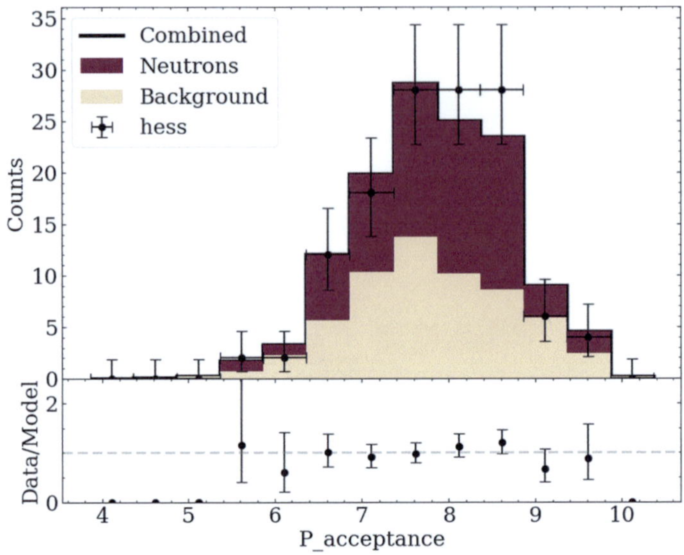

Fig. 8.49 $P_{Acceptance}$ distributions for cosmic neutron candidates classified by the trained BDT model compared to the data

To further build confidence with this classifier approach, the time between the putative muon event and the time of the predicted neutron event was fit with a simple exponential model, as shown in Fig. 8.50. The timing is only examined starting 3 microseconds after the initial putative muon event to focus on the characteristic exponential decay. This suffers from a notable lack of statistical power, but the results are taken in aggregate with the agreement shown in all the feature distributions. Despite the large uncertainty on the exponential fit to the neutron candidate data, there is clear qualitative agreement between this fit and the simulated neutron distribution (which was scaled by the number of predicted neutrons by the classifier). The maximum likelihood fit for the neutron capture time for the NuThor Science Run 2 detector geometry places the best fit value for τ and the 1σ uncertainty at 2962^{+994}_{-653} ns in agreement with the decay time of the simulation (Figs. 8.50 and 8.51).

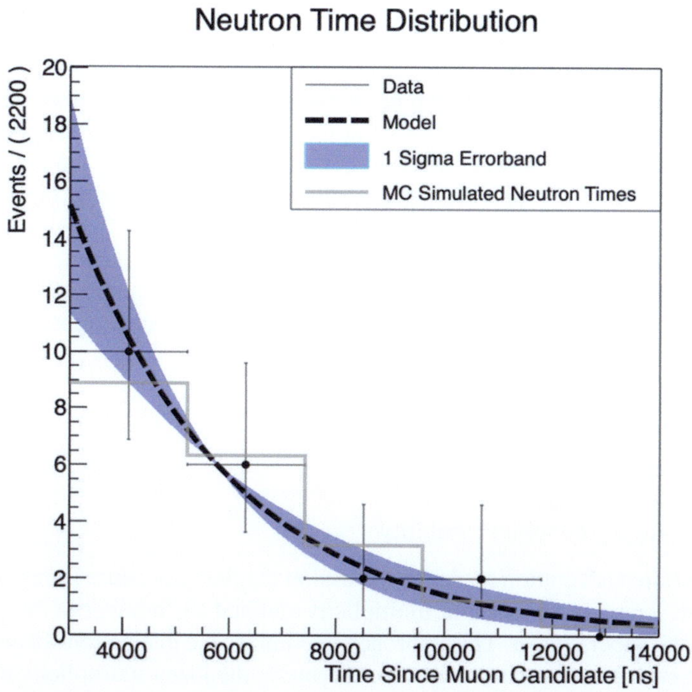

Fig. 8.50 The time between muon event and neutron candidate. An exponential model was fit to sparse data points. The result of the neutron simulation is overlaid and scaled by the number of predicted neutrons in the control sample. The fit is limited to the data beginning at 3 microseconds to view the exponential decay characteristic to the neutron capture time

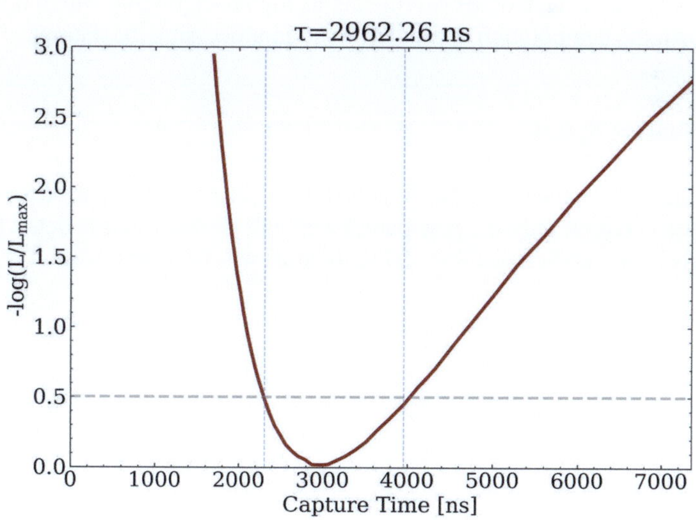

Fig. 8.51 The profile likelihood curve of the maximum likelihood fit performed on the muon-induced neutron selections data points in Fig. 8.50. The best fit τ and 1σ uncertainty band is 2962^{+994}_{-653} ns

8.2.5 *Fission Tagging*

The fission super clusters must then be evaluated for how "fission-like" in order to apply a fission tag. The approach is very similar to the process utilized in the neutron tagging approach taken in the previous section, but enhanced by the output of the neutron tags within each super cluster. Like the neutron tag, the fission tag is powered by a Boosted Decision Tree classification model.

The features input into the fission tag classifier are nearly identical to those in the neutron tag, but with a few additional parameters. The additional parameters are based on the collection of neutron-tag scores inside the super-cluster and are as follows:

1. Minimum neutron tag score
2. Maximum neutron tag score
3. Sum of neutron tag scores
4. Neutron tag score variation
5. Neutron Tag Likelihood Probability

An additional adaptation made for the fission tagging procedure is that the super-clusters are separated by neutron multiplicity–defined the number of N_{100} clusters present in the super-cluster. This approach maximizes the distinctions between populations at each neutron multiplicity. Furthermore, the lower multiplicity events are far more numerous than the higher ones, so training a classification model on all multiplicities could introduce bias toward the more common, lower multiplicity events. The neutron multiplicities considered range from 2 to 7, with multiplicities of 7 or greater grouped together as overflow. The $N = 1$ bin does not proceed further into the analysis due to its lack of sensitivity and its high background, which only serves to slow down the computation speeds involved in processing the dataset.

8.2.5.1 Feature Space

The input features for the fission tagging model are quite similar to the ones employed for the neutron tag, so only the new parameters will be described in detail here. For space's sake, only the Science Run 2 distributions will be shown here.

Maximum Neutron Tag Score

See Fig. 8.52.

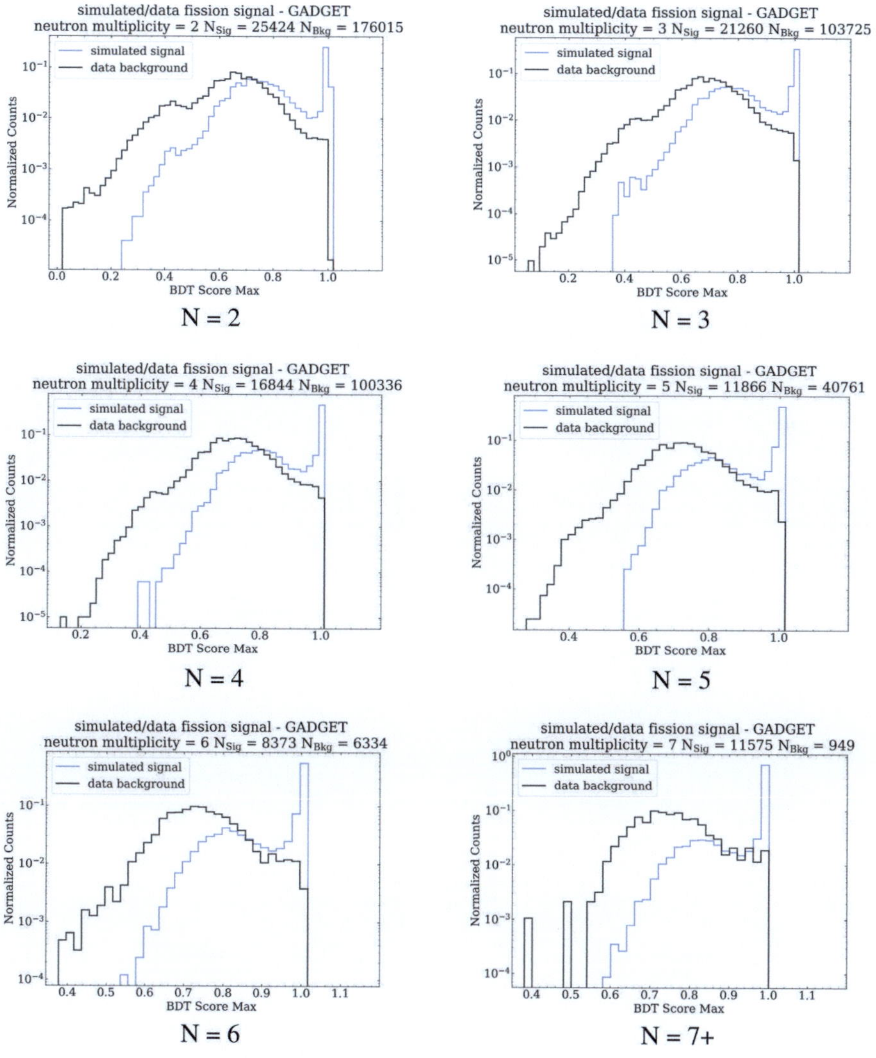

Fig. 8.52 Maximum neutron tag in super-cluster by neutron multiplicity

Minimum Neutron Tag Score

See Fig. 8.53.

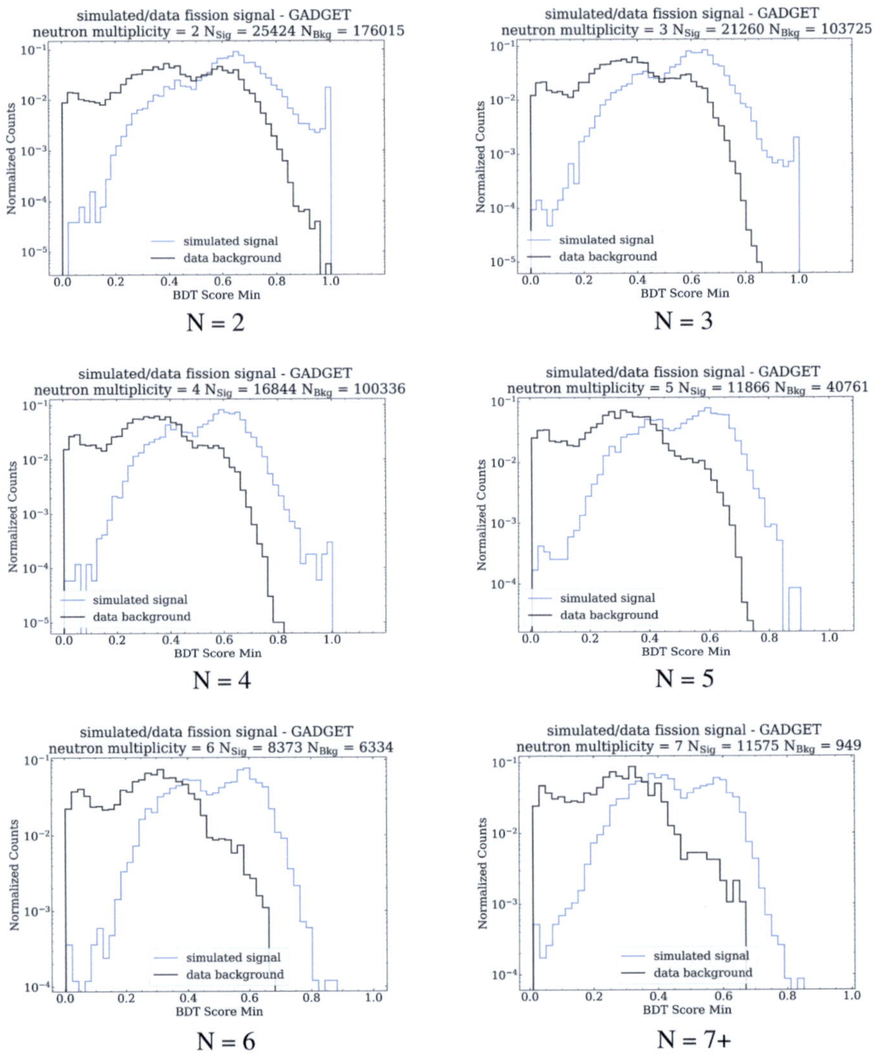

Fig. 8.53 Minimum neutron tag in super-cluster by neutron multiplicity

Sum of Neutron Tag Scores

See Fig. 8.54.

$$\sum_{i=0}^{N} \mathrm{N}_{\mathrm{tag}} \tag{8.22}$$

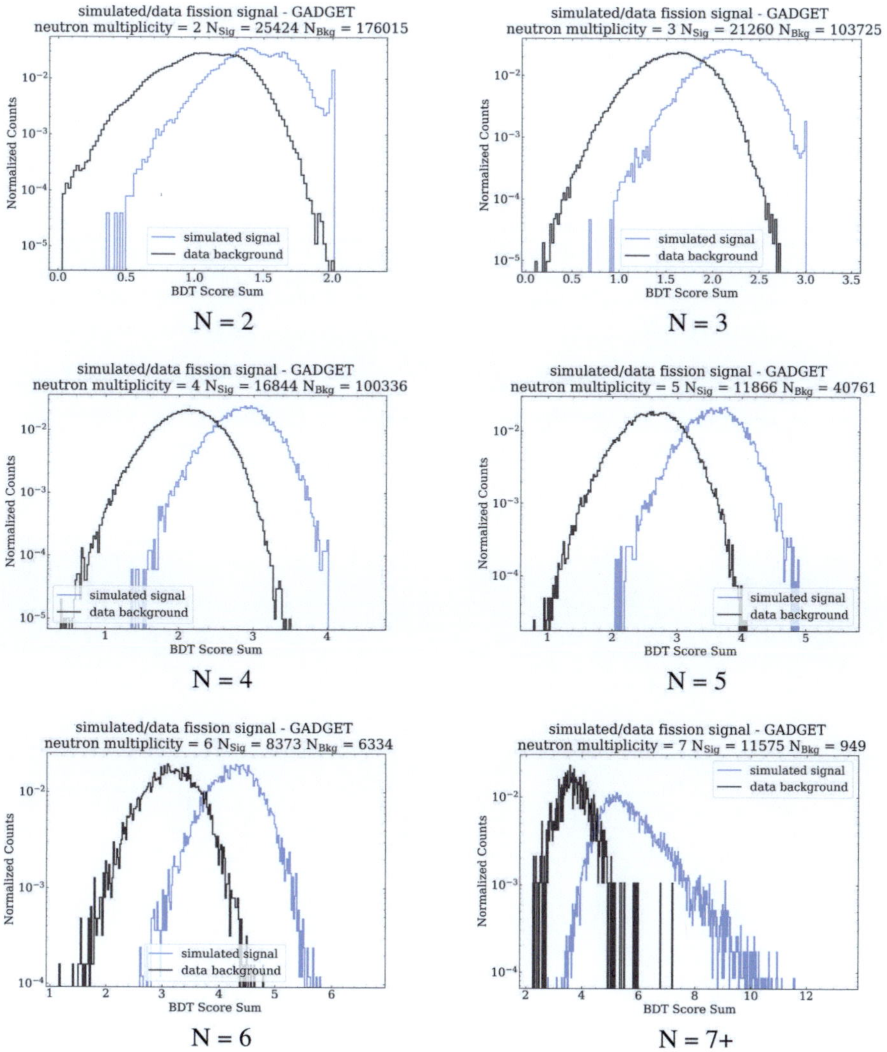

Fig. 8.54 Minimum neutron tag in super-cluster by neutron multiplicity

Variance of Neutron Tag Scores

See Fig. 8.55.

$$\frac{\sqrt{N_{tag} - \bar{N}_{tag}}}{N_{super}} \tag{8.23}$$

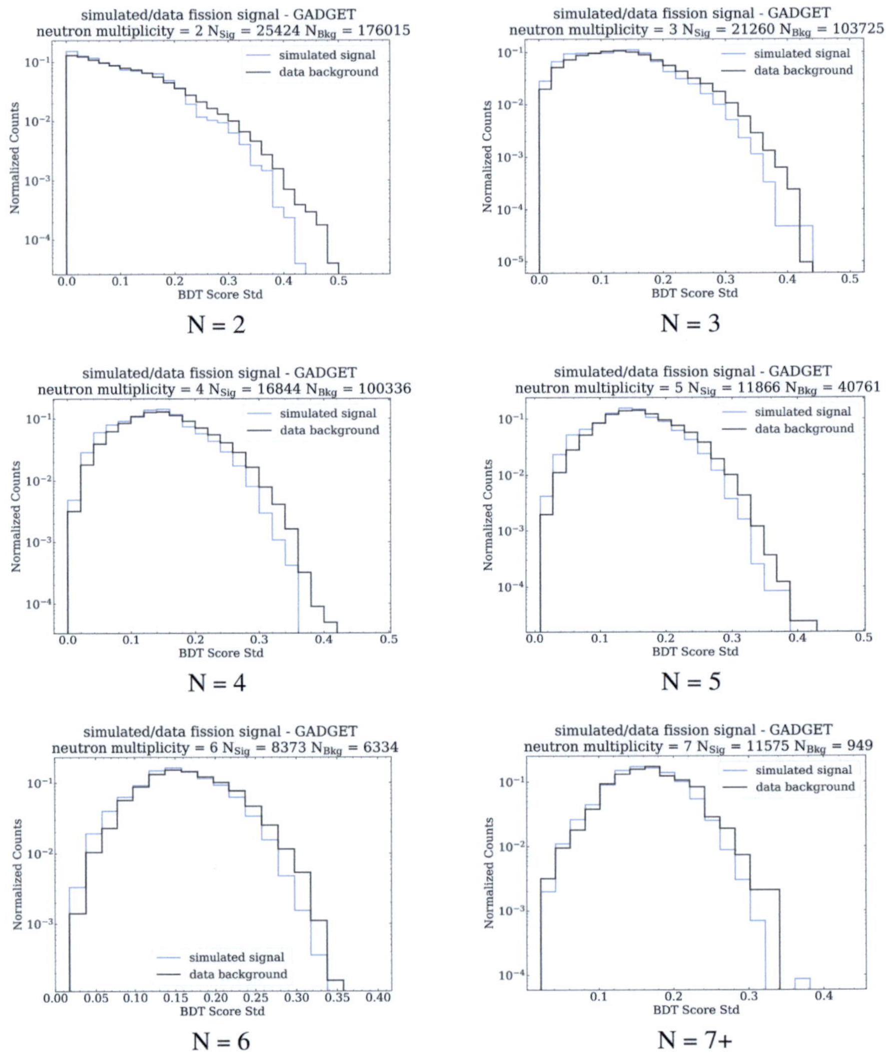

Fig. 8.55 Neutron tag variance in super-cluster by neutron multiplicity

Neutron Tag Likelihood Probability

The neutron tag likelihood probability is calculated by first forming a probability density function (PDF) of neutron tag BDT scores (See Fig. 8.56).

The likelihood is then found with the formula (Fig. 8.57)

$$L_{ntag} = \frac{\log\left(\prod_{i=1}^{N_{\text{super}}} P_i\right)}{N_{\text{super}}} \tag{8.24}$$

8.2.5.2 Over-Training Checks

The same procedure outlined for the neutron tag over-training was used to check the fission tag. For details on this procedure, see Sect. 8.2.2.3. The AUC over-training checks for the number and max depth of the trees is shown in Fig. 8.58.

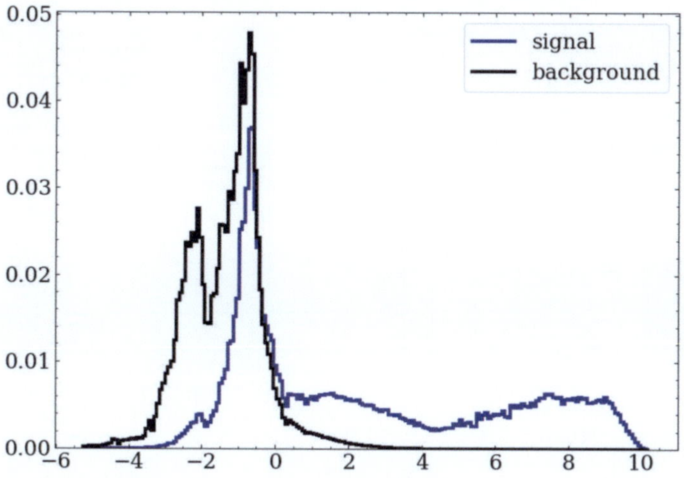

Fig. 8.56 Neutron tag distributions for signal and background

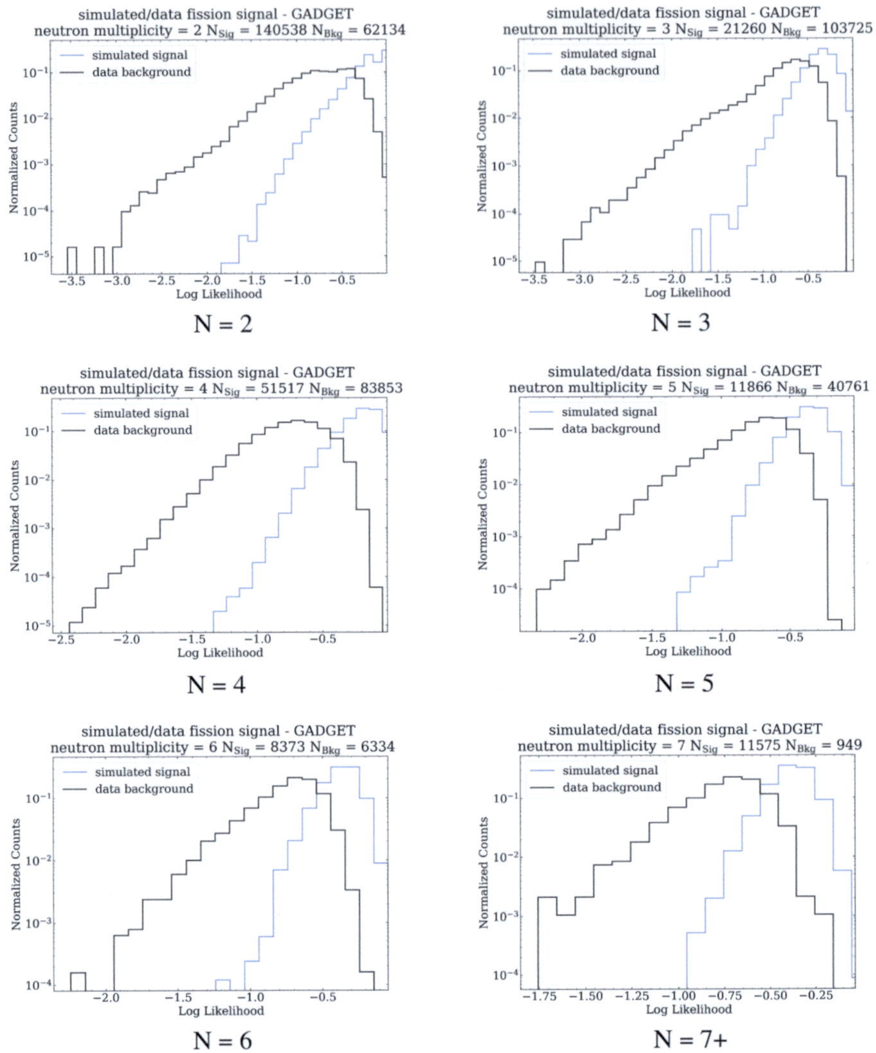

Fig. 8.57 Neutron tag likelihood probability in super-cluster by neutron multiplicity

8.2.6 *Fission Model Performance*

One of the metrics used to finalize the fission tag BDT models is the separation of the signal and background BDT scores. Where separation is quantified with the expression (Fig. 8.60):

$$\langle S^2 \rangle \frac{1}{2} \int \frac{(\hat{y}_S(y) - \hat{y}_B(y))^2}{\hat{y}_S(y) + \hat{y}_B(y)} dy \tag{8.25}$$

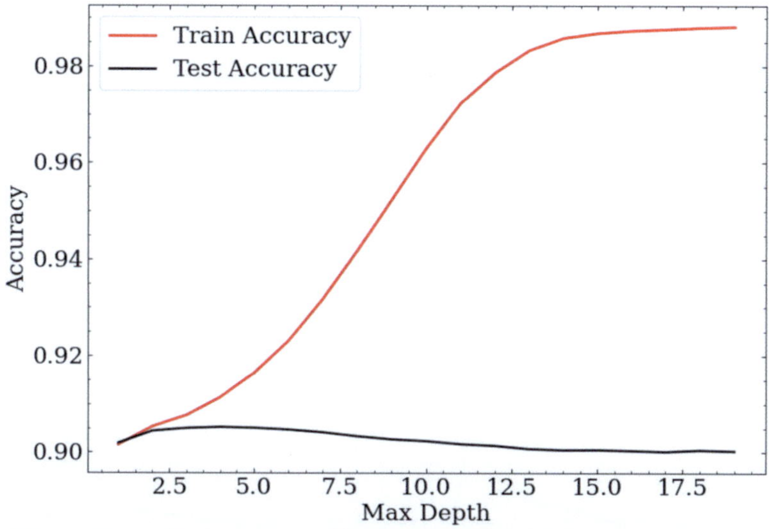

(a) After a max. depth of 5, there is both a divergence in the performances of the testing and training datasets and a reduction in the effectiveness in the evaluation of the testing dataset.

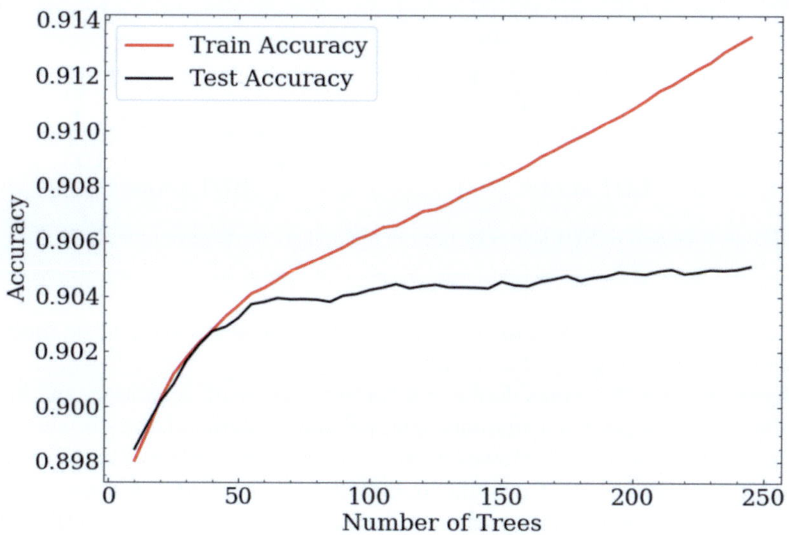

(b) There is a divergence after 100 trees where the test dataset's accuracy score no longer increases.

Fig. 8.58 Over-training checks for the fission-tag BDT model

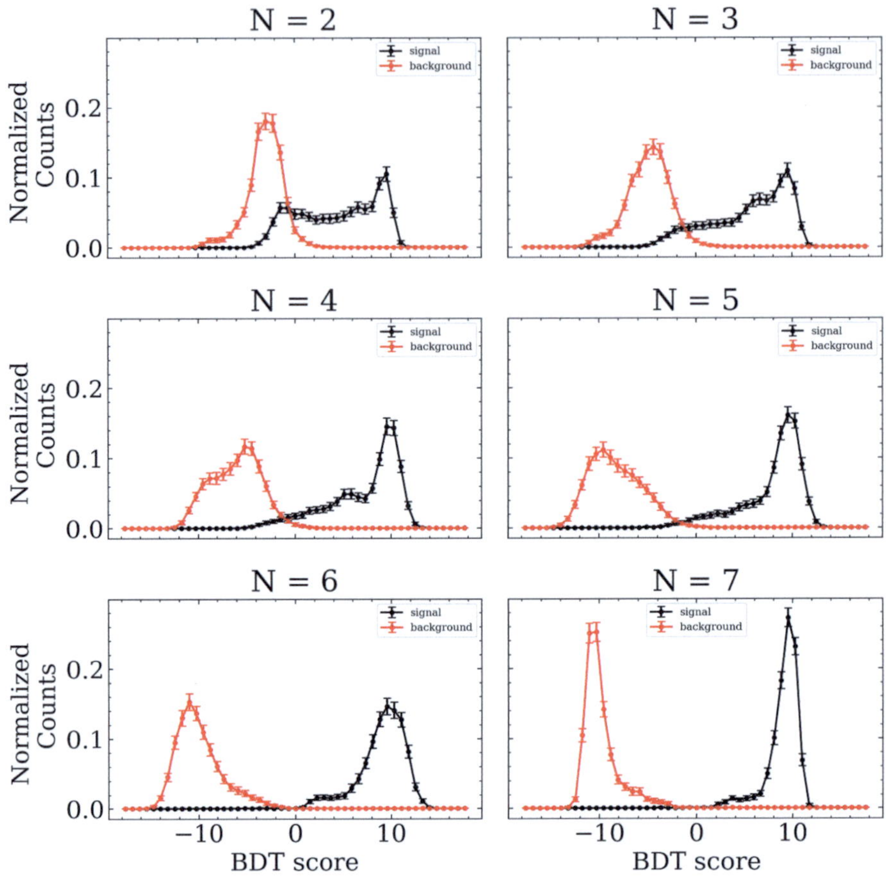

Fig. 8.59 The distributions of BDT scores by neutron multiplicity for Science Run 2

For Science Run 2, the separation continues to increase at greater neutron multiplicities as indicated in Figs. 8.59 and 8.60.

The shapes of the BDT scores demonstrate a high degree of separation, and in a scenario where there were equal amounts of signal and background, the story would end here quite straightforwardly. However, NuThor has many-fold more background than signal, thus when the BDT classification model was applied to the entirety of the beam anti-coincident dataset, a greater amount of overlap of these two BDT score curves—rarely events look more fission-like.

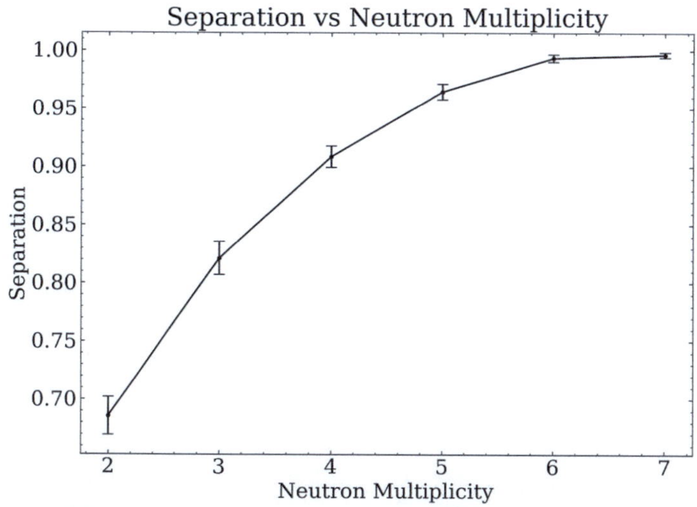

Fig. 8.60 The separation of the signal and background BDT score populations for each neutron multiplicity for Science Run 2

The status of the Science Run 1 BDT model is presented as well (Figs. 8.61 and 8.62).

The separation between signal and background BDT scores for Science Run 1 shares the trend of increasing rapidly with neutron multiplicity. However, the upswing is notably more gradual compared to Science Run 2. This is believed to arise from the increased background rate in Science Run 1 leading to higher random pile up of multiple thorium-originating γ-rays–producing more varietals of accidentally similar looking events.

The above BDT classification models were then applied to the entirety of the beam anti-coincident dataset for Science Run 2. The background BDT scores share key characteristics with the smaller-scale training/testing datasets, with the separation progressively improving (Fig. 8.63).

The fission-tag BDT score distributions for both the measured background and the simulated signal are shown below in Fig. 8.64. The signal distributions are scaled by the expected neutrino-induced nuclear fission events for that neutron multiplicity in Science Run 2. The background distribution is scaled to the expected exposure inside the beam coincident dataset for Science Run 2. This is tabulated by taking the entire Science Run 2 Golden Run List meta-information database and counting the number of Event 39s and number of Event 61s.

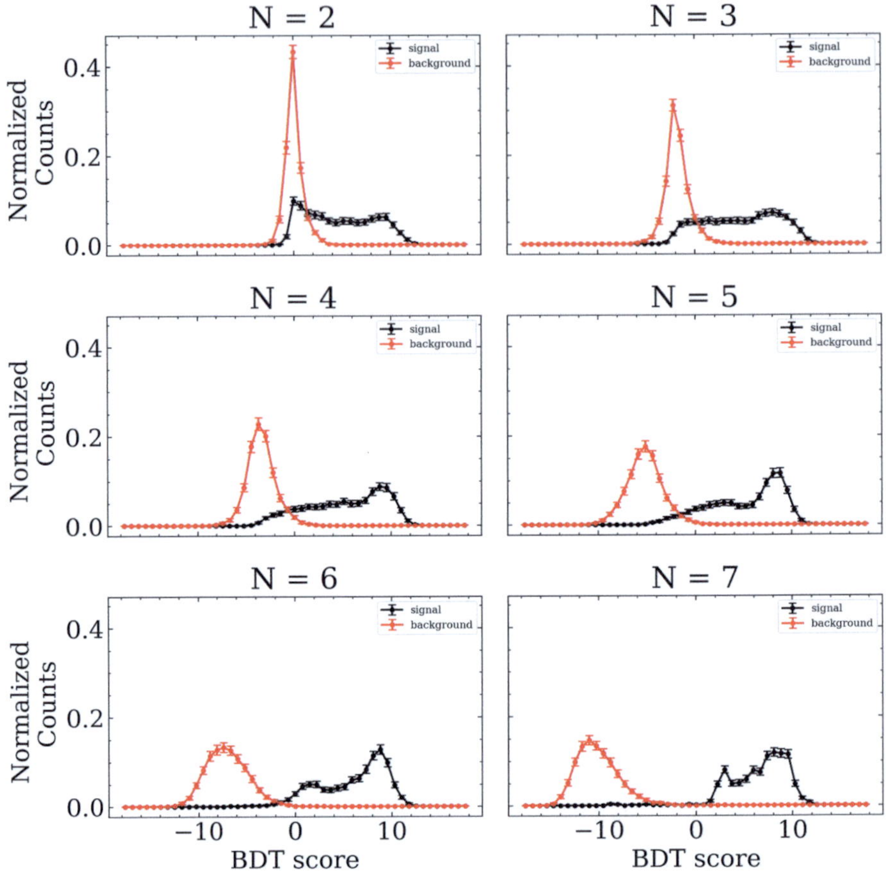

Fig. 8.61 The distributions of BDT scores by neutron multiplicity for Science Run 1

The background profiles encroached into the signal region, as expected. A closer examination of the overlapping BDT score region reveals a key insight: significant, exploitable differences still existed between the signal and background populations. The overlapping region of the neutron multiplicity = 3 populations is examined to better illustrate this behaviour. A sampling of fission-tag parameters for N = 3 in the overlapping BDT score region are displayed in Fig. 8.65. There are more model parameters this noticeably different shapes but for brevity's sake, only 4 are shown here.

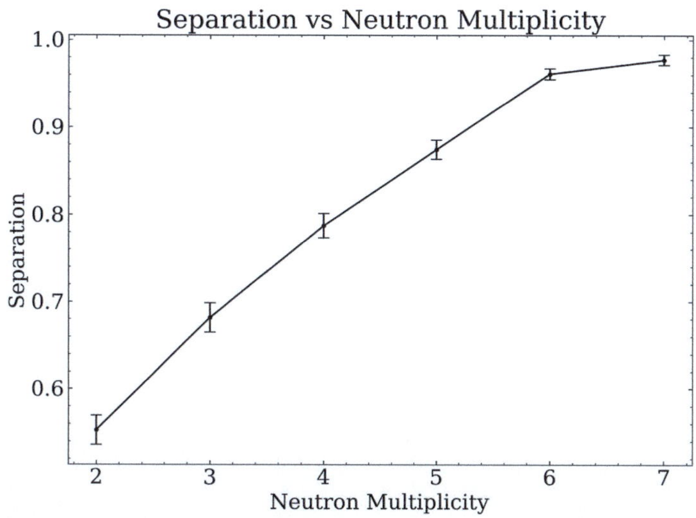

Fig. 8.62 The separation of the signal and background BDT score populations for each neutron multiplicity for Science Run 1

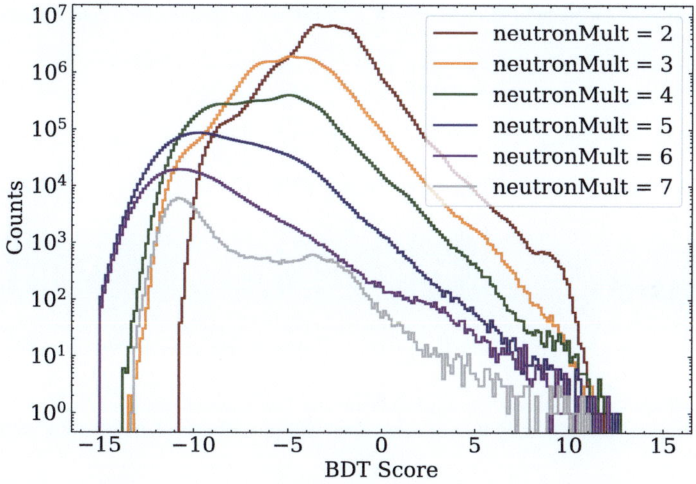

Fig. 8.63 The background BDT score distributions for each neutron multiplicity from the whole beam anti-coincident dataset

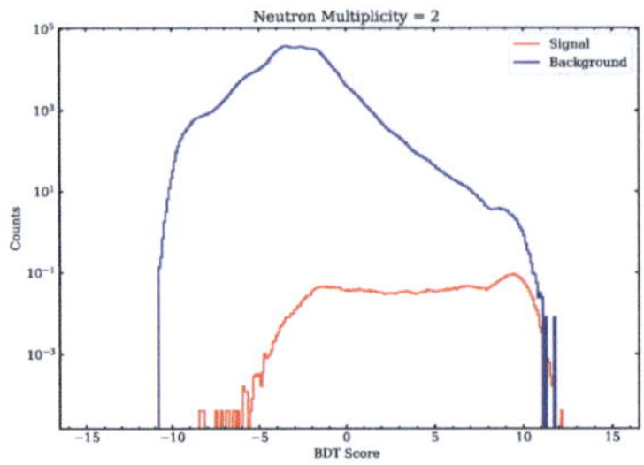

(a) Fission-tag score for Neutron Multiplicity = 2

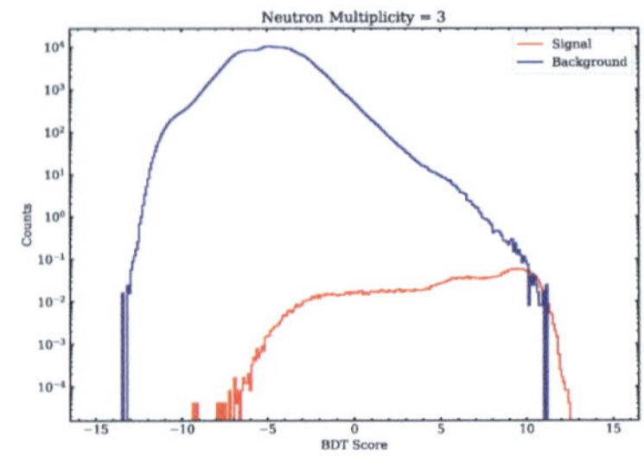

(b) Fission-tag score for Neutron Multiplicity = 3

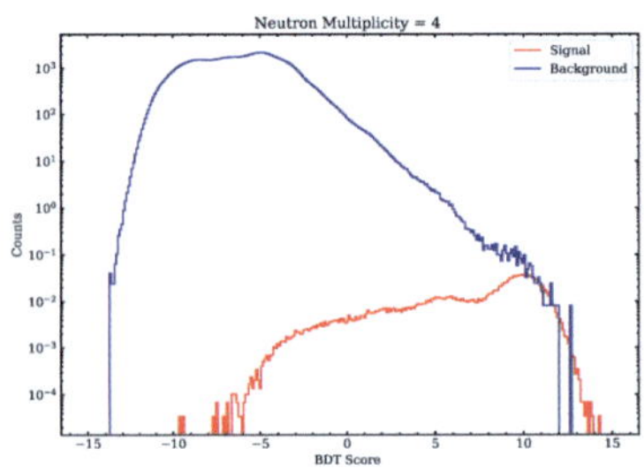

(c) Fission-tag score for Neutron Multiplicity = 4

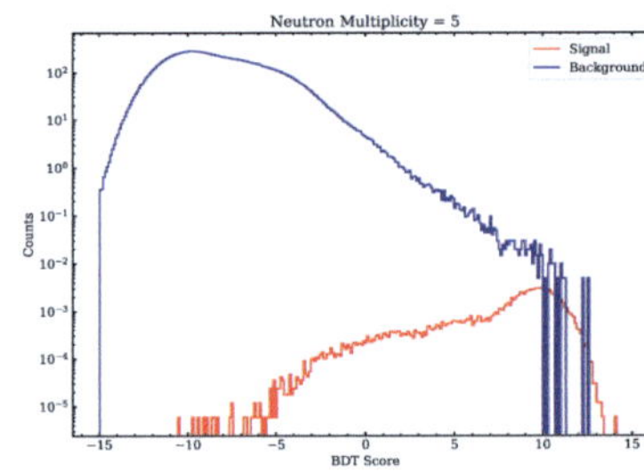

(d) Fission-tag score for Neutron Multiplicity = 5

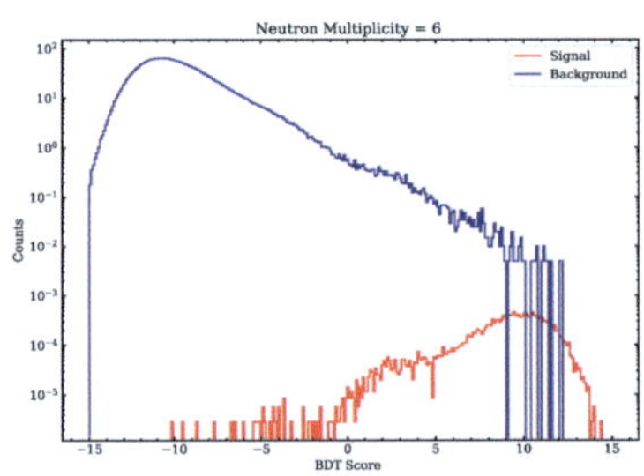

(e) Fission-tag score for Neutron Multiplicity = 6

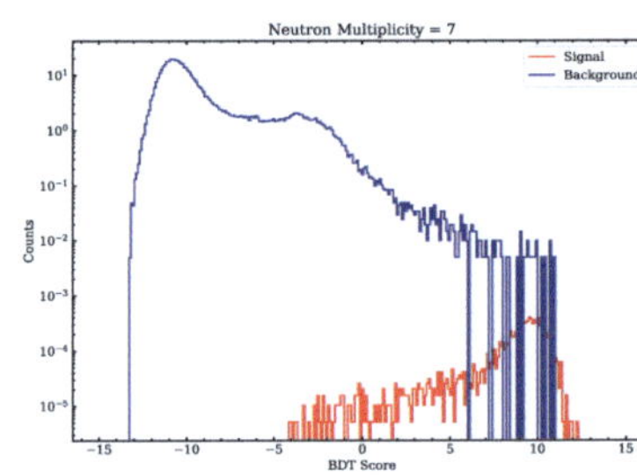

(f) Fission-tag score for Neutron Multiplicity = 7

Fig. 8.64 Fission-tag models applied the full beam anti-coincident dataset for NuThor Science Run 2. The distributions are scaled by the expected rates within the blinded, beam-coincident window

The fission-tag BDT effectively distinguished most labeled signal and background events, prioritizing the separation of the majority of data points into distinct categories. By applying a cut on the clearly differentiated events, the same BDT model was redeployed on the remaining post-cut events, without being hindered by known background events. This was the first and only BDT-related cut and it was designed as follows.

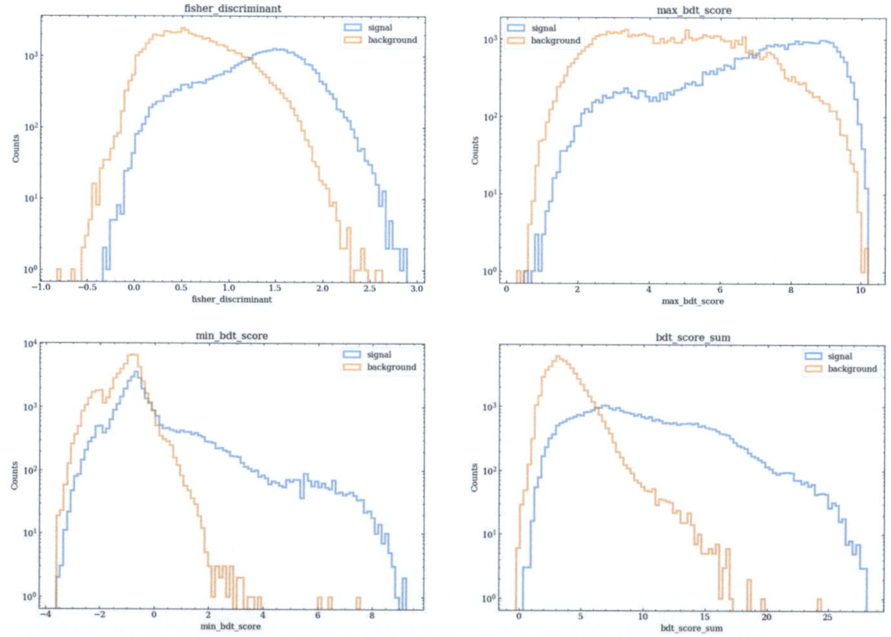

Fig. 8.65 Inspection of some of the BDT input features in the overlapping region of BDT Scores. Despite have overlapping BDT scores, there are plainly visible differences in the input parameters used to generate the classification score. This implied that the classifier can improve

8.2.7 Fission-Tag Score—Primary Cut

The topline priority for this cut was to preserve as much signal as possible while removing events the BDT classifier categorizes as obvious backgrounds. To start, BDT score thresholds accepting 95% of the signal BDT distributions were found (see Fig. 8.66). These initial BDT threshold values are adjusted to increase the sample size of background events in the training set. This allowed for sufficient training data and enough leftover events to construct a testing dataset (which is used as a PDF for subsequent fitting) (Fig. 8.67).

The fission-tag BDT score threshold values are shown on the background datasets in Fig. 8.68.

After applying these cuts to each of the BDT score distributions, the same BDT classifier model is fed 20% of the dataset for training. This final round of training was weighted by original BDT score value of each event. Weighting an event gives it greater importance to the model—a greater penalty is paid for misclassification. This practice is called "cascading".

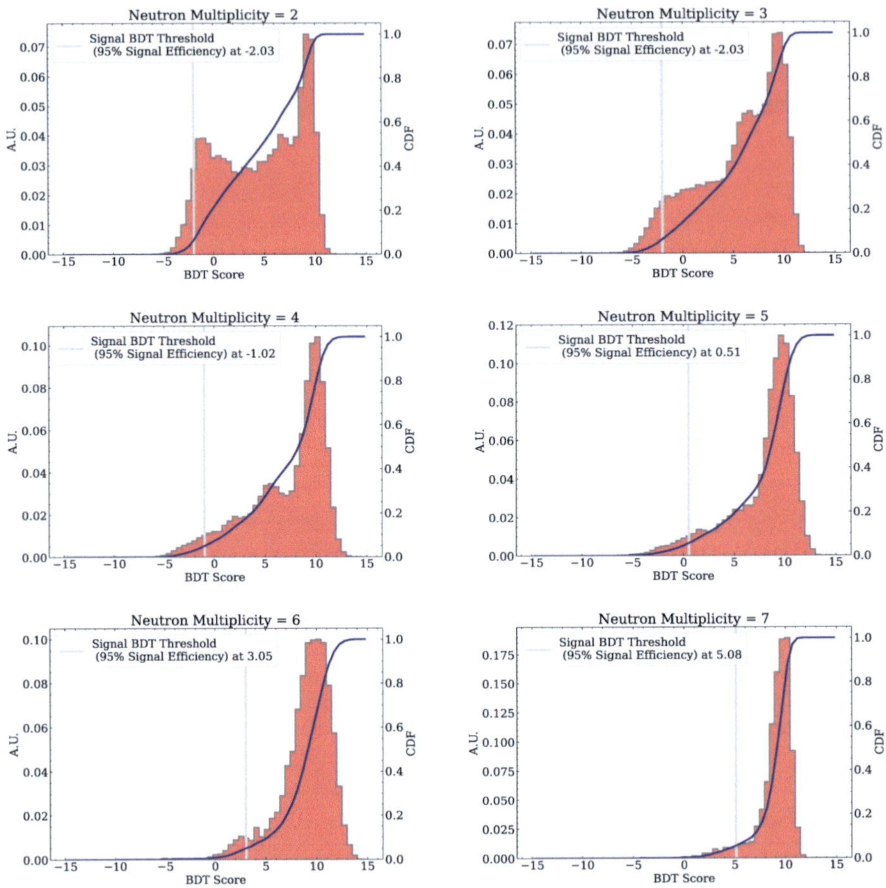

Fig. 8.66 Determining the thresholds in Fission-tag BDT Score for each neutron multiplicity to accept 95% of signal events

The results of this cascaded BDT classification score will be referred to as BDT Score* and each is depicted in Fig. 8.68a–f. The BDT Score Star signal values display an increased proportion of their total counts that are not overlapping with the background PDF. The distributions depicted here were then used in the fitting routine described in the proceeding chapter.

Following the two preselection cuts and the primary fission tag cut, the neutrino-induced nuclear fission signal acceptance at each neutron multiplicity was adapted (See Fig. 8.69).

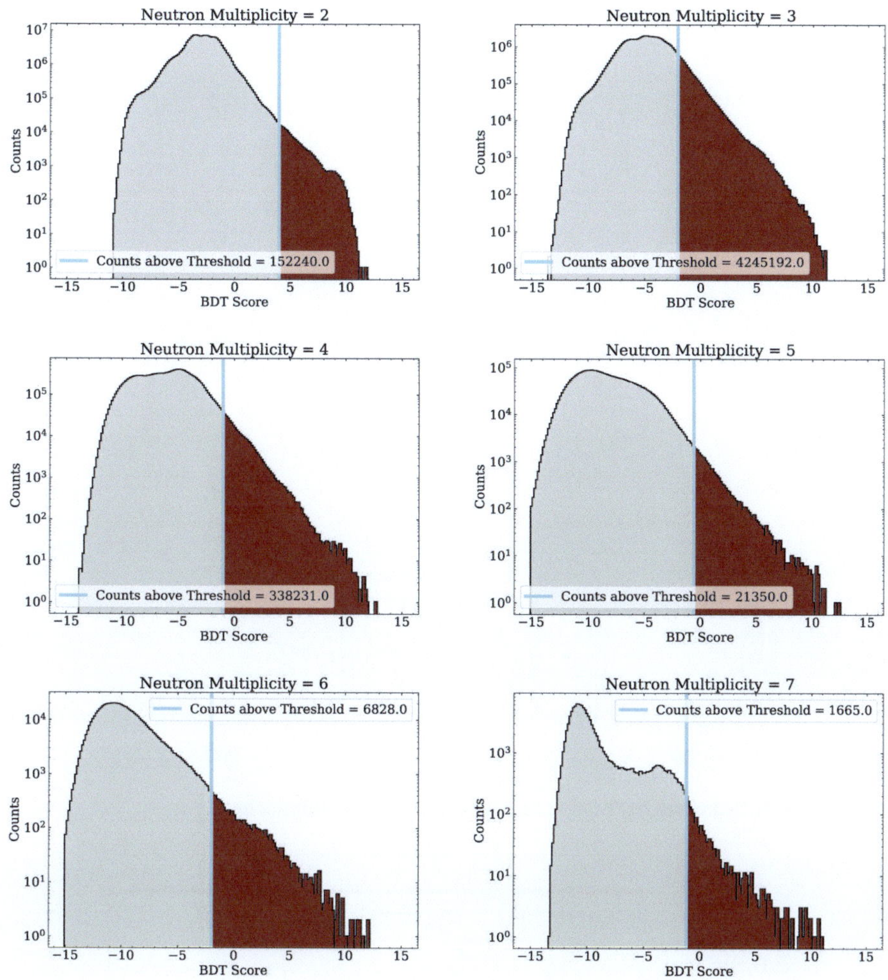

Fig. 8.67 The regions of BDT score space for each neutron multiplicity retained after the primary BDT score cut. These threshold in BDT score were selected to preserve a large fraction of the signal in each multiplicity bin

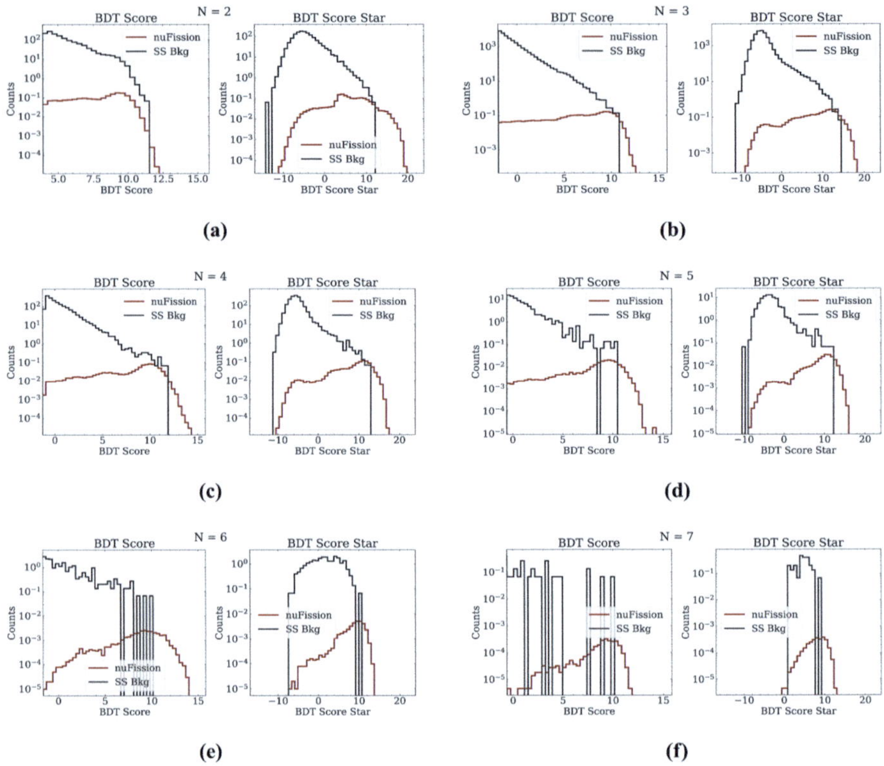

Fig. 8.68 BDT Score and BDT Score Star for each Neutron Multiplicity

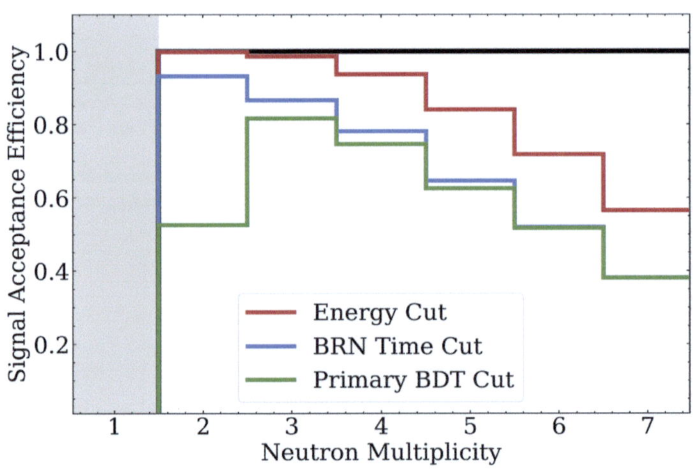

Fig. 8.69 Neutrino-induced nuclear fission Signal Acceptance after each of the preselection cuts and BDT Primary Cut

References

1. J. Therhaag, TMVA Toolkit for multivariate data analysis in ROOT, in *PoS ICHEP2010*, ed. by B. Pire et al., p. 510 (2010). https://doi.org/10.22323/1.120.0510
2. J.H. Friedman, Greedy function approximation: a gradient boosting machine. Ann. Stat. **29**(5), 1189–1232 (2001). ISSN: 00905364, 21688966. http://www.jstor.org/stable/2699986. Accessed 26 Apr 2025
3. W.Y. Loh, Classification and regression trees. WIREs Data Min. Knowl. Discov. **1**(1), 14–23 (2011). https://doi.org/10.1002/widm.8, https://wires.onlinelibrary.wiley.com/doi/abs/10.1002/widm.8
4. L. Breiman, Random forests. Mach. Learn. **45**(1), 5–32 (2001). ISSN: 1573-0565. https://doi.org/10.1023/A:1010933404324
5. F. Murtagh, Multilayer perceptrons for classification and regression. Neurocomputing **2**(5), 183–197 (1991). ISSN: 0925-2312. https://doi.org/10.1016/0925-2312(91)90023-5, https://www.sciencedirect.com/science/article/pii/0925231291900235
6. R.A. Fisher, The use of multiple measurements in taxonomic problems. Ann. Eugen. **7**(2), 179–188 (1936). https://doi.org/10.1111/j.1469-1809.1936.tb02137.x. eprint: https://onlinelibrary.wiley.com/doi/pdf/10.1111/j.1469-1809.1936.tb02137.x
7. S.D. Clarke et al., Verification and validation of the MCNPX-PoliMi code for simulations of neutron multiplicity counting systems. Nucl. Instrum. Methods Phys. Res. Sect. A: Accel. Spectrom. Detect. Assoc. Equip. **700**, 135–139 (2013). ISSN: 0168-9002. https://doi.org/10.1016/j.nima.2012.10.025. https://www.sciencedirect.com/science/article/pii/S0168900212011576
8. S. Agostinelli et al., Geant4—a simulation toolkit. Nucl. Instrum. Methods Phys. Res. Sect. A: Accel. Spectrom. Detect. Assoc. Equip. **506**(3), 250–303 (2003). ISSN: 0168-9002. https://doi.org/10.1016/S0168-9002(03)01368-8, https://www.sciencedirect.com/science/article/pii/S0168900203013688
9. P. An et al., Measurement of electron-neutrino charged-current cross sections on ^{127}I with the COHERENT NaIνE detector. Phys. Rev. Lett. **131**, 221801 (2023). https://doi.org/10.1103/PhysRevLett.131.221801
10. S. Gardiner, Simulating low-energy neutrino interactions with MARLEY. Comput. Phys. Commun. **269**, 108123 (2021). ISSN: 0010-4655. https://doi.org/10.1016/j.cpc.2021.108123. https://www.sciencedirect.com/science/article/pii/S0010465521002356
11. T. Chen, C. Guestrin, XGBoost: a scalable tree boosting system, in *Proceedings of the 22nd ACM SIGKDD International Conference on Knowledge Discovery and Data Mining (KDD '16)* (Association for Computing Machinery, San Francisco, California, USA, 2016), pp. 785–794. ISBN: 9781450342322. https://doi.org/10.1145/2939672.2939785
12. E.C. Norton, B.E. Dowd, Log odds and the interpretation of logit models. Health Serv. Res. **53**(2), 859–878 (2018). https://doi.org/10.1111/1475-6773.12712. eprint: https://onlinelibrary.wiley.com/doi/pdf/10.1111/1475-6773.12712
13. L. Reichhart et al., Measurement and simulation of the muon-induced neutron yield in lead. Astropart. Phys. **47**, 67–76 (2013). ISSN: 0927-6505. https://doi.org/10.1016/j.astropartphys.2013.06.002. https://www.sciencedirect.com/science/article/pii/S0927650513000972
14. P.S. Barbeau, Neutrino and astroparticle physics with P-type point contact high purity germanium detectors. English. Copyright—Database copyright ProQuest LLC; ProQuest does not claim copyright in the individual underlying works; Last updated 2023-11-29. PhD thesis, p. 291 (2009). ISBN: 978-1-109-52824-4. https://login.proxy.lib.duke.edu/login?url=https://www.proquest.com/dissertations-theses/neutrino-astroparticle-physics-with-p-type-point/docview/305051617/se-2

Chapter 9
Maximum Likelihood Fitting

Abstract After the NuThor Science Run 2 dataset is prepared, it is fit using a 2D binned maximum likelihood fitter implemented through RooFit. The fitter is evaluated for bias using a large collection of toy datasets. A high-energy sideband is used to constrain the possible contamination by beam-related neutrons. Lastly, the dataset is unblinded and fit using the aforementioned model. The best fit count rate for neutrino-induced nuclear fission events comes out to $17.32 + 12.26 - 10.00$. After evaluating 48,000 background-only toy datasets, the significance is estimated to be 2.4 sigma rejection of the null hypothesis of no neutrino-induced nuclear fission.

The NuThor Science Run 2 dataset underwent extensive preparation, including signal processing (Sect. 6.2), energy calibration (Sect. 6.9), beam correlation (Sect. 6.11.2), data cleaning, event reconstruction (Sect. 6.11.1.2), and classification (Sect. 8.2.5). After these studies, the data was evaluated in a search for neutrino-induced nuclear fission. The statistical analysis performed in this chapter employs a 2D-binned maximum likelihood fit to adjudicate the presence or absence of a neutrino-induced nuclear fission signal.

9.1 PDF Construction

The analysis of the search data first requires the construction of models for both the signal and the relevant background sources. Each model component is described with a probability density function (PDF) in two dimensions: neutron multiplicity and BDT score*. The total model is the sum of five PDFs describing neutrino-induced nuclear fission neutrons, thorium NINs, lead NINs, Beam-Related Neutrons, and Steady-State backgrounds.

T. Johnson, *The First Indication of Neutrino-Induced Nuclear Fission*,
https://doi.org/10.1007/978-3-032-01936-3_9

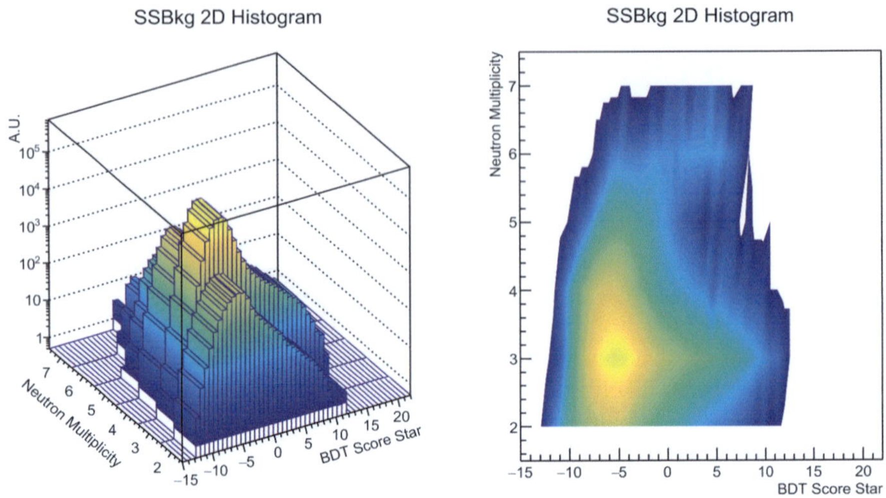

Fig. 9.1 2D steady-state background PDF in neutron multiplicity and BDT score*. This PDF was extracted from NuThor beam anti-coincident data. The data used to form this PDF was at no point used in the training of the classification models

9.1.1 Steady-State Backgrounds

The BDT score* distributions displayed in Fig. 8.68 are stacked together in the order of their neutron multiplicities from 2 to 7+, where 7+ is an overflow bin for any multiplicity equal to or greater than 7.[1] This stack of distributions forms the 2D PDF used in the analysis. Recall that the steady-state background events used to create this PDF come from the beam anti-coincident dataset and are isolated from the portion of anti-coincident data used in the classification training. Because four anti-coincident beam windows were recorded per one beam-coincident beam window in Science Run 2, the impact of statistical fluctuates on the PDF is partially mitigated. Future work will include the larger Science Run 3 dataset to further reduce statistical fluctuations. The steady-state background PDF utilized for this analysis is shown in Fig. 9.1.

9.1.2 Neutrino-Induced Nuclear Fission Neutrons

Similar to the steady-state backgrounds PDF, the neutrino-induced nuclear fission PDF stacks together the BDT score* distributions shown in Fig. 8.68. The BDT Score* distributions originate from simulations that are subjected to post-processing

[1] The neutron multiplicity = 1 bin was removed from the analysis because it is so fraught with background that nearly no sensitivity can be extracted. It's inclusion would immensely increase the computational burden of the analysis without much use to counter-balance the added effort.

that closely mimic the treatment of the real data. After applying the series of analysis cuts discussed in Sect. 8.2.5, the neutrino-induced nuclear fission PDF shape was obtained, shown in Fig. 9.2.

The relative amplitudes of the neutron multiplicity bins derive from the expected count rates calculated in Sect. 7.4, after the analysis cuts are applied. For the 1.57 GWHr beam exposure in Science Run 2, the expected counts in the dataset are summarized in Table 9.1

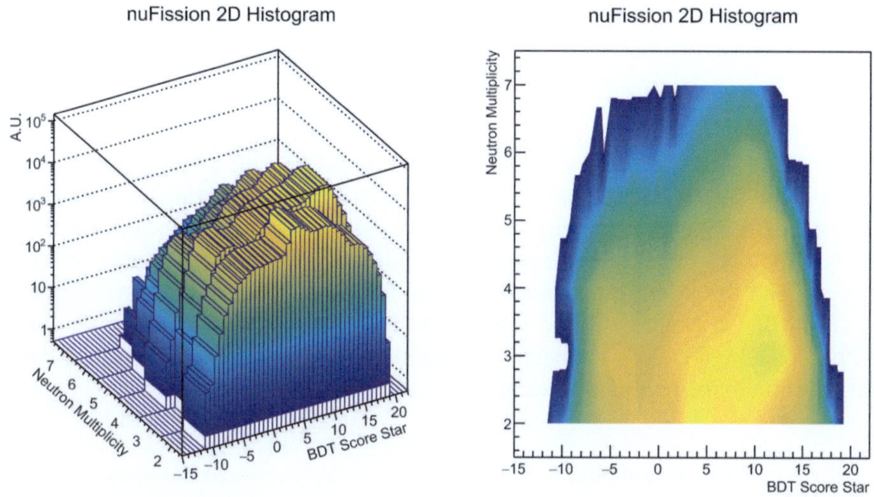

Fig. 9.2 2D neutrino-induced nuclear fission PDF in neutron multiplicity and BDT score*. The data used to form this PDF is generated from simulation and was subjected to all the same analysis cuts that the data was privy to

Table 9.1 Summary of expected neutrino-induced nuclear fission counts after the preselection and BDT cut for Science Run 2. The expected counts for the higher neutron multiplicities have a notably lower acceptance efficiency thanks to the preselection cuts applied to the data

Neutron multiplicity	Neutrino-induced nuclear fission
2	3.35
3	2.17
4	1.17
5	0.25
6	0.04
7	0.003
Total	7.98

9.1.3 Neutrino-Induced Neutrons

The same procedure used to generate the PDF for neutrino-induced nuclear fission is applied to the simulations of neutrino-induced neutron emission from lead and thorium. Due to their similarities in shape, the same PDF is utilized for both populations of neutrino-induced neutrons (See Fig. 9.3). After the full analysis pipeline, the total number of expected counts for lead and thorium neutron emission are 7.64 and 2.94, respectively.

9.1.4 Beam-Related Neutrons

The BRN PDF is generated from simulation and follows the same approach. The shape of the BRN PDF is illustrated in Fig. 9.4. However, estimating the normalization of the BRN population is challenging due to frequent changes in beam power and energy; the SNS upgrading to set world-record-breaking and unprecedented beam powers; and the poorly understood correlation between beam configurations and the BRN flux at Neutrino Alley. Rather than relying on an *a priori* estimate of the BRN flux, a high-energy sideband was unblinded to set an upper limit on the BRN contribution in the region of interest.

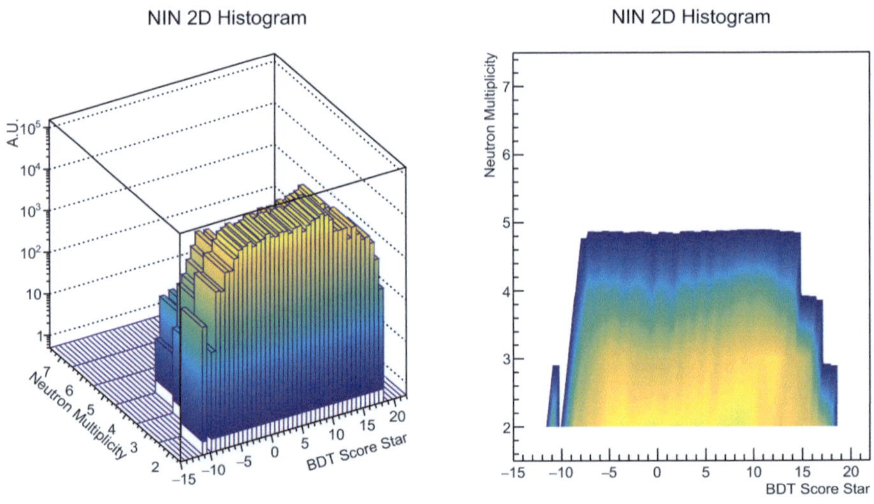

Fig. 9.3 2D NINs PDF in neutron multiplicity and BDT score*. This PDF shape is used for both the lead and thorium NINs but with differing amplitudes

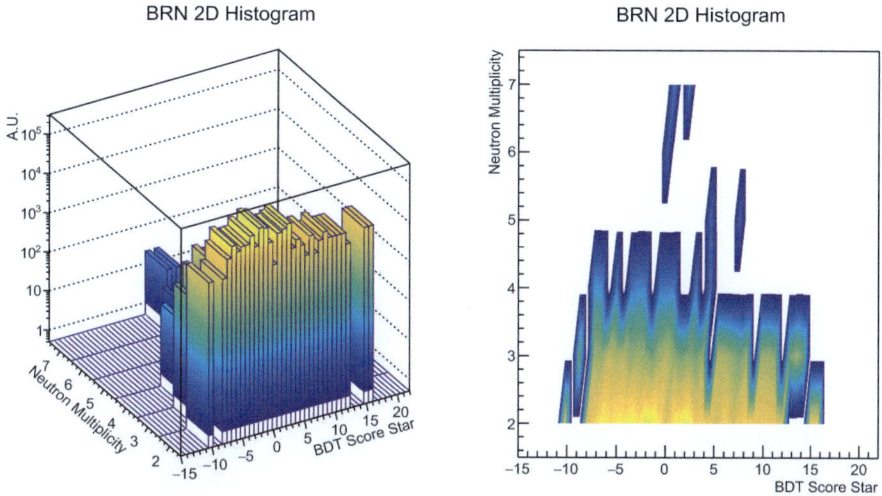

Fig. 9.4 2D BRN PDF in neutron multiplicity and BDT score*. This PDF was generated from simulation

High-Energy Sideband

For the entirety of the analysis preparation, all events coincident with the SNS beam were blinded and separated. In order to investigate the BRN flux and timing relative to the NuThor external triggers, a high-energy sideband was opened between 15 and 55 MeV. This is a safe region to study because no signal is expected to accumulate within it for Science Run 2. Approximately 19% of beam-related neutrons are estimated to be in this high energy sideband (See Fig. 7.25).

Previous neutron background measurement campaigns completed by MARS and the neutron scatter camera in or near NuThor's position have shown a lower BRN flux compared to other positions in the alley, where detectors were able to clearly identify an excess of events attributed to BRNs [1, 2]. Based on this, the BRN flux was expected to be considerably lower.

The arrival time of the background events in the high-energy sideband was modeled with an exponential plus uniform distribution (Fig. 9.5) then fit with an unbinned maximum likelihood routine using RooFit on the beam anti-coincidence dataset. The origin of the slight excess in the early portion of the beam window is hypothesized to be from contamination by pile-up of several thorium-related γ-rays. By Poisson statistics, the arrival time for these possible events would have a strong preference for the first part of the beam window. The best fit parameters were used to constrain the fit in the beam coincident sideband dataset.

The beam-coincident high-energy sideband dataset was also fit with an unbinned maximum likelihood approach using the background model previously discussed, as well as a Gaussian model to describe the arrive of the BRNs (Fig. 9.6).

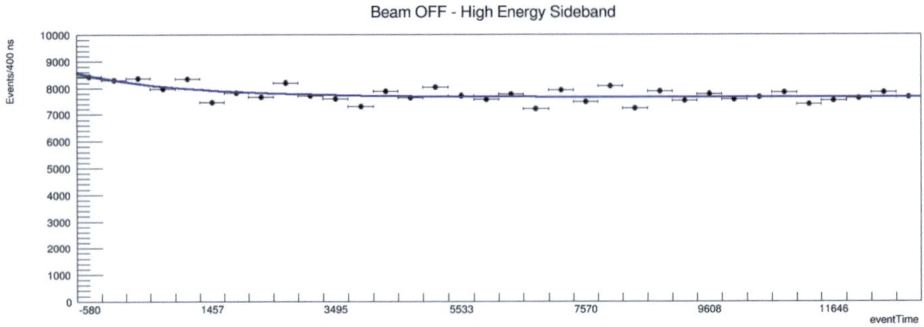

Fig. 9.5 Beam anti-coincident background fit

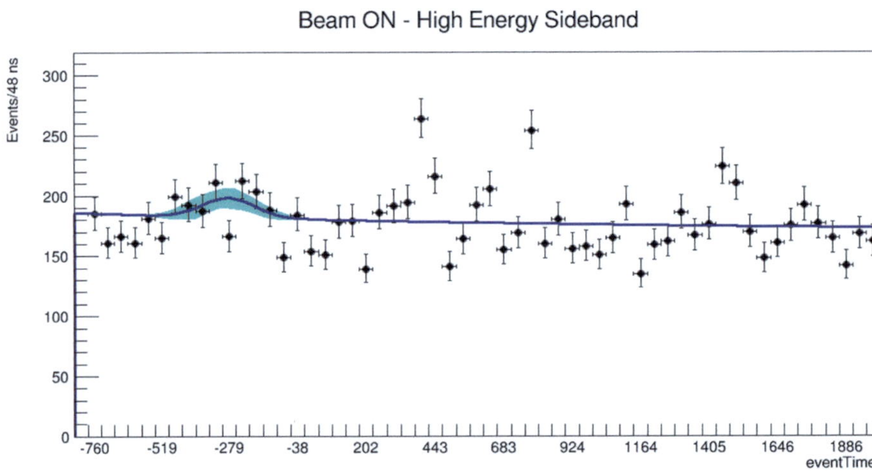

Fig. 9.6 Unbinned maximum likelihood fit in the beam ON high-energy sideband

The best fit parameters are shown in Fig. 9.7.

The excess of BRN counts is not as noticeable compared to the NaIvE-185 and Lead Neutrino Cube detectors previously deployed in Neutrino Alley, but that is possibly due to the different locations of each of these detector systems. The excess was small enough to motivate conducting a Monte Carlo evaluation of the fit results to establish the likelihood that this fit is picking up on statistics fluctuations from the background. 10,000 toy datasets were generated from the background model and passed through the same unbinned maximum likelihood fitting code used for the sideband fit. The ensemble of best fit values for the number of BRNs is plotted in Fig. 9.8. The resulted in a P-value of 0.0179. This was considered to be of slight statistical significance—enough to help set an upper bound on the BRN flux inside the NuThor shielding.

Using the best fit number of BRNs in the high-energy side (79.05 $\pm$40.0) and the estimated 19% of BRNs that manage to reach above this threshold, the estimated number of BRNs in the region-of-interest is 336.8 $\pm$ 170.5. The BRN time cut

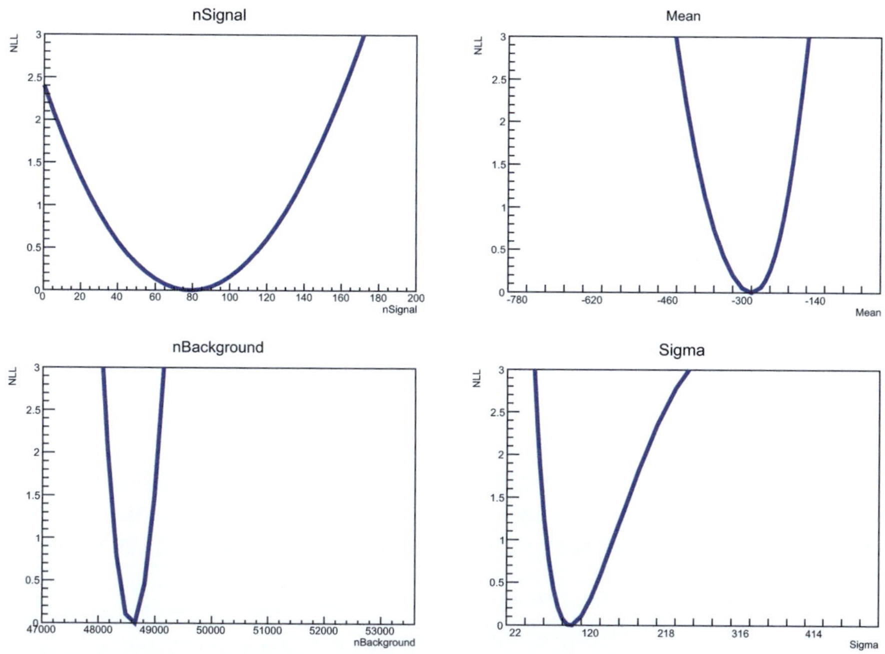

Fig. 9.7 BRN unbinned maximum likelihood profile likelihood curves

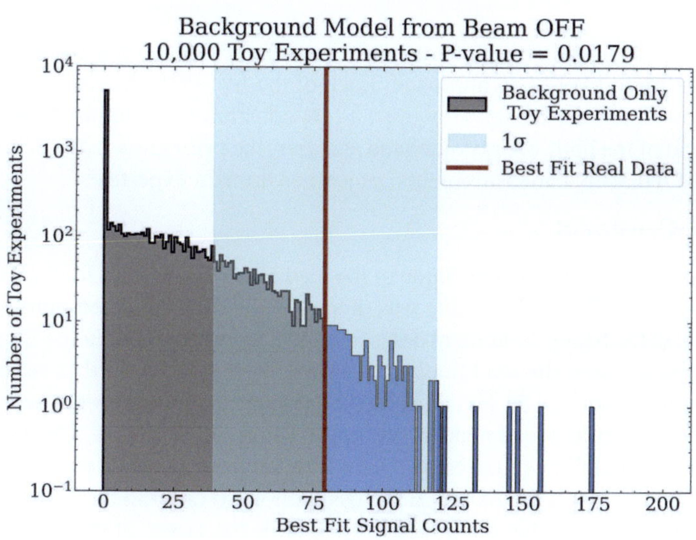

Fig. 9.8 An ensemble of toy Monte Carlo datasets generated with the BRN background model determined in the beam anti-coincident dataset. They were fit with the same unbinned maximum likelihood approach used in the beam-coincidence high energy sideband dataset. The best fit value from the beam-coincident dataset was 79 and is indicated as a red vertical line with its 1σ errorbars shown in light blue. The best fit value of the number of BRNs shows a P-value of 0.0179. This indicates statistical significance in rejecting the null hypothesis of no BRNs

Table 9.2 Summary of expected BRN counts after the preselection and BDT cut in Science Run 2

Neutron Multiplicity	BRNs
2	4.3
3	2.0
4	0.23
5	0.1
6	0.06
7	0.06
Total	6.75

discussed in Sect. 8.1 was designed specifically to reduce this background. After the analysis cuts are applied, the expected BRN counts region-of-interest is summarized in Table 9.2.

The BRN arrival time in the waveform trace was fit to 773.9 ± 68.8 nanoseconds into the waveform. Comparing this with the independent timing estimate of 732 nanoseconds used for the post-processing routine (See Fig. 7.13), they are seen to be in good agreement.

9.2 Binned Maximum Likelihood Fitter

The five PDFs comprising the total model are fit to the search data with a 2D-binned maximum likelihood approach. Each of the five PDFs has an amplitude that can be allowed to vary freely. However, a few constraints are applied based on the combination of the high-energy sideband analysis, the prior knowledge accrued from previous COHERENT measurements, as well as domain expertise.

Lead NIN Constraint

Rather than of allowing the amplitude of the lead NIN PDF to float freely, a Gaussian prior is imposed. The central value was determined by the expected number of lead NIN counts in the beam-coincident dataset, which takes into account the Science Run 2 beam exposure and the lead mass in NuThor. This value is further refined using a cross section from MARLEY, corrected by a suppression factor derived from the combined results of the lead neutrino cube and Eljen cell detectors. The 1σ width of the Gaussian constraint was based on the 1σ interval from the profile likelihood curve that was used to correct the MARLEY cross section. In order to be conservative, the profile likelihood curve has asymmetric error bars, the larger of the two is used for the Gaussian constraint.

BRN Constraint

A Gaussian constraint is applied to the amplitude of the BRN PDF, with the central value based on the total counts listed in Table 9.2. The sigma is determined by the

1σ fractional uncertainty from the best fit result for the number of BRNs in the high-energy sideband. The effect of relaxing the BRN constraint by increasing the width to double and triple its original value was negligible.

Thorium NIN Constraint

The amplitude of the thorium NIN PDF is set in a fixed ratio to the amplitude of the neutrino-induced nuclear fission PDF. This ratio was determined based on the fixed branching ratio between de-excitation via nuclear fission and neutron emission. This relationship is already accounted for by the de-excitation modeling used to estimate their respective event rates.

Systematic Uncertainties

There are several sources of uncertainty in this analysis, but many of them affect the normalization of the signal or background PDFs, which are allowed to float in the fit. A non-exhaustive list of the uncertainties follows:

Prediction Systematic Uncertainties

1. The neutrino flux normalization at the SNS has an uncertainty of 10% and stems from the fraction of protons-on-target that create a π^- [3, 4].[2]
2. Modeling of the nuclear de-excitation was done in the INCL/ABLA package [5] by varying the de-excitation model used and yields a 34% uncertainty. This was bench-marked with ENSDF database to compare the de-excitation of IAS decay following a charge-exchange reaction on ^{232}Th and the results agree relatively well [6]. The nuclear de-excitation model yields a fission probability of 0.29 ± 0.11 and the equivalent nuclear reaction experiment yields a fission probability of 0.19 ± 0.01.
3. The ^{127}I cross section is anchored to the uncertainty from the NaIvE-185 measurement of the exclusive 1+ n cross section. The exclusive cross section was reported as $2.2^{3.5}_{2.2} \times 10^{-40}$cm^2 [1].
4. The ^{nat}Pb NINs cross section suppression relative to the MARLEY prediction uncertainty is extracted from the combined Eljen Cell/Lead Neutrino Cube measurements[2] and was used to constrain the proceeding fit with a Gaussian prior about the central value.
5. The uncertainty of the BRN rate was extracted from the high-energy sideband (See the next section) where the best fit value for the number of BRNs is from the minimum of the profile likelihood and error bars come from the shape of the contour.
6. The distance to the SNS target is estimated at 20.9 ± 0.1. Because the distance enters the neutrino flux estimate quadratically, this results in a neutrino flux uncertainty of 0.7%.

[2] This uncertainty will be reduced in the future to as little as 2–3% by the heavy water detector deployed by the COHERENT collaboration.

Detector Systematic Uncertainties

1. The thorium mass is known to better than 0.01% precision because the plates are maintained by Oak Ridge National Laboratory's Nuclear Materica Control & Accountability (NMC&A) specialists as a special nuclear material.
2. The concentration of contaminants in the thorium plates was measured as 200 ppm by chemical analysis performed on the scraps leftover when they were first fabricated [7].
3. The variation in the mass of standard lead bricks used for the inner core is approximately 1.6%.
4. Neutron scatter cross sections on Thorium are known to 1% according to MCNP's cross section library references.

Toy Monte Carlo Fitting

In order to test for biases, the fitter was exercised by producing a large number of toy datasets from the combination of the five PDFs. For each toy Monte Carlo dataset, the total number of counts was equal to the sum of the expected counts for the signal and each individual background source. The fitter code was then applied to the toy dataset and a profile likelihood curve was generated by profiling over the parameters. In this context, profiling refers to the process of varying one parameter through a range of values and at each value, minimizing the negative log likelihood by allowing all the other parameters involved in the fit to vary.

Figure 9.9 depicts an example fit result from the fitter code when the true number of signal counts generated in the toy dataset was 8 and a non-zero N_{sig} is returned. That

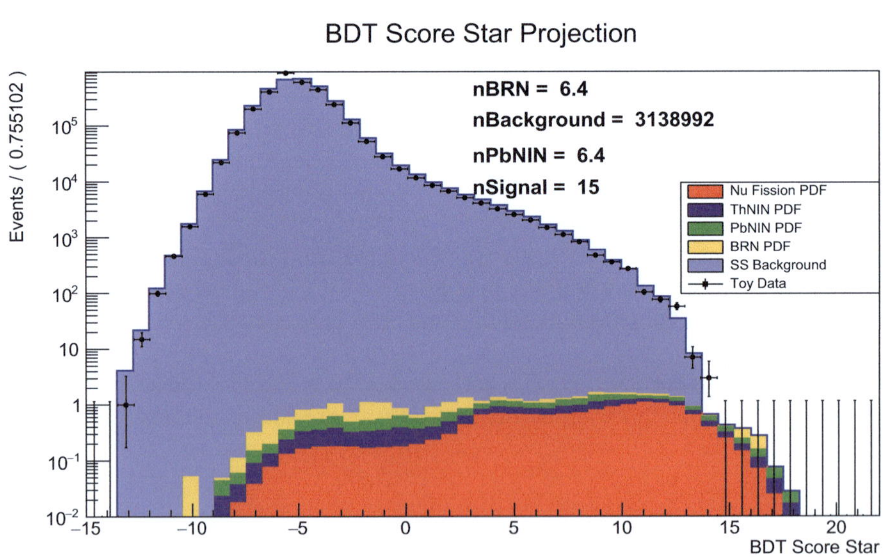

Fig. 9.9 The result of the fitter code applied to a toy dataset generated with the nominal number of signal counts. It is projected into BDT Score*

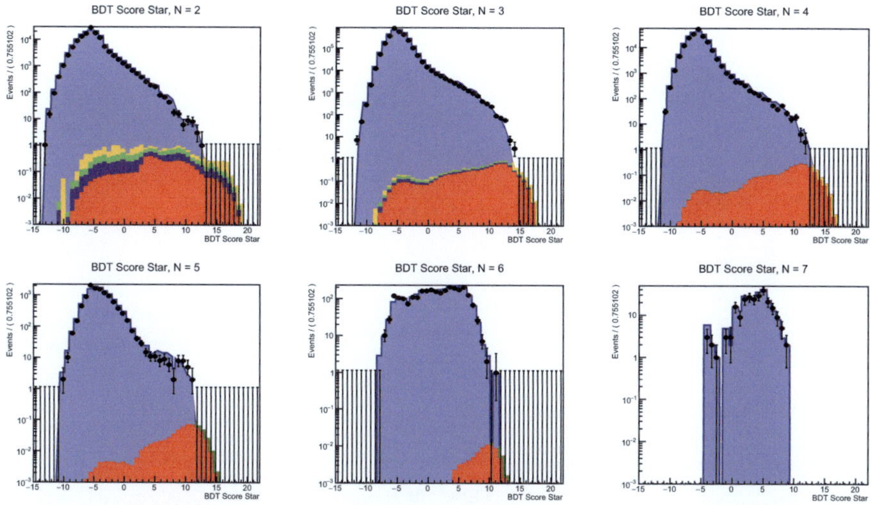

Fig. 9.10 Slices of the 2D fit for each neutron multiplicity for this toy data set. In the case of this toy dataset generated with the nominal, eight signal counts, the significance of the fit is driven by the outlying, higher BDT score* data points

Figure depicts the projection across all neutron multiplicities onto the BDT Score* axis. Figure 9.10 directly proceeding the projection shows the fit results in slices of neutron multiplicity. This is instructive towards understanding how a minority of data points seeded at high BDT Score* can pull the best fit amplitude to higher values.

Conversely, there is also the likelihood of all signal counts populating the region in BDT score* that is overridden with the steady-state background PDF. In those cases, the best fit value is very likely zero. The fit results for an ensemble of toy datasets with the nominal number of signal counts is shown in Fig. 9.11.

The profile likelihood curves are then shown in Fig. 9.12. The pertinent profile likelihood curve is the one which profiles over the N_{Signal} parameter, where N_{Signal} is the number of neutrino-induced nuclear fission counts. The best fit value is the minimum of the curve and the 1σ uncertainty is given by the location in the N_{Signal} axis where the profile NLL increases from the minimum by 0.5. The putative rejection of the null hypothesis (no signal) is the given by the y-intercept at $N_{Signal} = 0$.

The use of toy dataset reveals biases imparted by the fit itself. Also beneficial to understanding the performance of the fitting code is the fact that a true number of signal events can be injected into the toy dataset to demonstrate the fitter's response to these changes. This is depicted in Fig. 9.13, where toy datasets were generated with increasing numbers of true signal counts. For each increment, 18,000 datasets were produced. The green belt was constructed in horizontal slices but is meant to be read vertically. Based on the best fit number of signal counts, the green region indicates the 1σ confidence interval.

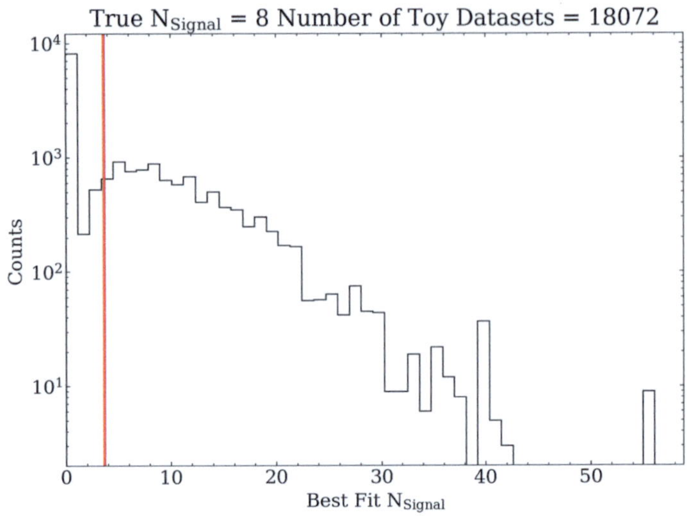

Fig. 9.11 The distribution of best fit N_{Signal} for 18,000 toy datasets generated with the nominal neutrino-induced nuclear fission expected count rate. The red vertical line indicates the median best fit value produced by the ensemble of toy datasets. The median is lower and the nominal number of neutrino-induced nuclear fission counts. This reflects the nature of the low count rate with highly overlapping PDFs. A large portion of the toy datasets generated featured signal counts populated in a region of BDT Score* that was inundated with stead-state backgrounds, thus returned a best fit value of zero. Conversely, many toy datasets exhibited the opposite behavior of returning a larger best fit value than the true number. This manifests from the toy data points populating high, outlying BDT Score*

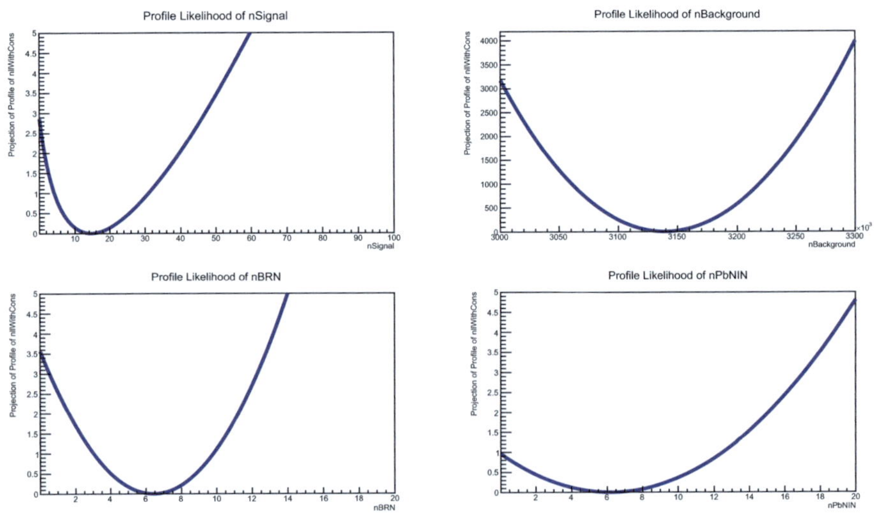

Fig. 9.12 The profile likelihood curves for the 4 parameters allowed to float when fitting a toy dataset with the nominal number of signal counts. The thorium NIN amplitude is not reported because it is fixed to the neutrino-induced nuclear fission amplitude

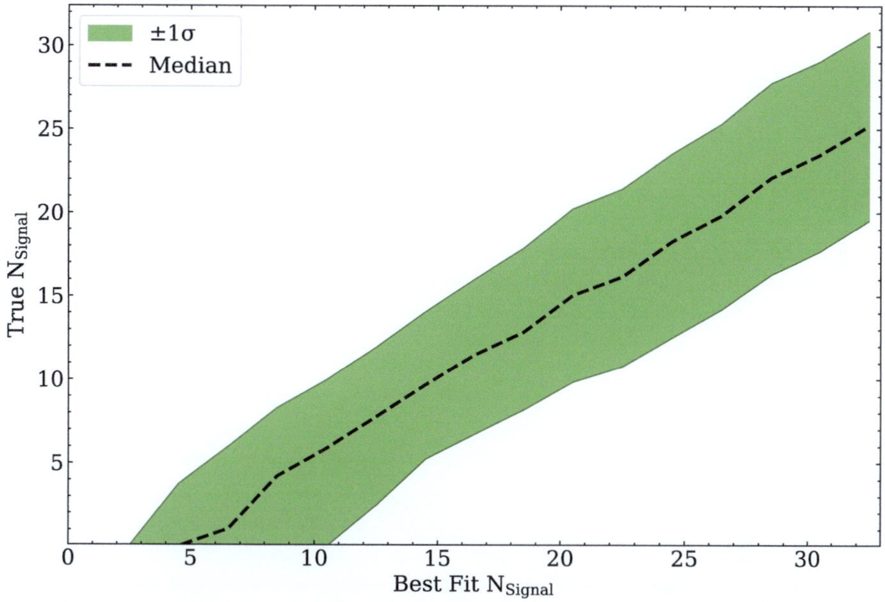

Fig. 9.13 Each slice of true N is composed of 18,000 toy datasets. The figure was constructed in horizontal slices by generating toy datasets with varying true number of signal counts. The diagram is then meant to be read vertically. Given a best fit value in the unblinded dataset, trace vertically to interpret the confidence interval. An open interval is expected for best fit values less than approximately 10. At increasing number of true signal counts, the median best fit value diverges from zero with increasing confidence. The central value (dashed line) was determined from the median true N_{Signal} for each best fit N_{Signal}. The 1σ confidence belt represents the range between the 33rd and 66th quantiles of true N_{Signal} for each best fit N_{Signal}

9.3 Unblinded Results

The beam-coincident data was unblinded and fit with the approach shown in the previous section. Figure 9.14 shows the profile likelihood curve resulting from the binned maximum likelihood fit. The best fit value for signal counts is $17.32^{+12.26}_{-10.00}$. The amplitude of the thorium NIN PDF was held in a fixed relationship with the neutrino-induced nuclear fission PDF. This results in the best fit number of thorium NIN counts of 6.38 ± 4.21. The lead NIN best fit amplitude returned as 7.34 ± 4.42.

A comparison of the estimated amplitudes for the number of neutrino-induced nuclear fission events, lead NINs and BRNs is depicted in Fig. 9.15. Both the lead NINs and BRN amplitudes receive Gaussian constraints in the fit and the effects induced by removing those constraints is studied in a proceeding section.

Looking at slices in neutron multiplicity is illuminating for identifying which data points may be pulling the fit result upward or downward. There appears to be considerable strength in the N = 4 and 5 neutron multiplicity bins from outlying data points in the signal region.

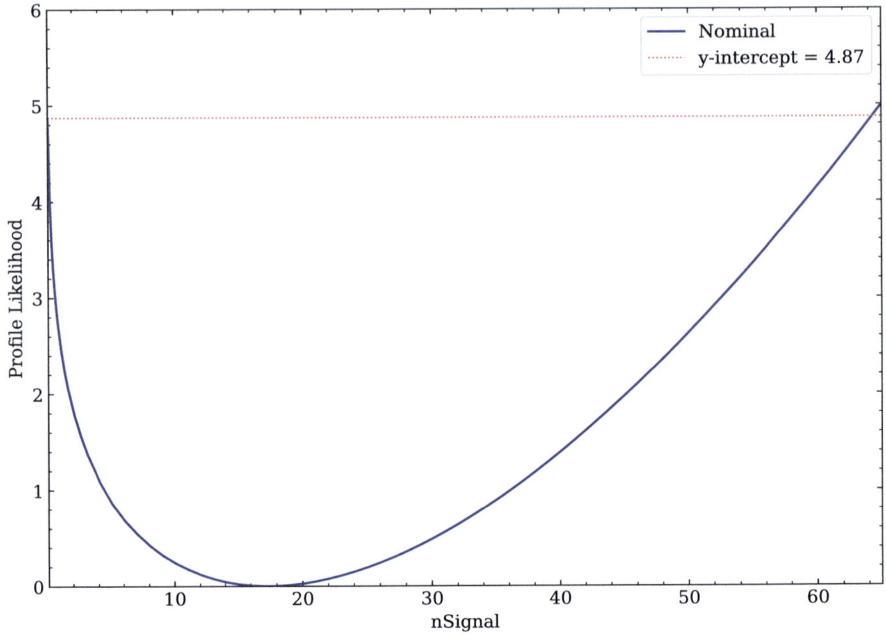

Fig. 9.14 Profile Likelihood curve resulting from the 2D binned maximum likelihood fit to the beam coincident data in Science Run 2. The minimum of the curve is shifted such that the best fit value is at zero. The best fit value for the number of signal counts is 17 ± 11. The best fit values of the nuisance parameters are listed in Table 9.3

Table 9.3 The best fit values for the number of neutrino-induced nuclear fission counts, thorium NIN counts, and lead NIN counts. † The components in the fit model that have a constraint imposed. The constraints are enumerated in Sect. 9.2

Parameter name	Best fit value
Neutrino-induced nuclear fission counts - N_{Signal}	17.32 ± 11.42
Thorium NIN counts†	6.38 ± 4.21
Lead NIN counts†	7.34 ± 4.42
Beam-related neutrons†	6.50 ± 2.44
Steady-state backgrounds	537706 ± 733

Often, the profile likelihood is used to deduce the significance associated with the measurement's ability to reject the null hypothesis (no signal). The significance derived from the profile curve is the y-intercept where, in this case, N_{Signal}=0. In the scenario that Wilk's theorem [8] is adhered to, the profile likelihood ratio follows a Chi-squared distribution and can be directly converted into a Z-score (σ) via the expression:

$$Z = \sqrt{2 \times \Delta\text{LL}} \tag{9.1}$$

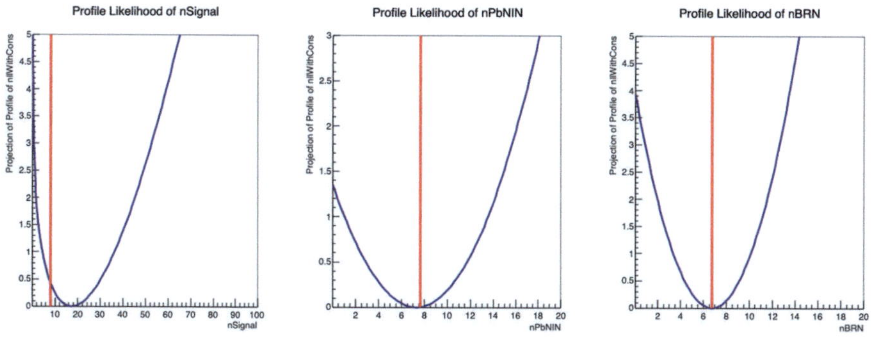

Fig. 9.15 The profile likelihood curves for the number of neutrino-induced nuclear fission events (left), lead NINs events (middle), and beam-related neutron events (right). The red vertical lines in each subplot indicates the expected values for that PDF's amplitude. Both lead NINs and BRNs amplitudes are nuisance parameters with Gaussian constraints applied to them in the fit

The y-intercept of the profile curve in Fig. 9.14 is 4.87, which would correspond to a Z-score of 3.12σ assuming the applicability of Wilk's theorem.

However, in the case here, where Wilk's theorem does not hold, an alternative approach to significance estimation is taken using an ensemble of toy datasets. In this exercise, a number of background-only toy datasets were produced using the PDFs applied in the maximum likelihood fit. In this context, "background-only" refers to the sum of steady-state backgrounds, beam-related neutrons, and lead NINs. Lead NINs are including in the null hypothesis because a non-zero lead NIN population is used as a prior in the fit. The components left out of the background-only toy datasets are the neutrino-induced nuclear fission and thorium NIN PDFs. A total of 48,000 background-only toy datasets were generated for this study. Each of the toy datasets was fit with the same fitting procedure used for the real dataset. The best fit value for N_{Signal} is recorded and reported in Fig. 9.16. The P-value derived from the background-only ensemble represents the proportion of background-only toy datasets that yield a best-fit value greater than the best-fit obtained from the real data. The result of this study returns a P-value of 0.01565 which corresponds to a Z-score of 2.41σ (Figs. 9.17 and 9.18).

A more intuitive data visualization is exemplified in Fig. 9.19 where the steady-state background is subtracted from the beam-coincident data. The data and PDFs are projected along the BDT Score* axis into the neutron multiplicity axis. Note that this is simply a visualization and is done after the fit ws performed. It is instructive in understanding how the profile likelihood determines there is significance when examining the region greater than BDT score* equals 11 where a large portion of the signal is expected.

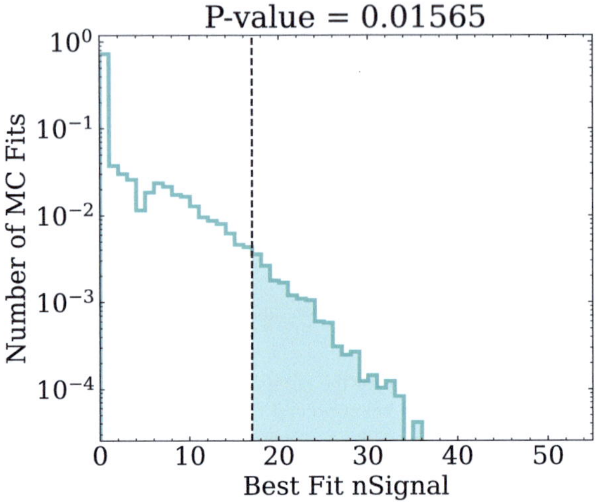

Fig. 9.16 The significance of the measurement was gauged by generating an ensemble of 48,000 background-only toy datasets with zero signal counts from the same PDFs used in the fitting. The toy datasets were then fit in the same approach used for the real data. The definition for a background-only dataset includes steady-state backgrounds, beam-related neutrons, and lead NINs. A P-value of 0.01565 is extracted from the study

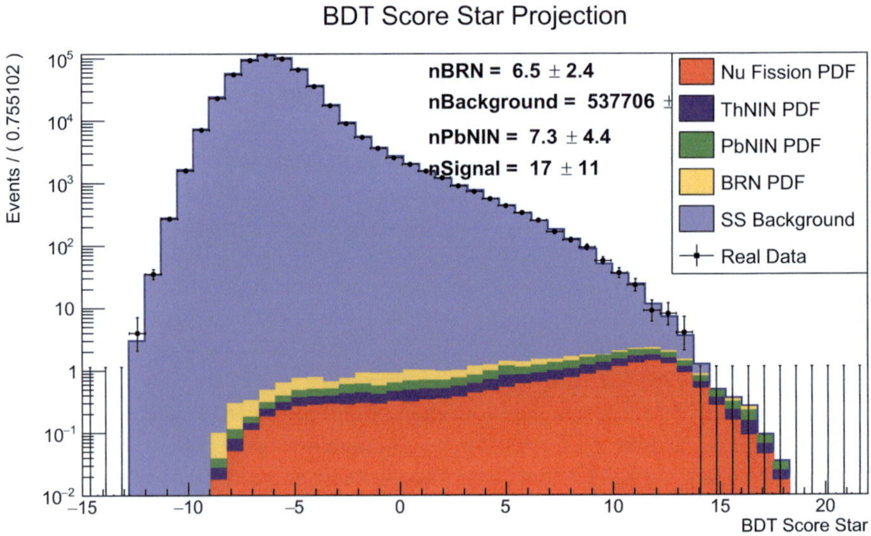

Fig. 9.17 The 2D PDFs projected into the BDT score* axis after fitting to the beam-coincident data. All neutron multiplicities are included in this projection. The individual components of the fit are indicated with separate colors

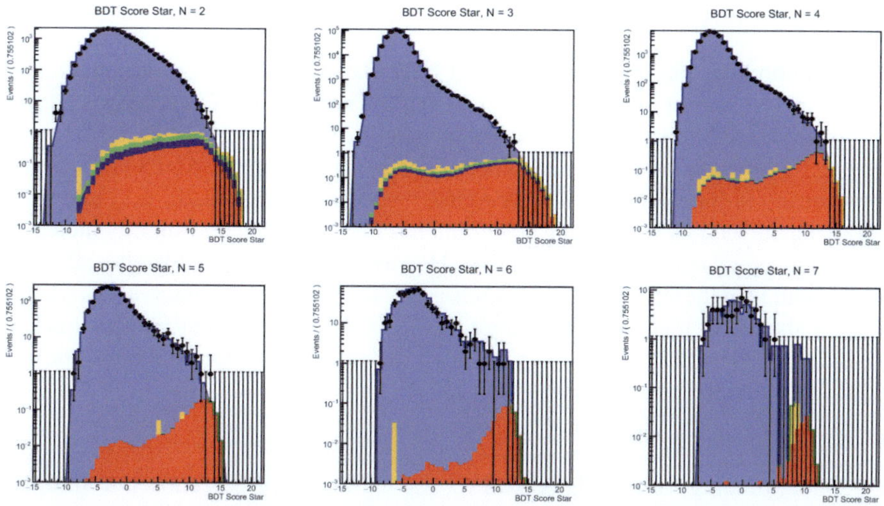

Fig. 9.18 The BDT score* of each neutron multiplicity after the fit. Much of the strength comes from events in the N = 4 and 5 neutron multiplicity bins

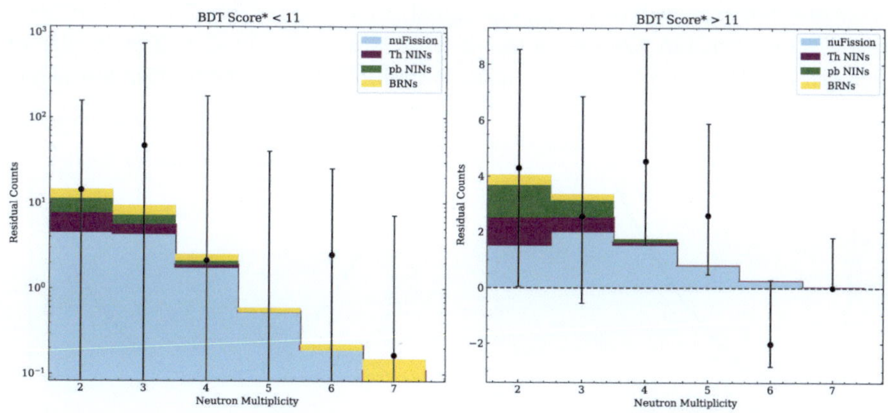

Fig. 9.19 The steady-state background subtracted residual data points projected into the neutron multiplicity axis. The post fit PDFs and residual data points are segmented along a threshold of 11 to illustrate the source of pull on the fit. This is simple a data visualization and should not be used for estimating significance. The left plot is in log scale because the error bars are much larger the the residual counts and it is difficult to see both the error bars and the residual simultaneously unless in log scale

Effect of Removing the Lead NIN Constraint

After unblinding, the impact of the lead NIN Gaussian constraint was studied to understand its impact on the best fit signal value. The purpose of this study is simply to inform about the potential bias that the constraint introduces. Particularly, the study is aimed at determining whether the Gaussian constraint on the lead NIN PDF inflates the neutrino-induced nuclear fission signal. As shown in Figs. 9.20, 9.21 and 9.22, removing the lead NIN Gaussian constraint has no such effect and slightly increases the best fit signal count to 19.58 ± 12.08. However, this had a considerable affect on the amplitude of the lead NINs population as it fit to 0 with an open interval. This helps assuage the worry that the Gaussian prior was biasing the neutrino-induced nuclear fission signal best fit value to an artificially high value as the opposite behavior was exhibited when the constraint was relaxed.

Effect of Bin Size in the Fit

The initial binning for the PDFs used for the unblinded fit was chosen to mitigate statistical fluctuations from sparsely population bins and to mostly eliminate the instances of zero counts in each PDF bin. The higher neutron multiplicity bins have a small portion of bins that exhibit some choppy behavior. Those bins (N = 6 and 7) are also not expected to contain any signal at the Science Run 2 beam exposure. The effect that the bin size has on the fit result was studied by doubling the size of each bin in BDT Score*. Because neutron multiplicity is discrete, it is left alone.

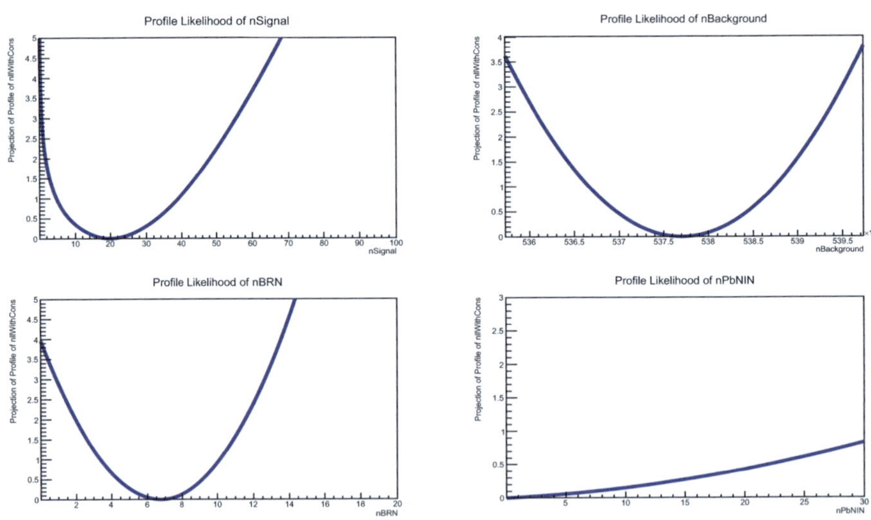

Fig. 9.20 The profile likelihood curve for the signal and nuisance parameters when the lead NINs Gaussian constraint is removed from the fit approach. This is an educational endeavor to better understand the effect that this constraint had on the fit results but the approach agreed upon before unblinding is the only valid, reportable result

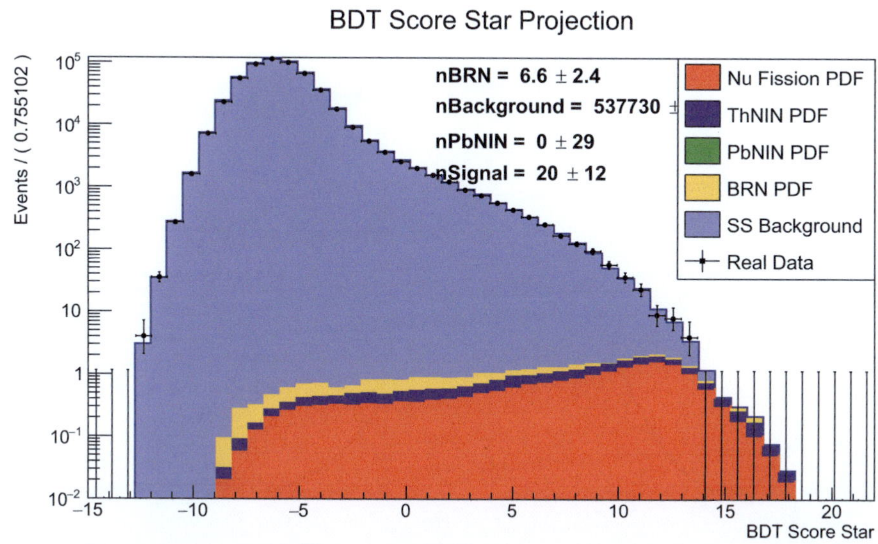

Fig. 9.21 The projection in the BDT score axis after the binned maximum likelihood fit is performed without the lead NINs Gaussian constraint applied

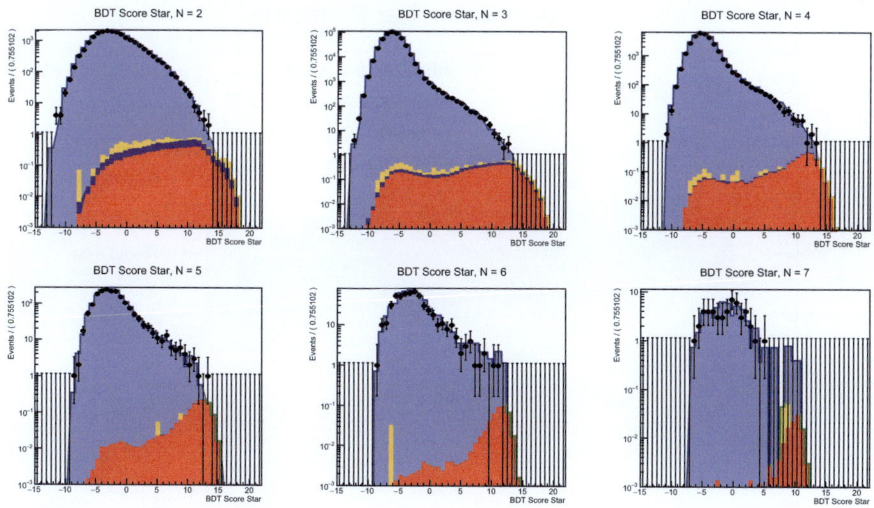

Fig. 9.22 The BDT score* distributions in slices of neutron multiplicity after the binned maximum likelihood fit is performed without the lead NINs Gaussian constraint applied

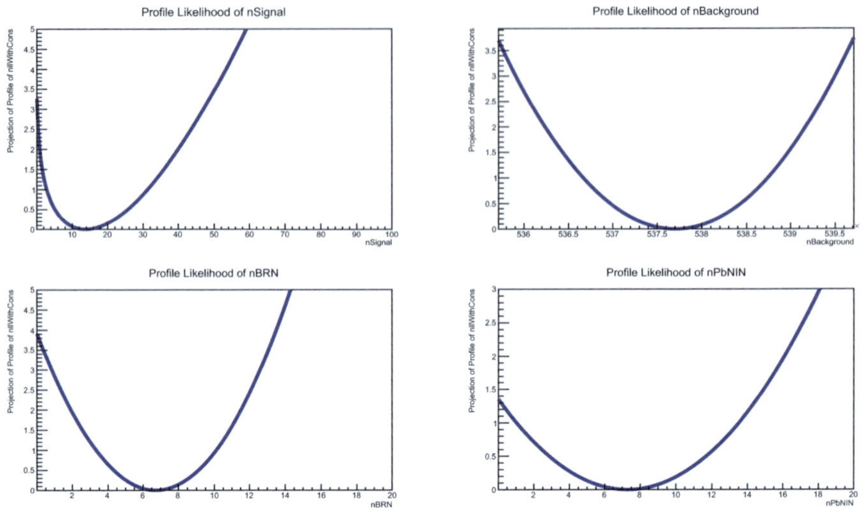

Fig. 9.23 The result of the binned maximum likelihood fit for PDFs with larger bins on the profile likelihood curves for the signal and nuisance parameters. The best fit value for signal counts slightly decreases from 17 to 13

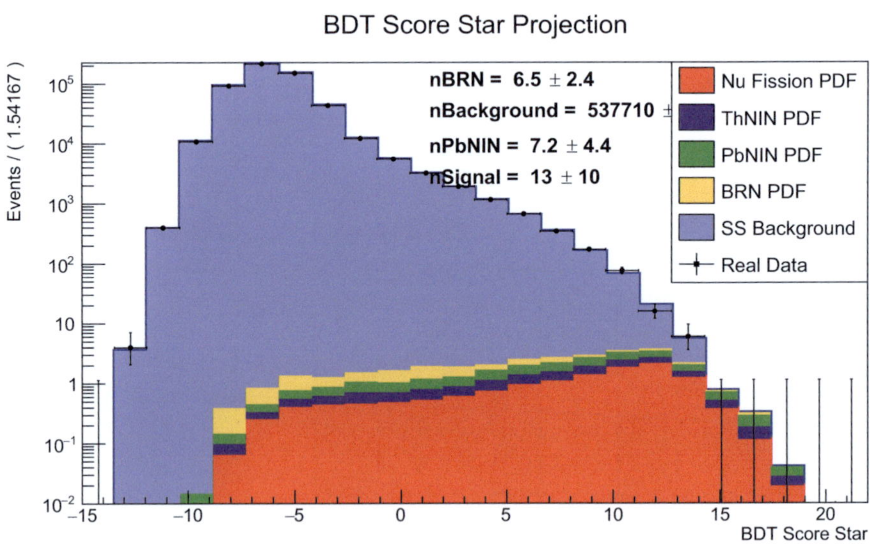

Fig. 9.24 The result of the binned maximum likelihood fit for PDFs with larger bins on the projection in BDT score*. The best fit value for the number of neutrino-induced nuclear fission events decreased slightly to 13 ± 10

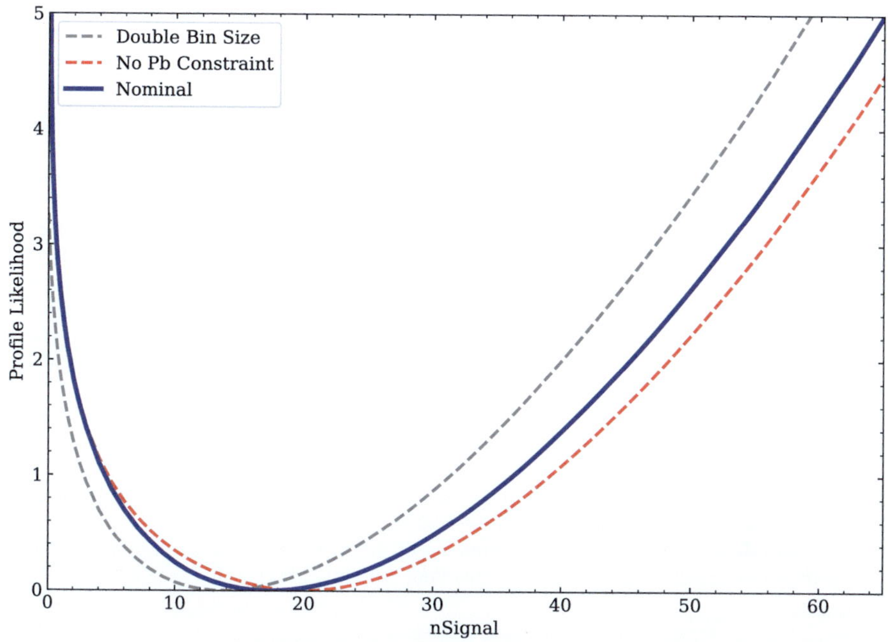

Fig. 9.25 A comparison of the profile likelihood curves for the nominal fit decided upon before unblinding (the only valid, ethical result) and the profile likelihood curves resulting from tests meant to better understanding the effects certain assumptions in the fitting approach had on the best fit value. Removing the Pb NINs constraint results in a best fit value for the number of neutrino-induced nuclear fission events to slightly increase. Conversely, doubling the BDT score* bin size results in a slight decrease

The purpose of this is not to revise the fitting approach that was agreed upon before unblinding—it is simply to get a sense of how that choice impacts the fit result. The PDFs were generated with larger bins and the same 2D binned maximum fit was performed. The result of the fit is shown in Figs. 9.23 and 9.24. The effect was modest and reduced the best fit value of the signal counts from 17 ± 11 to 13 $\pm$ 10

Because the binning and constraints employed in the maximum likelihood fit were set before unblinding, the initial conditions used in the fit are the only valid results reported here. Their effects are slight but worth comprehending. A direct comparison of the resulting profile likelihood curves is shown in Fig. 9.25. The significance (or lack thereof) of this result must be informed by coverage tests to determine if this fitting routine is appropriately coveraged. If not, a correction can then be applied. Because coverage testing is so computationally expensive, this work is still ongoing with hopes that is completed in the near future.

References

1. An P et al (2023) Measurement of electron-neutrino charged-current cross sections on ^{127}I with the coherent NaIνE detector. Phys Rev Lett 131:221801. https://doi.org/10.1103/PhysRevLett.131.221801
2. An P et al (2023) Measurement of ^{nat}Pb(ν_e, Xn) production with a stopped-pion neutrino source. Phys Rev D 108:072001. https://doi.org/10.1103/PhysRevD.108.072001
3. Akimov D et al (2022) Simulating the neutrino flux from the spallation neutron source for the coherent experiment. Phys Rev D 106:032003. https://doi.org/10.1103/PhysRevD.106.032003
4. Coherent collaboration et al (2021) A D2O detector for flux normalization of a pion decay-at-rest neutrino source. J Instrum 16(08):P08048. https://doi.org/10.1088/1748-0221/16/08/P08048
5. Heikkinen A, Kaitaniemi P, Boudard A (2008) Implementation of INCL cascade and ABLA evaporation codes in Geant4. J Phys: Conf Ser 119(3):032024. https://doi.org/10.1088/1742-6596/119/3/032024
6. Petit M et al (2004) Determination of the 233Pa(n,f) reaction cross section from 0.5 to 10 MeV neutron energy using the transfer reaction 232Th(3He,p)234Pa. Nucl Phys A **735**(3), 345–371. ISSN: 0375-9474. https://doi.org/10.1016/j.nuclphysa.2004.02.017
7. Olsen DK, Ingle RW, Portney JL (1980) Precise measurement and analysis of neutron transmission through 232Th. Technical report CONF-800942–8. United States
8. Wilks SS (1938) The large-sample distribution of the likelihood ratio for testing composite hypotheses. Ann Math Stat 9(1):60–62. https://doi.org/10.1214/aoms/1177732360

Chapter 10
Conclusion

Abstract In this conclusion, the results of the thesis are summarized. Additionally, the future direction of the neutrino-induced nuclear fission campaign is laid out with emphasis places on the import of this measurement on the field of neutrino-induced nuclear reactions.

Neutrino-induced nuclear fission was first proposed in 1971 [1]. For the first time, a search for this rare process was undertaken. The initial results are promising and yielded a significance of 2.41σ based on Monte Carlo toy datasets. The future NuThor program has strong prospects for conclusively putting the half-century old mystery to rest with the analysis and inclusion of more data. The Science Run 2 dataset analysis in this work had the best sensitivity of the two campaigns completed at the time of this writing, but both Science Run 1 and Science 3 will be instrumental.

Thorium is primed to join the small number of low-energy neutrino-nucleus measurements [2]. As the largest nucleus tested to date, this places the thorium measurement with a pivotal role in the describing how neutrino interactions map across ranges of nucleus sizes. Recall that the measured cross sections for both ^{127}I [3] and ^{nat}Pb [4] were lower than their respective theoretical predictions. It is perhaps too early to determine whether thorium follows or deviates from this downward trend. The complete dataset is essential to address this question and extend a non-zero detection of neutrino-thorium interactions to a cross-section estimate. While the best fit number of neutrino-induced nuclear fission counts (17 ± 11) is consistent with the prediction within one sigma, it is worth considering what physical mechanisms could contribute to an increased cross section, particularly if there is a broader suppression trend for large nuclei. As was discussed in Chap. 4, the predicted event rate includes only allowed charged-current interactions, excluding all neutral-current interactions and forbidden transitions from both the charged-current and neutral-current. Nuclei with significant neutron excesses are more prone to forbidden transitions, which could contribute to an increased cross section. At the time of this writing, no calculation exists for neutrino-induced nuclear fission partial cross section on thorium. However, such a computation would clarify whether a significant portion of the total neutrino-induced nuclear fission rate derives from that interaction channel, and was thus overlooked in this work's predictions.

T. Johnson, *The First Indication of Neutrino-Induced Nuclear Fission*,
https://doi.org/10.1007/978-3-032-01936-3_10

In terms of future work with both the NuThor detector and neutrino-induced nuclear fission more broadly, there is a great deal of additional study that is possible and warranted. The current incarnation of the NuThor design would benefit from additional validation of the simulation and event selection process using a time-tagged ^{252}Cf sources available at Oak Ridge National Laboratory. With a time-tagged ^{252}Cf source, NuThor's neutron detection efficiency can be calibrated with the source's precisely known neutron multiplicity. This calibration effort did not take place in the time span of this thesis because it is likely optimal to conduct this calibration without the thorium mass present, which requires a sizeable effort from SNS staff, radiation control staff and experimenters.

A more distant prospect is an upgrading NuThor detector that is less sensitive to the thorium's γ-ray backgrounds. There are early designs for a ^{3}He-based neutron multiplicity meter detector at Neutrino Alley. The ability to discriminate γ-rays while maintaining a high neutron capture efficiency makes this approach very enticing as a fortuitous path forward.

In addition to the neutrino-induced nuclear fission interaction channel, there are many more neutrino-nucleus interactions that the NuThor detector can examine. Without the thorium present in the inner core, this neutron detector can be an excellent neutrino-induced neutron detector with the ability to swap out target materials.

References

1. Andryushin VI, Bilen'fkii SM, Gershtein SS (1971) Possible method of measuring the electromagnetic form factor of the neutrino. ZhETF Pis Red
2. Formaggio JA, Zeller GP (2012) From eV to EeV: neutrino cross sections across energy scales. Rev Mod Phys 84:1307–1341. https://doi.org/10.1103/RevModPhys.84.1307
3. An P et al (2023) Measurement of electron-neutrino charged-current cross sections on ^{127}I with the coherent Naive detector. Phys Rev Lett 131:221801. https://doi.org/10.1103/PhysRevLett.131.221801
4. An P et al (2023) Measurement of $^{nat}Pb(v_e, Xn)$ production with a stopped-pion neutrino source. Phys Rev D 108:072001. https://doi.org/10.1103/PhysRevD.108.072001

Appendix A
Alternative NuThor Designs

Abstract This appendix presents several of the previous iterations of the NuThor detector's design. There were many avenues this measurement campaign could have taken, however, a combination of scientific and logistical reasons came together to make the detector geometry presenting in the body of this thesis optimal.

There were several iterations on the design for the NuThor detector before converging on its final form. Before gadolinium-loaded water bricks were determined to be an effective and cheap effective neutron capture scheme, a top contender was layers of plastic scintillator and gadolinium paint. The following figures illustrate early mock-up designs for a gadolinium paint and plastic scintillator style apparatus.

The design featured in Figs. A.1 and A.2 consists of a thin lead inner shielding around the thorium plates. At the time, the rate of radioactive decays was heavily underestimated, thus the thickness of this lead shielding was increased substantially. Outside of the thorium and lead is several inches of plastic scintillator followed by alternating layers of gadolinium paint. This design has a similar approach to detecting neutrons as the original; neutron moderate on the hydrogen in the plastic scintillator then capture on either the gadolinium in the paint or on a hydrogen nucleus. Those materials are all then surrounded by a layer of NaI crystals to detect the resultant γ-rays from the neutron capture.

Although there were several other incarnation of NuThor designs, this Gd-Paint style detector is emblematic of the approach where a prompt signal would be observed from the elastic scattering of neutrons as they moderated to thermal energies. This was an enticing prospect because the prompt elastic scattering would provide an additional coincidence condition for rejected backgrounds.

This design was simulated in MCNP-Polimi to adjudicate its efficiency and practicality as a final design—See Fig. A.3. A flaw in this approach was the increase in the fraction of neutrons that capture on hydrogen, which emits a single γ-ray at 2.2 MeV rather than the more noticable $\sim$8 MeV worth of γ-rays after a neutron capture on gadolinium.

T. Johnson, *The First Indication of Neutrino-Induced Nuclear Fission*,
https://doi.org/10.1007/978-3-032-01936-3

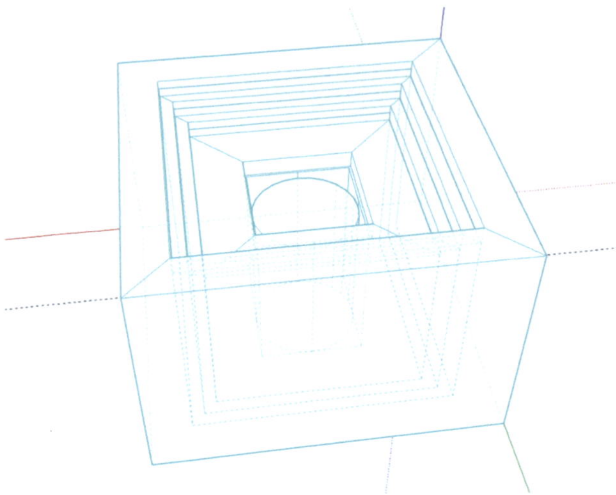

Fig. A.1 3D CAD mockup of a gadolinium paint version of NuThor

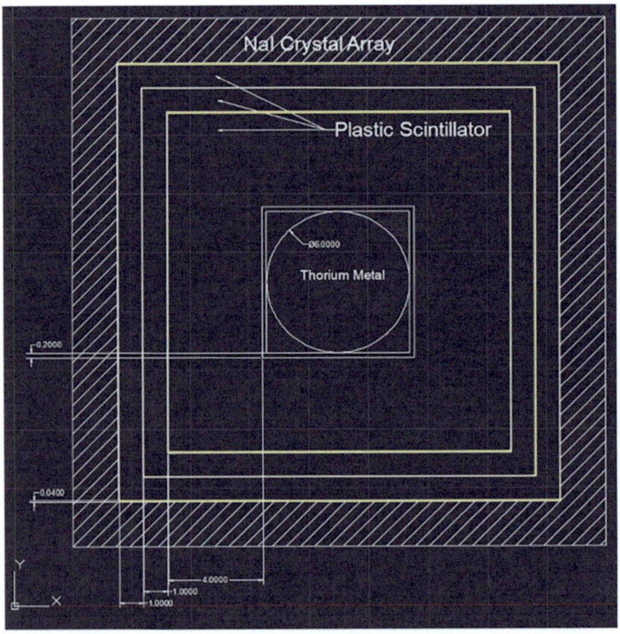

Fig. A.2 CAD Drawing of a gadolinium paint version of NuThor

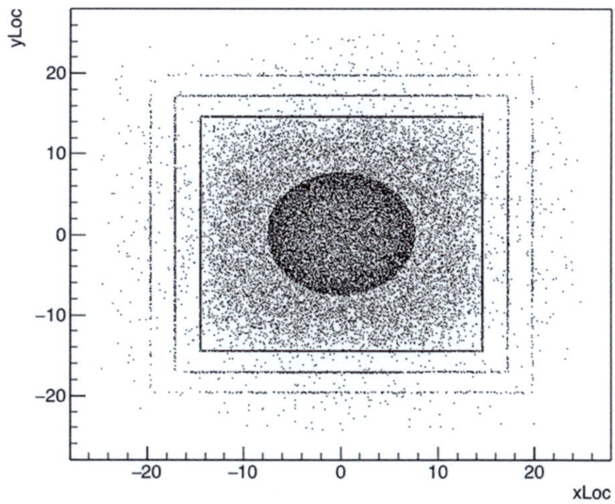

Fig. A.3 MCNP simulation of neutron propagation in a Gd-Paint based NuThor

Other designs considered featured gadolinium-loaded scintillator from Eljen Technologies instead of the layers of plastic scintillator and Gd-Paint.[1] This options was judged to be too costly.

[1] https://eljentechnology.com/products/liquid-scintillators/ej-331-ej-335.

Appendix B
Seismic Stability

Abstract This appendix presents the study of the NuThor detector geometry's stability to seismic activity in addition to its own gravitational load. The finite element analysis (FEA) was carried out in Autodesk Inventor and collaboratively examined by Oak Ridge National Laboratory engineering professionals.

The components of the detector are analyzed to ensure they will not fail under gravitation and seismic loads. These analyses are performed using Autodesk Inventor's Stress Calculation functionality. This requires fixing points in place on the detector and applying loads. Wall Anchors are used as fixed points. Bolts are separately analyzed to ensure they can withstand the tangential or axial forces that could be placed on them, using the Inventor Bolted Connection Component Generator. For the stress calculations, a Safety Factor is, in part, used to determine whether the component can meet the seismic requirements. This factor is the ratio of the maximum allowable stress to the Von Mises stress. A minimum safety factor of 1.0 is required, which ensures no permanent deformation from applied loads. This criterion is met with all calculations contained within this document. Additionally, the planar and shear stresses are enumerated and judged not to exceed allowable limits as described in the AISC's "Specification for Structural Steel Buildings" (2016) and the Aluminum Association's "Aluminum Design Manual." The tensile yield strength and shear strength of hot-rolled ASTM A36 steel (which I acquired from Benco Steel Inc. in Hickory, NC) is 36 ksi and 22 ksi, respectively, according to Chapter A Section 3 subsection 1a of the text. Additionally, the maximum yield for aluminum alloy 6015-T5 is 35 ksi according to 80/20's "Deflection Calculator" and its maximum allowable shear stress is 24 ksi according to the Aluminum Design Manual (Table B.1).

T. Johnson, *The First Indication of Neutrino-Induced Nuclear Fission*,
https://doi.org/10.1007/978-3-032-01936-3

Table B.1 Summary of loads associated with each detector component for NuThor at the time of deployment in June 2022

Detector component	Item	Weight [kg]
Inner core	Lead Bricks	536
	Thorium Plates	52
	Aluminum Sheets	7.4
	80/20 Frame	13.1
	Steel Angle Irons	1.2
	Subtotal	611
Neutron multiplicity meter	NaI[Tl] Scintillators	276.5
	80/20 Frame	53.2
	Steel Plates	172.7
	Water Bricks	240.3
	Borated Polyethylene	412
	Corner Brackets & Screws	5.98
	Subtotal	1160.7
	Total	1771.7

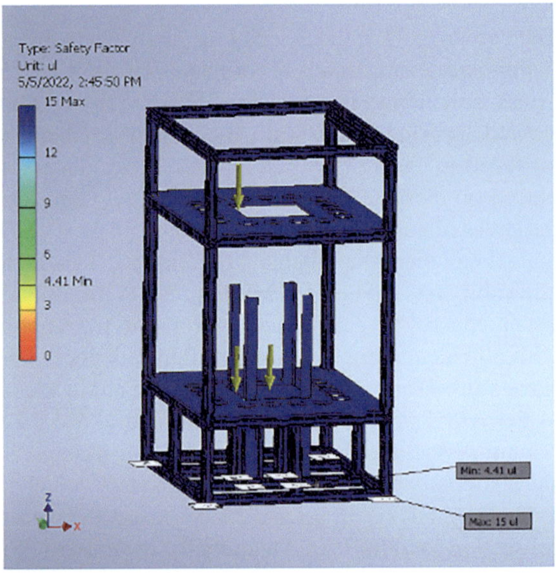

Fig. B.1 Gravitational force applied to each tier of the Neutron Multiplicity Meter. 2,250 N is applied to the top tray. 8,600 N is applied to the bottom tray. 6,000 N is applied to the inner core. Minimum Safety Factor of 4.4

The NaI[Tl] detectors, lead, thorium, water bricks, and borated polyethylene themselves are not structural components, and their geometry is not included in the seismic analyses due to the large memory usage required in stress calculations when present. However, their weight is included in calculated loads. For reference, a table is provided to hold all the proceeding results shown in Figs. B.1, B.2, B.3, B.4, and B.5.

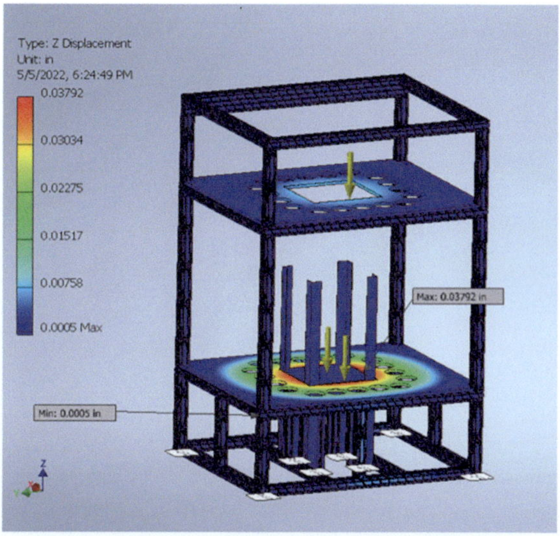

Fig. B.2 The Z-direction has a maximum stress of 9.075 ksi, or 25.9% of the allowable maximum for aluminum

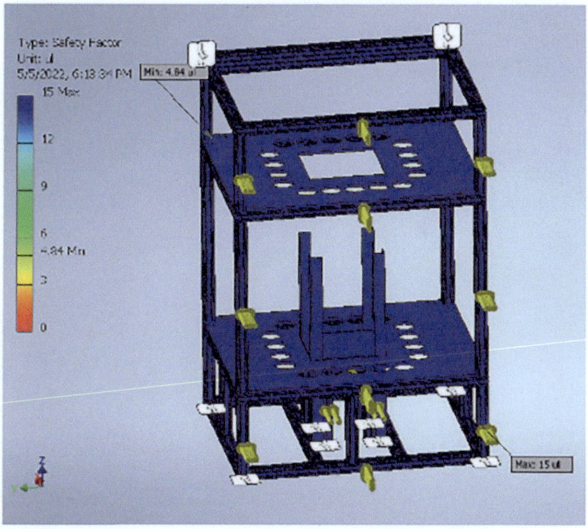

Fig. B.3 Applying 0.4g horizontal acceleration along one perpendicular axis, with a minimum Safety Factor of 4.84

B.0.1 Gravitational Load

First, the detector structure is examined to ensure that it can handle the gravitational force of its contents. Because the Neutron Multiplicity Meter consists of two steel trays with differing loads, they are treated separately with two unique loads/forces applied in Inventor's Stress Analysis platform. The inner core then has its own independent load from the lead and thorium. The bottom tier of the Neutron Multiplicity Meter experiences a load of 8597 N but is simulated with a load of 8,600 N of down-

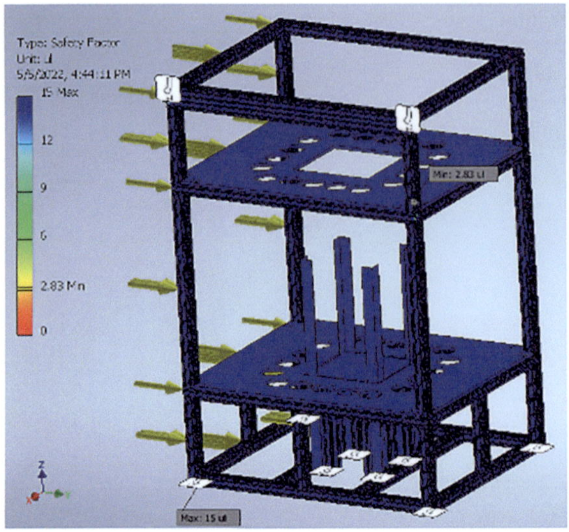

Fig. B.4 Applying 0.4g horizontal acceleration along a second perpendicular axis, with a minimum Safety Factor (with Von Mises Stress) of 2.83

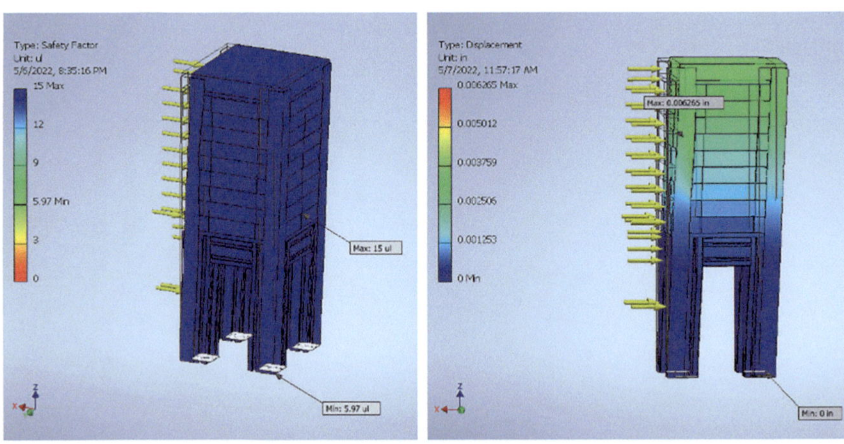

(a) The Safety Factor of the Inner Core on its own with a lead and thorium load which is allowed to slide in reaction to the applied force. A minimum SF of 5.97

(b) the maximum displacement experienced by the inner core when influenced by a 0.4g horizontal force is 0.006 in.

Fig. B.5 The FEA for both the gravitational and seismic loads on the inner core ward force. The upper tier of the Neutron Multiplicity Meter experiences a load of 2206.6 N but is simulated with a load of 2250 N of downward force. The inner core experiences a load of 5762 N but is simulated with a load of 6000 N of downward force to account for a slight uncertainty in the final amount of thorium which will be utilized. The minimum Safety Factor (based on the Von Mises Stress) for the Neutron Multiplicity Meter design is 4.41 which is greater than the 1.0 requirement (Table B.2).

B.0.2 Seismic Analysis

Next, the detector was subjected to 0.4g of horizontal acceleration. It is worth emphasizing that the 0.4g horizontal acceleration applied in this work is notably larger than the minimum required by ASCE 7-16 13.3-3:

$$F_P = 0.3 S_{DS} I_P W_P = 0.218 g W_P \tag{B.1}$$

where S_{DS} is the spectrum acceleration and determined by ASCE/SEI 7.16 and is taken to be 0.485 k. I_P is the component importance factor and is taken to be 1.5. By passing the criteria of this analysis at 04 g, one substantiates the safety of this apparatus in a seismic event all the way down to the minimum required acceleration and below. Two other neutrino detectors deployed to the same space (the 8700 Basement TA-B158) demonstrated their seismic analysis using 0.25 g and 0.3 g.

The weight of the neutron multiplicity meter is 1160.7 kg (2553.5 lb) so a horizontal load of 0.4 g corresponds to 4,549.9 N but is simulated to be 4.550 N. The inner core bears a load of 6,000, thus receives a horizontal force of 2,400 N parallel to the force applied on the neutron multiplicity meter. These forces are applied along the two primary horixontal axes, parallel and perpendicular to the length of the SNS basement hallway. The floor and wall anchors are assumed to be fixed points of contact in this analysis. The safety factors for the two analyses are 4.84 and 2.83.

Table B.2 Summary of stress analysis for different conditions and directions, showing the material with the maximum value, the maximum calculated stress, and the maximum allowable stress

Analysis type	Stress type	Material with maximum value	Maximum calculated stress (ksi)	Max. allowable stress (ksi)
Gravitational (Von Mises safety factor = 4.4)				
	Planar XX	Aluminum	6.532	35
	Planar YY	Aluminum	5.147	35
	Planar ZZ	Aluminum	9.075	35
	Shear XY	Steel	1.513	22
	Shear XZ	Aluminum	4.085	24
	Shear YZ	Aluminum	3.674	24
	Maximum X-Displacement		0.03 inches	
	Maximum Y-Displacement		0.004 inches	
	Maximum Z-Displacement		0.038 inches	
Seismic: X-Direction (Von Mises safety factor = 4.84)				
	Planar XX	Aluminum	8.913	35
	Planar YY	Aluminum	3.757	35
	Planar ZZ	Aluminum	8.758	35
	Shear XY	Aluminum	1.86	24
	Shear XZ	Aluminum	4.206	24
	Shear YZ	Aluminum	1.891	24
	Maximum X-Displacement		0.025 inches	
	Maximum Y-Displacement		0.003 inches	
	Maximum Z-Displacement		0.003 inches	
Seismic: Y-Direction (Von Mises safety factor = 2.83)				
	Planar XX	Aluminum	12.16	35
	Planar YY	Aluminum	6.4	35
	Planar ZZ	Aluminum	14.29	35
	Shear XY	Aluminum	2.563	24
	Shear XZ	Aluminum	6.66	24
	Shear YZ	Aluminum	2.713	24
	Maximum X-Displacement		0.015 inches	
	Maximum Y-Displacement		0.053 inches	
	Maximum Z-Displacement		0.005 inches	
Inner core only Seismic: Z-Direction (Von Mises safety factor = 5.97)				
	Planar XX	Steel	4.946	36
	Planar YY	Steel	2.207	36
	Planar ZZ	Aluminum	3.9	35
	Shear XZ	Aluminum	3.264	24
	Shear YZ	Aluminum	2.667	24
	Maximum X-Displacement		0.006 inches	
	Maximum Y-Displacement		0.003 inches	
	Maximum Z-Displacement		0.0003 inches	

Appendix C
Nuclear Transitions in Thorium

Abstract Conclude

For $^{232}_{90}\text{Th}_{126}$, the protons fill the $1h_{9/2}$ sub-shell while the neutrons fill up to the $2d_{5/2}$ sub-shell. A cartoon of Thorium's shell structure is shown in Fig. C.1.

The transformation of one nucleon into another via neutrino capture begs the question of what nuclear sub-shell the new nucleon would be created in. Answering this question requires considering angular momentum and parity selection rules. The most likely transitions from the neutron energy states to the proton energy states are the "allowed" transitions in which the orbital angular momentum does not change—any change to the total angular momentum occurs from the spin of the two final state particles. The final state particles are the electron (positron) and up (down) quark for (anti)neutrino capture. Additionally, because the orbital angular momentum does not change, the parity, $(-1)^l$ [1], so the allowed transition selection rules are:

$$\Delta I = 0, 1 \quad \Delta\pi\,(\text{parity change}) = \text{no} \tag{C.1}$$

when total S = 1, it is a Gamow-Teller transition and when total S = 0, it is a Fermi transition.

Further, each transition has a corresponding strength (or probability of moving from the initial state to the final state). The transition strength is approximated as:

$$|M_{GT}|^2 = \frac{(n_{p,n})_i (n_h)_f}{2J_f + 1} |M_{GT}^{SP}|^2 \tag{C.2}$$

where J_f is the angular momentum of the final state nucleon, $(n_{p,n})_i$ is the number of filled spots in the initial sub-shell, $(n_h)_f$ is the number of unfilled nucleons/number of holes in the final sub-shell, and $|M_{GT}^{SP}|^2$ is the single-particle Gamow-Teller strength

T. Johnson, *The First Indication of Neutrino-Induced Nuclear Fission*,
https://doi.org/10.1007/978-3-032-01936-3

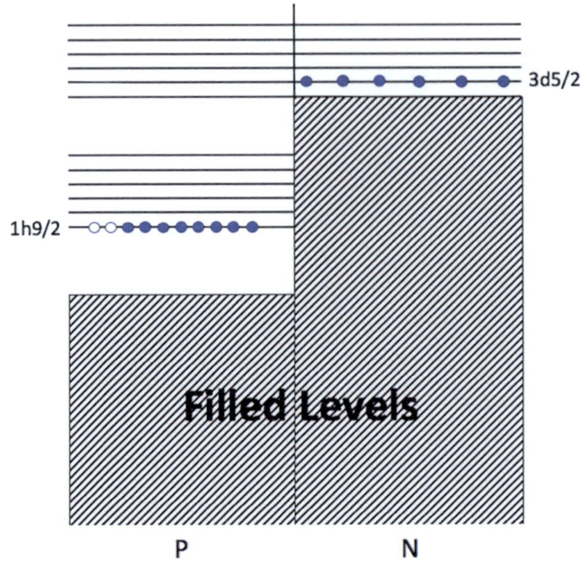

Fig. C.1 The shell structure of ^{232}Th for both protons and neutrons

Table C.1 Single-particle GT strengths

j_i	j_f: $l+\frac{1}{2}$	j_f: $l-\frac{1}{2}$
$l+\frac{1}{2}$	$\frac{j_i+1}{j_i}$	$\frac{2j_i-1}{j_i}$
$l-\frac{1}{2}$	$\frac{2j_i+3}{j_i+1}$	$\frac{j_i}{j_i+1}$

given by Table C.2. Notably, the term $(n_h)_f/(2J_f+1)$ is the fraction of the final state sub-shell that is unfilled (Table C.1).

$l \pm \frac{1}{2}$ refers to which of the spin-orbit doublets said state is. For $^{232}_{90}\mathrm{Th}_{142}$, the model space for calculations is $3d_{5/2}$, $3d_{3/2}$, $2g_{9/2}$, $2g_{7/2}$, $3p_{1/2}$, $3p_{3/2}$, $1i_{13/2}$, $1i_{11/2}$, $2f_{5/2}$, $2f_{7/2}$,$1h_{9/2}$ for proton orbits and almost the same for neutron orbits but with the addition of $1h_{11/2}$. As such, the allowed transitions and associated GT strengths are as follows.

Notice that the sum of the strength calculated in the allowed Gamow-Teller transitions above exactly equal the value predicted by the Ikeda Sum rule [2] shown in Eq. C.3:

$$|M_{GT}|^2 = 3(N-Z) \tag{C.3}$$

Table C.2 Single-particle GT strengths

^{232}Th(ν_e,e$^-$)^{232}Pa Allowed Gamow-Teller transitions

Shell model transition	$l_{i,f}$	$(n_n)_i$	$\frac{(n_h)_f}{2j_f+1}$	$\lvert M_{GT}^{sp}\rvert^2$	$\lvert M_{GT}\rvert^2$ ($\log ft$)
$3d^+_{5/2} \rightarrow 3d^+_{5/2}$	2	6	1	7/5	8.40 (1.47)
$3d^+_{5/2} \rightarrow 3d^+_{3/2}$	2	6	1	8/5	9.60 (1.33)
$2g^+_{9/2} \rightarrow 2g^+_{9/2}$	4	10	1	110/9	12.22 (1.09)
$2g^+_{9/2} \rightarrow 2g^+_{7/2}$	4	10	1	16/9	17.77 (0.72)
$3p^-_{1/2} \rightarrow 3p^-_{1/2}$	1	2	1	1/3	0.67 (4.00)
$3p^-_{1/2} \rightarrow 3p^-_{3/2}$	1	2	1	8/3	5.33 (1.92)
$1i^+_{13/2} \rightarrow 1i^+_{13/2}$	6	14	1	15/13	16.15 (1.92)
$1i^+_{13/2} \rightarrow 1i^+_{11/2}$	6	14	1	24/13	25.85 (0.34)
$3p^-_{3/2} \rightarrow 3p^-_{3/2}$	1	4	1	5/3	6.67 (1.70)
$3p^-_{3/2} \rightarrow 3p^-_{1/2}$	1	4	1	4/3	5.33 (1.92)
$2f^-_{5/2} \rightarrow 2f^-_{5/2}$	3	6	1	5/7	4.29 (2.14)
$2f^-_{5/2} \rightarrow 2f^-_{7/2}$	3	6	1	16/7	13.71 (0.98)
$2f^-_{7/2} \rightarrow 2f^-_{7/2}$	3	8	1	9/7	10.29 (1.26)
$2f^-_{7/2} \rightarrow 2f^-_{5/2}$	3	8	1	12/7	13.71 (2.14)
$1h^-_{9/2} \rightarrow 1h^-_{9/2}$	5	10	2/10	9/11	1.64 (3.10)
$1h^-_{11/2} \rightarrow 1h^-_{9/2}$	5	12	1/5	20/11	4.36 (2.12)
Total GT strength					156.00

References

1. Osterfeld F (1992) Nuclear spin and isospin excitations. Rev Mod Phys 64:491–557. https://doi.org/10.1103/RevModPhys.64.491
2. Krmpotić F et al (1997) Ikeda sum rule, self-consistency and double-beta decay in the renormalized quasiparticle random phase approximation. Nucl Phys A 612(2):223–238. ISSN: 0375-9474. https://doi.org/10.1016/S0375-9474(97)80010-3, https://www.sciencedirect.com/science/article/pii/S0375947497800103

Curriculum Vitae

Tyler Johnson e-mail: Tyler.e.johnson@duke.edu (913) 575-9490
201 Munson Street Apt 360 New Haven, CT

Research Focus

My research is focused on developing novel detector technologies to search for rare physics events related to neutrinos, dark matter and nuclear reactions. Currently, my primary concentration is on the development of novel quantum sensors for detecting dark matter candidate particles called "axions" and the search for the ultra rare process called "neutrino-less double beta decay." My doctoral work was completing the first search for neutrino-induced nuclear fission—a process named "nuFission."

Education

Doctor of Philosophy, Nuclear Physics, Duke University, Fall 2024

- Dissertation: "The First Indication of Neutrino-Induced Nuclear Fission"

Master of Arts, Physics, Duke University, Fall 2021
Physics, Bachelor of Arts, University of Chicago, 2018

- Thesis: "Cosmic Ray Tracker Monte Carlo Simulation and Analysis for the Short Baseline Near Detector (SBND) at Fermilab"

T. Johnson, *The First Indication of Neutrino-Induced Nuclear Fission*,
https://doi.org/10.1007/978-3-032-01936-3

Research Experience

Nov 2024-ongoing **Yale University** New Haven, CT
Wright Laboratory
Postdoctoral Researcher

- Develop quantum sensing instruments for the RAY experiment to probe for the dark matter candidate particle called an axion
- Lead of the Cryogenics and Facilities working group coordinating the creation of the lab space for the plasma metamaterial haloscope experiment, ALPHA.

Nuclear Particle Astrophysics (NPA) Seminar Series Co-coordinator

- Organizing and hosting the Yale NPA seminar series to cover the cutting-edge developments in nuclear, particle and astrophysics.

Jul 2018–Oct 2024 **Duke University** Durham, NC
Experimental Neutrino Physics Laboratory
PhD Candidate and Research Assistant

- Developed, built and analyzed the world's first neutrino-induced nuclear fission experiment
- Characterize detector response to low energy nuclear recoils (quenching factors) as a certified operator of the Tandem Van De Graaff Accelerator

Jun 2019–Sept 2024 **National Nuclear Security Administration** Ann Arbor, MI
Consortium on Monitoring, Technology and Verification
Applied Neutrino PhD Fellow

- Conduct particle/nuclear physics research to advance the American technological portfolio for nuclear non-proliferation and international arms control treaty verification

Aug 2018–Aug 2019 **Stanford University** Palo Alto, CA
Stanford U.S. —Russia Forum
Arms Control Fellow

- Among a small delegation of US and Russia scholars (5% selectivity) working on policy matters to better US-Russia relations in nuclear non-proliferation and international arms control

Sept 2016–Jul 2018 **University of Chicago** Durham, NC
Experimental Neutrino Physics
Undergraduate Research Assistant

- Created a simulation of neutrino-induced muons steaming through the SBND facility before the LArTPC is installed to gain understanding of the beam-correlated muon background

Mar 2015– Jul 2017 **Ultra-Cold Atomic Physics**
Undergraduate Research Assistant

- Produced precision electrical devices for the Quantum Matter Synthesizer (QMS) experiment to ultra-cool atoms to near absolute zero

Summer 2016 **Fermi National Accelerator Laboratory** Batavia, IL
Neutrino Division, Deep Underground Neutrino Experiment (DUNE)
Research Intern

- Assisted in the design and evaluation of the focusing and targeting system for what will be the world's most intense neutrino beam

Summer 2015 **Enrico Fermi Institute of Physics** New Chicago, IL
Undergraduate Research Fellow

- Fabricated the design for a microwave frequency synthesizer device, wrote the C++ control software and created the physical circuitry to construct 2 functioning instruments

Honors and Awards

Fellowships and Scholarships

Jun 2024 **Henry W. Newson Fellowship from Duke Physics** Competitive departmental award granted for final year of graduate study

Jun 2019 **Applied Neutrino Fellowship from the Consortium on Monitoring, Technology and Verification** Selected for the competitive five-year doctoral fellowship to conduct research on nuclear non-proliferation

Apr 2019 **Honorable Mention for the National Science Foundation Graduate Research Fellowship**Conferred honorable mention for my research proposal, outreach efforts and personal statement

Jul 2019 **Robert C. Richardson Fellowship from Duke Physics** Competitive departmental fellowship for summer research funding

Aug 2018 **Goshaw Family Fellowship from Duke Physics** Departmental award allotting funding the first year of research

Jul 2018 **Minority PhD Scholarship from the Alfred Sloan Foundation** Five-year scholarship to support minority PhD students

May 2016 **Mellon Mays Undergraduate Fellowship from the Andrew Mellon Foundation** Selected for a small cohort of under-represented scholars to aid in pursuing doctoral education and subsequent professorship

Awards and Prizes

Apr 2025	**Forever Duke Leadership Award** Awarded this leadership award for distinguished service to the university
Jul 2023	**NextProf Nexus 2023 Cohort** Intensive program geared towards diversifying the American professorate in science (5% acceptance rate)
Mar 2022	**Best Poster Prize at the 2022 Consortium for Monitoring, Technology and Verification Workshop** Awarded best poster presentation among 40 posters presented
Oct 2021	**Best Talk Prize at the Defense Nuclear Nonproliferation R&D Conference** Selected as best talk from 70 talks delivered through the conference
Mar 2021	**Best Talk Prize at the 2021 Consortium for Monitoring, Technology and Verification Workshop** Selected as best talk from among 60 talks delivered by PhD Students
Feb 2017	**2nd Place Presentation in Physical Science at the Yale Undergraduate Research Conference** Placed second among all posters presented for the physical sciences and engineering

Professional Service

Institutional Service

2021–2023	**Finance Committee for the Duke University Board of Trustees**
2023–present	**Capital Subcommittee for the Duke University Board of Trustees**
2018–2022	**Advisory Board for the University Center of Exemplary Mentoring**
2021	**Search Committee for the Dean of Arts and Sciences at Duke University**
2018–2020	**President's Advisory Committee on Investment Responsibility at Duke University**
2019	**President's Council for Black Affairs at Duke University**

Departmental Service

2019–2020	**Physics Conduct and Accountability Committee at Duke University**
2019–2020	**Graduate Student Mentor at Duke Physics**
2018–2020	**Physics Representative at the Graduate and Professional Student Government**
2017–2018	**University of Chicago Physics Chair's Advisory Panel**

Mentoring Experience

Graduate Students

- Ana Zaalishvili, Ph.D. Student, Duke University
- Natalie Jones, Ph.D. Student, Duke University
- Aitor Bracho, Ph.D. Student, Duke University
- Nixon Ogoi, Master's Student, North Carolina Central University
- Godwin Smart, Master's Student, North Carolina Central University

Undergraduate Students

- Ingrid Rodriguez Vila, Undergraduate Student, Yale University
- Erem Ujah, Undergraduate Student, North Carolina Central University
- Will Henson, Undergraduate visiting student from University of Massachusetts Amherst
- Patrick Jacobs, Undergraduate Student, North Carolina Central University
- Gavi Fischer, Undergraduate Student, William and Mary College
- Yiwei Liu, Undergraduate Student, Duke University

High School Students

- Fin Densmore, High School Student, North Carolina School of Science & Math
- Elliott Kaufman, High School Student, North Carolina School of Science & Math

Talks and Presentations

Invited Seminars

Apr 2025	**Virginia Tech Center for Neutrino Physics Seminar**, Blacksburg, VA, USA
Apr 2024	**Yale University—Wright Lab Seminar**, New Haven, CT, USA
Feb 2024	**Mitchell Institute of Physics, High Energy Phenomenology, Experiment and Cosmology Seminar**, College Station, TX, USA
Nov 2023	**MIT Lincoln Nuclear Science Seminar**, Cambridge, MA, USA
Sept 2023	**Oak Ridge National Laboratory Nuclear Non-Proliferation Seminar**, Oak Ridge, TN, USA
Apr 2023	**University of Tennessee at Knoxville HEP Seminar**, Knoxville, TN, USA
Feb 2023	**Sandia National Laboratory Radiation and Nuclear Detection Seminar**, Virtual Format
Oct 2023	**University of Virginia HEP Seminar**, Charlottesville, VA, USA

Conferences and Workshops

Jun 2024 **The Magnificent CEvNS International Workshop**, University-Enterprise Foundation of the University of Valencia (ADEIT), Valencia, Spain

Mar 2024 **Consortium for Monitoring Technology and Verification Workshop 2024**, Ann Arbor, MI

Nov 2023 **APS DNP US-Japan Joint Nuclear Physics Conference 2023**, Waikoloa, Hawaii, USA

Jun 2023 **Defense Nuclear Non-Proliferation R&D Conference 2023**, University of California—Berkeley, Berkeley, CA, USA

Mar 2023 **The Magnificent CEvNS International Workshop**, Technical University of Munich, Munich, Germany

Mar 2023 **Consortium for Monitoring Technology and Verification Workshop 2023**, University of Michigan—Ann Arbor, Ann Arbor, MI

Mar 2023 **Workshop on Neutrino Interactions Measurements for Supernova Neutrino Detection**, Oak Ridge National Laboratory, Oak Ridge, TN

Jun 2022 **Defense Nuclear Non-Proliferation R&D Conference 2022**, University of Michigan—Ann Arbor, Ann Arbor, MI

Oct 2021 **Defense Nuclear Non-Proliferation R&D Conference 2021**, Awarded Best Talk Prize, Georgia Institute of Technology, Atlanta, GA

Apr 2021 **Stanford US-Russia Forum Research Summit**, Virtual Format

Mar 2021 **Consortium for Monitoring Technology and Verification Workshop 2021**, Awarded Best Talk Prize, Virtual Format

Jun 2020 **Consortium for Monitoring Technology and Verification Research Symposium**, Virtual Format

Mar 2020 **Consortium for Monitoring Technology and Verification Workshop 2020**, University of Michigan—Ann Arbor, Ann Arbor, MI, USA

Feb 2020 **Sloan Research Symposium**, Duke University, Durham, NC, USA

Nov 2017 **Naval Academy Science and Engineering Conference**, U.S. Naval Academy, Annapolis, MD, USA

Oct 2017 **Mellon Mays Fellowship Conference 2017**, Carleton College, Northfield, MN, USA

Oct 2016 **Mellon Mays Fellowship Conference 2016**, Washington University in St. Louis, St. Louis, MO, USA

Collaboration Meetings

May 2024 **COHERENT Collaboration Meeting 2024**, Oak Ridge National Lab, Oak Ridge, TN

May 2023 **10th Anniversary COHERENT Collaboration Meeting**, Oak Ridge National Lab, Oak Ridge, TN

Dec 2023 **COHERENT Collaboration Meeting 2023**, North Carolina State University, Raleigh, NC

May 2022 **COHERENT Collaboration Meeting 2022**, Oak Ridge National Lab, Oak Ridge, TN

Dec 2021 **COHERENT Collaboration Meeting 2021**, Oak Ridge National Lab, Oak Ridge, TN
Sept 2017 **Fermilab SBND Collaboration Meeting 2017**, Fermilab, Batavia, IL
Sept 2016 **Fermilab Neutrino Monte Carlo Group Meeting** 2016, Fermilab, Batavia, IL
July 2016 **Fermilab DUNE Collaboration Meeting 2016**, Fermilab, Batavia, IL

Lectures and Outreach

Sept 2021 **Summer High School Physics Lecture Series**, Duke University, Durham, NC
May 2021 **Research Experience for Undergraduate Lunch Talk**, Triangle Universities Nuclear Laboratory, Durham, NC

Posters

Jun 2024 **XXXI International Conference on Neutrino Physics and Astrophysics**, University of Milano—Bicocca, Milano, IT
May 2024 **Defense Nuclear Non-Proliferation R&D Conference 2024**, Texas A&M—College Station, TX, USA
Mar 2022 **Consortium for Monitoring Technology and Verification Workshop 2022**, Awarded Best Poster Prize, University of Michigan—Ann Arbor, Ann Arbor, MI
Feb 2017 **Yale University Undergraduate Research Conference**, Awarded 2nd Place, Yale University, New Haven, CT, USA
Jan 2017 **Harvard University National Collegiate Research Conference**, Harvard University, Cambridge, MA, USA
Oct 2016 **3rd Annual Chicago Undergraduate Research Symposium**, University of Chicago, Chicago, IL, USA

Publications

- S. Adamski., et al "Evidence of Coherent Elastic Neutrino-Nuclear Scattering with COHERENT's Germanium Array" Phys. Rev. Lett. 134 231801 https://doi.org/10.1103/PhysRevLett.134.231801
- Barbeau, P.S., et al "Accessing new physics with an undoped, cryogenic CsI CEvNS detector for COHERENT at the SNS" Phys. Rev. D 109, 092005 https://doi.org/10.1103/PhysRevD.109.092005
- An, P., et al "Measurement of the inclusive electron-neutrino charged-current cross section on ^{127}I with the COHERENT NaIνE detector", Phys. Rev. Lett. 131, 221801 https://doi.org/10.1103/PhysRevLett.131.221801
- An, P., et al "Measurement of ^{nat}Pb (ν_e,Xn) Production with a Stopped-Pion Neutrino Source", Phy. Rev. D 108 (2023) 7, 072001 https://doi.org/10.1103/PhysRevD.108.072001
- Akimov, D., et al "The COHERENT Experimental Program", arXiv:2204.04575 [hep-ex]. https://doi.org/10.48550/arXiv.2204.04575

- Akimov, D., et al "Constraint on Leptophobic Dark Matter from COHERENT using CsI[Na] Data", Phys. Rev. D 106 (2022) 5, 052004 https://doi.org/10.1103/PhysRevD.106.052004
- Akimov, D., et al "Monitoring the SNS basement neutron background with MARS detector", 2022 JINST 17 P03021 https://doi.org/10.1088/1748-0221/17/03/P03021
- Akimov, D., et al "A D2O Detector for Flux Normalization of a Pion Decay-at-rest Neutrino Source" 2021 JINST 16 P08048 https://doi.org/10.1088/1748-0221/16/08/P08048
- Akimov, D., et al "Measurement of Scintillation Response of CsI[Na] to Low-energy Nuclear Recoils by COHERENT", 2022 JINST 17 P10034 https://doi.org/10.1088/1748-0221/17/10/P10034
- Akimov, D., et al "First Probe of Sub-GeV Dark Matter Beyond the Cosmological Expectation with the COHERENT CsI Detector at the SNS", Phys. Rev. Lett. 130 (2022) no. 5, 051803. https://doi.org/10.1103/PhysRevLett.130.051803
- Akimov, D., et al. "Measurement of the Coherent Elastic Neutrino-Nucleus Scattering Cross Section on CsI by COHERENT", Phy. Rev. Lett. 129 (2022) no. 8, 081801 https://doi.org/10.1103/PhysRevLett.129.081801
- Akimov, D., et al. "Simulating the neutrino flux from the Spallation Neutron Source for the COHERENT experiment", Phy. Rev. D 106 (2022) no. 3, 032003 https://doi.org/10.1103/PhysRevD.106.032003
- Awe, C., Barbeau, P. S., Haghighat, A., Hedges, S., Johnson, T., Li, S., et al. "Measurement of Proton Quenching in a Plastic Scintillator Detector", 2021 JINST 16 P02035 https://doi.org/10.1088/1748-0221/16/02/P02035
- Akimov, D., et al. "Development of a ^{83m}Kr source for the calibration of the CENNS-10 Liquid Argon Detector", 2021 JINST 16 P04002 https://doi.org/10.1088/1748-0221/16/04/P04002
- Akimov, D., et al. "Sensitivity of the COHERENT Experiment to Accelerator-Produced Dark Matter" Phy. Rev. D. 102 (2020) no. 5, 052007 https://doi.org/10.1103/PhysRevD.102.052007
- Akimov, D., et al. "First Measurement of Coherent Elastic Neutrino-Nucleus Scattering on Argon" Phys. Rev. Lett. 126 (2021) no. 1, 012002. https://doi.org/10.1103/PhysRevLett.126.012002
- Akimov, D., et al. "COHERENT Collaboration data release from the first detection of coherent elastic neutrino-nucleus scattering on argon" arXiv:2006.12659[nucl-ex] https://doi.org/10.48550/arXiv.2006.12659
- Akimov, D., "First Constraint on Coherent Elastic Neutrino-Nucleus Scattering in Argon", Phys. Rev. D 100 (2019) no. 11, 115020 https://doi.org/10.1103/PhysRevD.100.115020

Professional Affiliations

- American Physical Society
- National Society of Black Physicists
- Institute of Teaching and Mentoring
- Southern Regional Institute of Teaching and Mentoring
- Mellon Mays Graduate Initiative
- Institute of Physics
- Duke Society of Fellows

MIX
Papier aus verantwortungsvollen Quellen
Paper from responsible sources
FSC® C105338

FSC
www.fsc.org

If you have any concerns about our products,
you can contact us on
ProductSafety@springernature.com

In case Publisher is established outside the EU,
the EU authorized representative is:
Springer Nature Customer Service Center GmbH
Europaplatz 3, 69115 Heidelberg, Germany

Printed by Libri Plureos GmbH
in Hamburg, Germany